Retail Supply Chain Management

Early Titles in the

INTERNATIONAL SERIES IN OPERATIONS RESEARCH & MANAGEMENT SCIENCE

Frederick S. Hillier, Series Editor, *Stanford University*

Saigal/ *A MODERN APPROACH TO LINEAR PROGRAMMING*
Nagurney/ *PROJECTED DYNAMICAL SYSTEMS & VARIATIONAL INEQUALITIES WITH APPLICATIONS*
Padberg & Rijal/ *LOCATION, SCHEDULING, DESIGN AND INTEGER PROGRAMMING*
Vanderbei/ *LINEAR PROGRAMMING*
Jaiswal/ *MILITARY OPERATIONS RESEARCH*
Gal & Greenberg/ *ADVANCES IN SENSITIVITY ANALYSIS & PARAMETRIC PROGRAMMING*
Prabhu/ *FOUNDATIONS OF QUEUEING THEORY*
Fang, Rajasekera & Tsao/ *ENTROPY OPTIMIZATION & MATHEMATICAL PROGRAMMING*
Yu/ *OR IN THE AIRLINE INDUSTRY*
Ho & Tang/ *PRODUCT VARIETY MANAGEMENT*
El-Taha & Stidham/ *SAMPLE-PATH ANALYSIS OF QUEUEING SYSTEMS*
Miettinen/ *NONLINEAR MULTIOBJECTIVE OPTIMIZATION*
Chao & Huntington/ *DESIGNING COMPETITIVE ELECTRICITY MARKETS*
Weglarz/ *PROJECT SCHEDULING: RECENT TRENDS & RESULTS*
Sahin & Polatoglu/ *QUALITY, WARRANTY AND PREVENTIVE MAINTENANCE*
Tavares/ *ADVANCES MODELS FOR PROJECT MANAGEMENT*
Tayur, Ganeshan & Magazine/ *QUANTITATIVE MODELS FOR SUPPLY CHAIN MANAGEMENT*
Weyant, J./ *ENERGY AND ENVIRONMENTAL POLICY MODELING*
Shanthikumar, J.G. & Sumita, U./ *APPLIED PROBABILITY AND STOCHASTIC PROCESSES*
Liu, B. & Esogbue, A.O./ *DECISION CRITERIA AND OPTIMAL INVENTORY PROCESSES*
Gal, T., Stewart, T.J., Hanne, T. / *MULTICRITERIA DECISION MAKING: Advances in MCDM Models, Algorithms, Theory, and Applications*
Fox, B.L. / *STRATEGIES FOR QUASI-MONTE CARLO*
Hall, R.W. / *HANDBOOK OF TRANSPORTATION SCIENCE*
Grassman, W.K. / *COMPUTATIONAL PROBABILITY*
Pomerol, J-C. & Barba-Romero, S. / *MULTICRITERION DECISION IN MANAGEMENT*

(*continued after index*)

Narendra Agrawal · Stephen A. Smith
Editors

Retail Supply Chain Management

Quantitative Models and Empirical Studies

Foreword by Hau L. Lee

Editors
Narendra Agrawal
Santa Clara University
CA, USA
nagrawal@scu.edu

Stephen A. Smith
Santa Clara University
CA, USA
ssmith@scu.edu

ISBN: 978-0-387-78902-6 e-ISBN: 978-0-387-78903-3
DOI 10.1007/978-0-387-78902-6

Library of Congress Control Number: 2008925269

Printed on acid-free paper

springer.com

Foreword

It is with great pleasure that I can write a foreword for the book *Retail Supply Chain Management: Quantitative Models and Empirical Studies*. I want to congratulate the editors, Narendra Agrawal and Stephen Smith, for compiling this impressive volume. This, no doubt, will be a book that provides a solid reference on research on retail supply chains, and inspires new research on this subject.

Retailing forms the part of the supply chain that interfaces between the ultimate consumers and the rest of the supply chain. As such, it is often viewed as the part of the supply chain where the real demands of the consumers first show up. Whether we are talking about a physical retail store or a virtual store, the consumer demands that occur here drive the demands in the rest of the supply chain. So in that sense, it is like the frontier of all supply chains.

I remember that, when I started working on research on the "bullwhip" effect (demand information distortion) in supply chains, industry practitioners all recognized the importance of having the demand information at the retail level under control, or else it would be hopeless to dampen the bullwhip. In the famous beer game, we have witnessed many examples in which, once the retailer started ordering beer with anxiety or nervousness, then the rest of the supply chain would be in chaos.

It is therefore gratifying to see Naren and Steve focusing their volume on retail supply chains. The innovations, lessons in practice, and new technological solutions in managing retail supply chains are not just important in retailing, but crucial in the ultimate effective management of the complete supply chain.

There are two distinguishing features in the research of retail supply chains, which the current volume captures well. First, retail supply chains are loaded with a lot of empirical data. This is an area that has traditionally been rich in data, which provides fertile grounds for us to pursue empirical research. Second, research on retail supply chains naturally intersects with research in marketing in two ways – category management and pricing. Of course, category management and pricing have traditionally been key areas in the marketing literature. But what the current volume has added is the dimension of supply chain management to these marketing approaches. Integrating category management with inventory planning, and coordinating price optimization with supply chain management are unique dimensions that distinguish this book.

I am sure that the readers will share my great enthusiasm for this book as a wonderful addition to the emerging literature on retail supply chain management.

Stanford, CA

Hau L. Lee
Thoma Professor of Operations,
Information and Technology
Graduate School of Business
Stanford University
Stanford, CA 94305
USA

Preface

We began working in retail supply chain management through the retail research program of the Retail Management Institute (RMI) at Santa Clara University. RMI was founded in 1980 by its current Executive Director, Dale Achabal, who is the L. J. Skaggs Distinguished Professor of Marketing at Santa Clara University. Research at RMI has focused on marketing and supply chain decisions in department store chains and specialty retailers. Over 30 major retail chains have participated in our research by providing data and problem descriptions and by sponsoring projects. The goal of our research has always been to develop new analytical tools for supporting the operational and planning decisions that retailers face. The sponsoring organizations saw the potential benefit from developing new analytical methodologies that could take advantage of the capabilities offered by emerging information technologies in retailing. Consequently, a number of the decision support prototypes developed at RMI were later converted into operational software systems by consulting organizations, and application software products by independent vendors. In this sense, the research done at RMI, as well as the research by other authors of chapters in this volume, has led to an array of retailing applications that constitute a great success story for management science, and for supply chain management in particular.

We are grateful to all authors who have contributed their research to this endeavor, and thank them for their patience as we went through multiple rounds of the review process for their submissions. We are indebted to our colleagues who painstakingly reviewed the various revisions of the submissions, adhering to standards typical of professional journals. These reviewers include Goker Aydin (University of Michigan), Gerard Cachon (University of Pennsylvania), Nicole DeHoratius (University of Chicago), Vishal Gaur (Cornell University), Warren Hausman (Stanford University), Kirthi Kalyanam (Santa Clara University), Steven Nahmias (Santa Clara University), Andy Tsay (Santa Clara University), and Jin Whang (Stanford University). Finally, we would like to thank Gary Folven, our original editor with Kluwer and later with Springer Publishing, who encouraged us to undertake this project and supported our efforts.

We wish to thank our colleagues from the Marketing Department, Dale Achabal, Shelby McIntyre and Kirthi Kalyanam, for collaborating with us on a wide range of projects. Many retail executives from sponsoring companies have contributed immensely to our research. There are simply too many for us to acknowledge individually, but we are very grateful for their continued support. And we are especially grateful to our wives, Niti Agrawal and Karen Graul for graciously supporting our efforts during the time it took to complete this volume.

Santa Clara, CA Narendra Agrawal
Santa Clara, CA Stephen A. Smith

Contents

Contributors

Narendra Agrawal
Department of Operations and MIS, Leavey School of Business, Santa Clara University, Santa Clara, CA 95053, USA, nagrawal@scu.edu

Ravi Anupindi
Associate Professor of Operations Management, Department of Operations and Management Sciences, Stephen M. Ross School of Business, University of Michigan, 701Tappan St., Ann Arbor, MI 48109-1234, anupindi@umich.edu

Goker Aydin
Assistant Professor, Department of Industrial and Operations Engineering, 1205 Beal Avenue, The University of Michigan, Ann Arbor, MI 48109-2117, ayding@engin.umich.edu

Nicole DeHoratius
University of Chicago, Graduate School of Business, 5807 S. Woodlawn Ave., Chicago IL 60637, USA, Nicole.DeHoratius@ChicagoGSB.edu

Marshall L. Fisher
Professor of Operations and Information Management, The Wharton School, University of Pennsylvania, 500 Jon M. Huntsman Hall, 3730 Walnut Street, Philadelphia, PA 19104-5340, USA, fisher@wharton.upenn.edu

Vishal Gaur
Johnson Graduate School of Management, Cornell University, Sage Hall, Ithaca, NY 14853, USA, vg77@cornell.edu

Sachin Gupta
Professor of Marketing, Johnson Graduate School of Management, 369 Sage Hall, Cornell University, Ithaca, NY 14853-6201, USA, Sg248@cornell.edu

A. Gurhan Kok
Fuqua School of Business, 1 Towerview Drive, Duke University, Durham, NC 27708, USA, Gurhan.kok@duke.edu

Saravanan Kesavan
Kenan-Flagler Business School, University of North Carolina at Chapel Hill, Chapel Hill, NC 27599, USA, skesavan@unc.edu

Mümin Kurtuluş
Owen Graduate School of Management, Vanderbilt University, 401 21st, Avenue South, Nashville, TN 37203, USA, mumin.kurtulus@owen.vanderbilt.edu

Evan L. Porteus
Professor of Management Science, Graduate School of Business, Stanford University, Stanford, CA 94305, USA, eporteus@stanford.edu

Stephen A. Smith
Department of Operations and MIS, Leavey School of Business, Santa Clara, University, Santa Clara, CA 95053, USA, ssmith@scu.edu

L. Beril Toktay
Associate Professor Operations Management, College of Management, Georgia Institute of Technology, 800 West Peachtree Street NW, Atlanta, GA 30308-0520, USA, beril.toktay@mgt.gatech.edu

Zeynep Ton
Associate Professor of Business Administration, Harvard Business School, Morgan Hall 425, Boston, MA 02163, USA, zton@hbs.edu

Ramnath Vaidyanathan
The Wharton School, University of Pennsylvania, 3730 Walnut Street, Philadelphia, PA 19104-5340, USA, ramnathv@wharton.upenn.edu

M.A. Venkataramanan
Kelley School of Business, 1275 E. 10th Street, Indiana University, Bloomington, IN 47405, USA, venkatar@indiana.edu

Seungjin Whang
Graduate School of Business, Stanford University, Stanford, CA 94305, USA, Whang_jin@gsb.stanford.edu

Chapter 1
OVERVIEW OF CHAPTERS

Narendra Agrawal and Stephen A. Smith
Department of Operations and MIS, Leavey School of Business, Santa Clara University, Santa Clara, CA 95053, USA

1. BACKGROUND

The retail industry has emerged as a fascinating choice for researchers in the field of supply chain management. It presents a vast array of stimulating challenges that have long provided the context of much of the research in the area of operations research and inventory management. However, in recent years, advances in computing capabilities and information technologies, hyper-competition in the retail industry, emergence of multiple retail formats and distribution channels, an ever increasing trend towards a globally dispersed retail network, and a better understanding of the importance of collaboration in the extended supply chain have led to a surge in academic research on topics in retail supply chain management. Many supply chain innovations (e.g., vendor managed inventory) were first conceived and successfully validated in this industry, and have since been adopted in others. Conversely, many retailers have been quick to adopt cutting edge practices that first originated in other industries.

However, for every example of leading edge progressive thinking among retailers, there are numerous examples of archaic systems and planning processes. Moreover, there continue to be a host of open problems facing practitioners and academics. All of this is, of course, good news for academics engaged in research in retail supply chain management. The recent past has witnessed exciting new research – theoretical as well as applied – aimed at addressing some of the retail industry's many pressing challenges. This book is an attempt to summarize some of this research and present a perspective on what new applications may lie ahead.

The past twenty years have seen a revolution in retailer's computing capabilities. Circa 1990, retailers' information systems tracked and stored dollar receipts for their merchandise, but often retained only cumulative sales data, as opposed to the selling patterns for individual stock keeping units (SKUs) by time period. Merchandise planners had access to various kinds of product level financial and inventory count information through computer terminals connected to the corporate data base systems. But there was no computing

N. Agrawal, S.A. Smith (eds.), *Retail Supply Chain Management*,
DOI: 10.1007/978-0-387-78902-6_1,

technology capable of applying quantitative forecasting and inventory management methods to evaluate alternative strategies, analyze market sensitivity to assumptions or optimize buying, promotions and clearance markdown decisions.

Since that time, the technology required to implement these methodologies has become widely available to buyers and inventory control analysts, as retailers have greatly expanded the information captured in their data bases and have distributed networked PCs to their professional employees. Retailers today can choose from a variety of commercial products that perform sales forecasting, pricing and inventory management functions, integrated as modules in their corporate information systems. Networked personal computers allow access to detailed sales and financial information, as well as offering localized processing power to analyze certain types of decisions. While the analytical methods imbedded in today's commercial offerings may appear to be fairly simple by academic standards, retailers' increasing investment and reliance on these systems indicates that they are providing value to retail supply chain operations today.

There is a natural development path for academic research in supply chain management to find its way into general use by major retailers. A number of the authors of the chapters in this volume have been instrumental in the successful implementation of methodologies for retailers. For purposes of illustration, let us consider the typical steps leading up to the implementation of a new methodology developed at the Retail Workbench at Santa Clara University. First, working with a sponsoring retailer, a decision support prototype is designed and developed for testing by buyers or other analysts in the merchandise planning cycle. Successful decision support prototypes were then adapted into an operational system by a third party software company or consulting organization, that works in cooperation with the sponsoring retailer. Finally, if market demand is perceived to be large enough, the one of a kind operational system is transformed into a commercial software product to be sold by an independent software vendor. We hope that many of the methodologies presented in this volume will find their way into mainstream retail practice through such a process as well.

2. THE FOCUS OF ACADEMIC RESEARCH IN THIS VOLUME

Despite the advances in analytical applications discussed in the preceding section, retailers today face many important unsolved problems in supply chain management. The chapters in this book focus on three crucial areas of retail supply chain management in which academic researchers have been very active recently: (1) empirical studies of retail supply chain practices, (2) assortment and inventory planning and (3) integrating price optimization into retail supply chain decisions. There are clearly other important research areas related to retail supply chain management, but in these three areas, recent research has successfully addressed some problems, while significant challenges remain.

Empirical studies of retail supply chain practices
Chapter 2 (Agrawal and Smith) begins with a description of supply chain practices and processes observed at two retailers in the home furnishing sector. Because of the large number of SKUs, the inter- relationships among the SKUs, as well as use of multiple store formats and marketing channels targeted to different customer segments, home furnishings is one of the most complex retail sectors. In addition to documenting the complex flows of materials and information in such multi-channel environments, they present details of key supply chain planning processes: product design and assortment planning, sourcing and vendor selection, logistics planning, distribution planning and inventory management, clearance and markdown optimization, and cross-channel optimization. Due to its complexity, the assortment selection and supply chain management decisions for this sector pose many challenging problems, whose solutions extend beyond the current state of the art. At the same time, the challenges in this sector are relevant to many other retail sectors as well. Thus, we hope that documenting the practices for these supply chains will provide a foundation for future methodological research, some of which are identified in the chapter.

Product level inventory management has been the subject of numerous papers in the area of supply chain management. More recently, researchers have begun to evaluate empirical evidence regarding the relationship between inventory management and overall firm performance. Some past research shows that inventory turnover varies substantially across firms as well as over time. Gaur *et al.* (2005) demonstrate that a significant portion of this variation can be explained by gross margin, capital intensity, and sales surprise (the ratio of actual sales to expected sales for the year). Using additional data, in Chapter 3, Gaur and Kesavan confirm these previously published results. Extending the findings of Gaur *et al.* (2005), they investigate the effects of firm size and sales growth rate on inventory turnover using data for 353 public listed US retailers for the period 1985–2003. With respect to size, they find strong evidence of diminishing returns to size: inventory turnover increases with size at a slower rate for large firms than for small firms. With respect to sales growth rate, they find that inventory turnover increases with sales growth rate, but its rate of increase depends on firm size and on whether sales growth rate is positive or negative. Their results are useful in (i) helping managers make aggregate-level inventory decisions by showing how inventory turnover changes with size and sales growth, (ii) employing inventory turnover in performance analysis, benchmarking and working capital management, and (iii) identifying the causes of performance differences among firms and over time.

In Chapter 4, DeHoratius and Ton direct attention to store level performance. In order to ensure product availability in retail settings, most existing research in this area has focused on two factors – poor assortment and poor inventory planning. The authors' research with several retailers during the last few years highlights a third factor, poor execution, or the failure to carry-out an operational plan. Poor store execution leads to stock outs and distorts sales and inventory data that are important inputs to assortment and inventory planning.

In this chapter they focus on two common execution problems – inventory record inaccuracy and misplaced products. Drawing on well-researched case studies, they describe the magnitude and root causes of these problems. They also describe the findings of empirical studies that have identified factors that exacerbate the occurrence of these problems. These factors include product variety, inventory levels, employee turnover and training, employee workload and employee effort. They describe the effect of inventory record inaccuracy and misplaced products on inventory planning and summarize how researchers have incorporated these problems into existing inventory models. They also discuss future research opportunities for studying the impact of store execution on product availability, in particular, and on retail supply chains, in general.

In addition to scientific inventory management and keen attention to execution of operational policies, leading edge retailers are resorting to other innovative management practices. In Chapter 5, Kurtulus and Toktay discuss one interesting example from the consumer goods sector, called category captainship. It is a form of manufacturer-retailer collaboration in which retailers rely on a leading manufacturer for management of items in a given category. There are reported success stories about category captainship, but also a growing debate about its potential for creating anti-competitive practices by category captains. The goal of this chapter is to provide an overview of the existing research on category captainship.

Despite a decade of implementation, there is limited academic research concerning category captainship. The existing research on category captainship can be grouped into four broad categories (1) emergence of category captainship; (2) delegation of pricing decisions; (3) delegation of assortment decisions; and (4) anti-competitive issues related to category captainship practices. The limited research in this field is due to challenges arising from the broad scope of implementation of category captainship programs. This chapter reviews the current research on category captainship and proposes some avenues for future research that could potentially overcome these challenges and improve our understanding of category captainship practices. The chapter also sheds light on how category captainship practices could potentially change the nature of the manufacturer-retailer relationships and the landscape in the retail industry.

Assortment and inventory planning

The assortment a retailer carries has a significant impact on sales, margins and customer traffic. Therefore, assortment planning has received high priority from retailers, consultants and software providers. The academic literature on assortment planning from an operations perspective is relatively new, but quickly growing. The basic assortment planning problem focuses on choosing the optimal set of products to be carried and the inventory level of each product. Decisions for products are interdependent and complex, due to considerations such as shelf space availability, substitutability between products, and brand management by vendors.

Kok, Fisher and Ramnath present an in depth review of the research on this topic in Chapter 6. This chapter is composed of four main parts. In the first part they discuss empirical results on consumer substitution behavior and present three demand models used in assortment planning: the multinomial logit, exogenous demand and locational choice models. In the second part, they describe optimization based assortment planning research. In the third part, they discuss demand and substitution estimation methodologies. In the fourth part, they present industry approaches to assortment planning by describing the assortment planning process at four prominent retailers. The authors conclude by providing a critical comparison of the academic and industry approaches and identifying research opportunities to bridge the gap between the two approaches.

In Chapter 7, Anupindi, Gupta and Venkatraman present a specific optimization methodology for the rationalization of retail assortment and stocking decisions for retail category management. They assume that consumers are heterogeneous in their intrinsic preferences for items and are willing to substitute less preferred items to a limited extent if their preferred items are not available. The authors propose an objective function for a far-sighted retailer that includes not only short-term profits but also a penalty for disutility incurred by consumers who do not find their preferred items in the available assortment. The retailer's problem is formulated as a constrained integer programming problem. They demonstrate an empirical application of their proposed model using household scanner panel data for eight items in the canned tuna category. Their results indicate that the inclusion of the penalty for disutility in the retailer's objective function is informative in terms of choosing an assortment to carry. They find that customer disutility can be significantly reduced at the cost of a small reduction in short term profits. They also find that the optimal assortment behaves non-monotonically as the weight on customer disutility in the retailer's objective function is increased.

Smith, in Chapter 8, considers an assortment planning model for retailers who sell multi-featured products such as consumer electronics and must tailor their assortments to appeal to a diverse set of customer tastes. The assortment decision affects both the probability that customers choose a particular retailer and the demands for the various products in the retailer's assortment. By explicitly including diverse customer segments, this paper develops an operational methodology for optimizing retail assortments for heterogeneous product preferences. A multinomial logit model is used for computing customers' joint probabilities of retailer choice and product choice. An optimization problem is then formulated for determining the assortment that maximizes the retailer's expected profit. The relationship between the optimal assortment and the retailer's competitive strength is also analyzed. Limiting properties of the relationship are derived for the special cases of a monopoly retailer and perfect competition among retailers. A commercial data base of consumer preferences for DVD players is used to illustrate the assortment optimization methodology and the sensitivity to various input assumptions. It was found that including customer heterogeneity in the choice model had a significant impact on expected profits for this data set.

The assortment planning decision is tightly connected to the inventory planning decision, about which there is extensive literature in the field of operations management. However, much of this literature assumes that the assortment has already been specified, and focuses solely on the inventory management decision. In Chapter 9, Agrawal and Smith provide a review of some recent research that is related to retail supply chain management.

In order for the review to be meaningful, it is restricted in scope in a number of ways. First, the focus is on papers that model multi-level inventory systems, since virtually all retail supply chains are multi-level. Second, attention is restricted to papers after 1993, and the reader is referred to the reviews in other papers for articles prior to 1993. For example, Axsater (1993), Federgruen (1993), and Nahmias and Smith (1993) contain excellent reviews of the work up to that point. Third, certain model formulations that are not typical of retail inventory management are also excluded, such as serial systems, since they are not representative of typical retail chains, and are a special case of general multi-location multi-echelon systems. Also excluded are papers that assume deterministic demand, since demand uncertainty is a key aspect of most retail systems.

Finally, the primary focus is on periodic review systems. Most retail chains today employ technologies such as point-of-sale (POS) scanner systems that provide real time access to sales and inventory data. Consequently, in principle, continuous review models could be an appropriate construct for these retail systems. However, two issues limit the practical applicability of this assumption. First, due to contracts with vendors and shipping companies, shipments occur primarily on a pre-specified schedule, and often a variety of items are delivered simultaneously. Second, despite the real time access to sales information, the ERP databases and inventory allocation algorithms are typically updated periodically. Thus, strictly speaking, inventory decisions must be made by planners according to predefined cycles. Thus, periodic review systems are a better representation of the inventory management systems used by most retailers. They conclude with suggestions for future research in this area.

Integrating price optimization into retail supply chain decisions

In addition to more efficient operational decisions, recent research has shown that better designed incentive systems can also be very effective in improving the operational and financial performance of supply chains. These incentive systems are captured in the supply chain contracts that define the relationship between buyers and suppliers. Reviews of some of the supply chain literature that focuses on the design of these contracts are contained in Tsay et al. (1999) and Cachon (2003).

In Chapter 10, Aydin and Porteus study the effect of the type of rebate offered to customers on the performance of the supply chain, and on the preference of the manufacturer and the retailer for such rebates. Starting with a newsvendor model (single-product, single-period, stochastic demand), they analyze a single-retailer, single-manufacturer supply chain with endogenous manufacturer rebates and retail pricing. The demand uncertainty is multiplicative, and the expected demand depends on the effective (retail) price of the product. A retailer rebate goes from the manufacturer to the retailer for each unit it sells. A consumer rebate goes from the manufacturer to the consumers for each unit they buy. Each consumer's response to consumer rebates is characterized by two exogenous parameters: α, the effective fraction of the consumer rebate that the consumer values, leading to the lower effective retail price perceived by the consumer, and, β, the probability that a consumer rebate will be redeemed. The type(s) of rebate(s) allowed and the unit wholesale price are given exogenously. Simultaneously, the manufacturer sets the size of the rebate(s) and the retailer sets the retail price. The retailer then decides how many units of the product to stock and the manufacturer delivers that amount by the beginning of the selling season. Compared to no rebates, an equilibrium retailer rebate leads to a lower effective price (hence, higher sales volume) and higher profits for both the supply chain and the retailer. An equilibrium consumer rebate also leads to a lower effective price and higher profits for the retailer, but not necessarily for the chain. Under their assumptions, such a consumer rebate (with or without a retailer rebate) allocates a fixed fraction of the (expected) supply chain profits to each player: The retailer gets $\alpha/(\alpha+\beta)$ and the manufacturer gets the rest, leading to interesting consequences. However, both firms prefer a higher α and a lower β, even though the manufacturer gets a smaller share of the chain profits, the total amount received is higher. Neither the retailer nor the manufacturer always prefers one particular kind of rebate to the other. In addition, contrary to popular belief, it is possible for both firms to prefer consumer rebates even when all such rebates are redeemed.

Another important aspect of pricing that has received some attention in the operations management literature is markdown planning, i.e., the price charged by the retailer at the end of the season to clear leftover inventory. This is important financially for retailers, since studies by the National Retail Federation have found that over one third of merchandise is sold on markdowns in some retail chains. Clearance markdowns are the focus of Chapter 11 by Smith. In the basic newsvendor model, the salvage value (which is related to the markdown price) is assumed to be given, but, in practice, this will depend upon the retailer's markdown pricing strategy. As the season draws to a close, sales rates depend upon price, seasonal effects and the remaining assortment of items available to customers. There is little time to react to observed sales, and pricing errors result in either loss of potential revenue or excess inventory to be liquidated. This chapter develops optimal clearance prices and inventory management policies that take into account the impact of reduced assortment and seasonal changes on sales rates. Versions of these policies have been implemented and tested at a number of major retail chains and these results are summarized and discussed.

Finally, in Chapter 12, Whang extends the markdown strategy discussion by including the element of retailer competition, using a stylized model of markdown competition. He considers two retailers who compete in a market with a fixed level of initial inventory. The initial inventory level is known to one retailer, but not to the other. To maximize the profit, each retailer marks down at a time of his individual choice. The model assumes deterministic demands, a single chance of price change, and a prefixed set of prices. He considers a two-parameter strategy set where a retailer chooses the timing of markdown as a function of the current time, his inventory level and the other retailer's actions so far. The paper characterizes the equilibrium of the game and derives managerial insights.

Retail supply chain management is a relatively new, but very exciting field of research. Fortunately, there is a substantial body of research in the areas of traditional inventory management, multi-echelon systems, channel coordination and pricing that the field of retailing can rely upon. The challenge, of course, is to develop and adapt methodologies that most accurately reflect the realities and constraints faced by retailers. As the practice of retailing evolves at increasing speed because of changes in the global competitive landscape, technology, and consumer expectations, we expect the array of research challenges in front of academics and practitioners to expand as well. We hope that this book will serve as a useful reference for these colleagues, and look ahead to the evolution of this field with much anticipation.

REFERENCES

Axsater, S. 1993. Continuous review policies for multi-level inventory systems with stochastic demand. *Handbooks in Operations Research and Management Science*, (Eds.) S.C. Graves, A.H.G. Rinnooy Kan and P.H. Zipkin, Volume 4 (Logistics of Production and Inventory), Elsevier Science Publishing Company B.V., Amsterdam, The Netherlands. 175–197.

Cachon, G. 2003. Supply chain coordination with contracts. *Handbooks in Operations Research and Management Science: Supply Chain Management*. (Eds.) S.C. Graves and A.G. De Kok. Kluwer Academic Publishers, The Netherlands. 229–340.

Federgruen, A. 1993. Centralized planning models for multi-echelon inventory systems under uncertainty. *Logistics of production and inventory. Handbooks in Operations Research and Management Science*, (Eds.) S.C. Graves, A.H.G. Rinnooy Kan, P.H. Zipkin, vol. 4, ch. 3. Elsevier, Amsterdam. 133–173.

Gaur, V., M.L. Fisher, A. Raman. 2005. An econometric analysis of inventory turnover performance in retail services. *Management Science*. **51**(2), 181–194.

Nahmias, S., S.A. Smith. 1993. Mathematical models of retailer inventory systems: A review. *Perspectives in Operations Management*, (Ed.) R.K. Sarin, Kluwer Academic Publishers, MA. 249–278.

Tsay, A.A., S. Nahmias and N. Agrawal. 1999. Modeling supply chain contracts: A review. *Quantitative Models for Supply Chain Management*, (Eds.) S. Tayur, and R. Ganeshan, and M. Magazine, Kluwer Academic Publishers, Norwell, MA. 299–336.

Chapter 2
SUPPLY CHAIN PLANNING PROCESSES FOR TWO MAJOR RETAILERS

Narendra Agrawal and Stephen A. Smith
Department of Operations and MIS, Leavey School of Business, Santa Clara University, Santa Clara, CA 95053, USA

1. INTRODUCTION

This chapter provides descriptions of the supply chain structures and planning processes of two major retailers in the home furnishings sector. These descriptions are based on a series of interviews with senior executives at these two retailers. Our objective is not to provide a comprehensive survey of such retail firms, but rather to describe the structures and planning processes commonly found in this sector and the corresponding implications for supply chain management based on these two case studies.

Home furnishings is one of the most complex areas in retailing, because of the large number of SKUs, the inter-relationships among the SKUs, as well as use of multiple brands and multiple marketing channels targeted to different customer segments. Due to its complexity, we believe that the assortment selection and supply chain management decisions for this sector pose many challenging problems, whose solutions extend beyond the current state of the art. Thus, we hope that documenting the practices for these supply chains will provide a foundation for future methodological research.

Since both companies requested that we not reveal their identities, we will refer to them as Companies A and B. A number of our observations about planning processes were similar at the two retailers. Also, as described later, Company A has a more complex supply chain because it is a multi-channel retailer. Thus, its structure and planning process are more general than Company B. Therefore, rather than presenting two separate case studies, we will discuss them simultaneously, focusing primarily on Company A, while highlighting the differences at Company B.

Company A, with revenues of about $3.5 Billion per year, consists of six different retail brands or "concepts," with a total of the nearly 600 stores in over 40 states in the US. Each brand sells products through its own distinct set of retail stores. For example, while one brand focuses on casual home furnishings, another focuses on cookware essentials, and a third focuses on children's

N. Agrawal, S.A. Smith (eds.), *Retail Supply Chain Management*,
DOI: 10.1007/978-0-387-78902-6_2, © Springer Science+Business Media, LLC 2009

furnishings. In addition, Company A also operates direct-to-consumer channels, with eight different brands of catalogs and six different web sites. A true multi-channel retailer, this firm generates nearly 40% of its revenues from its direct-to-consumer marketing channels.

Company B has yearly revenues of approximately $1 Billion, and operates roughly 300 stores, selling products in the casual home furnishings, housewares, gifts, decorative accessories categories. In contrast to Company A, this retailer is primarily a single channel retailer, selling mostly through stores. Its Internet channel was initiated very recently, and it does not have a catalog channel. Also, the great majority of its products are branded merchandise. Therefore, its supply chain structure is much simpler than Company A's. However, Company B generates a significant fraction of its revenue from foods and beverages, which present special challenges due to the perishable nature of these products.

The number of different SKUs is quite large for both retailers. Within their largest brand, Company A offers roughly 70,000 different SKUs at a given point in time. Company B operates smaller stores (about 18,000 square feet), with approximately 36,000 SKUs at each store. The SKUs are partitioned into categories, such as furniture, home accessories, table top accessories, food and decorative accessories. Within a category, strong demand interactions across SKUs could be expected to occur, e.g., many SKUs may complement or substitute for each other. SKUs across different categories would have weaker and less specific demand interactions. The products vary significantly in their physical characteristics, prices, perishability, seasonality, procurement lead times and country of origin.

The assortment must address two key marketing objectives (1) providing customers with as complete an assortment as possible and (2) providing an assortment that creates attractive presentations. Since stores carry manufacturers' name brands, it is important to provide a comprehensive selection of related items within a given brand, e.g., Sheffield cutlery. Both retailers emphasized that "presentation drives demand" in each of the channels. Therefore, products are often displayed as they might actually appear in a customer's home for maximum advertising impact. In fact, some customers will purchase an entire room as displayed in the store, or will purchase the complete set of items in a tabletop display. In addition, the best types of items to feature in the catalog or Internet presentations may differ from those in the ideal store presentation. For example, a completely furnished room works well in a store, but would be difficult to capture photographically for a catalog. A large assortment of wall hangings shows well in a catalog, but would require too much wall space in a store.

The merchandise featured in each channel's presentation is, of course, only a small subset of the available merchandise. Store and catalog presentations are modified as frequently as every thirty days depending on the seasons of the year. The products offered in the assortments change much less frequently than the presentations, with the majority of the SKUs continuing for at least six months or more. One rapidly changing type of SKU, known as "ornamentation," is seasonal and fashion driven, and thus the ornamentation assortment tends to change with the presentation. Also, some products may be discontinued in their original sales channel, but still continue to be offered through the outlet stores or Internet and catalog channels. Therefore, the presentation requirements lead to additional constraints on both the assortment planning process and the management of the supply chain.

Neither retailer optimizes supply chain costs as part of the product design and assortment selection process. Instead sourcing costs and financial outcomes are viewed as constraints, rather than primary objectives. Supply chain decisions are handled by a sourcing team, which is separate from the design and assortment selection team. In general, the sourcing team is responsible for managing the supply chain as effectively as possible for whatever assortment is chosen. If problems arise, the sourcing team does have some power to initiate assortment modifications later in the planning process, as we discuss in the next section. It is generally recognized that this partitioning of responsibilities is suboptimal, but the problem persists because of the complexity of the decisions.

We note that some of these characteristics of home furnishings supply chains are common to retailers in other areas, which indicates that the structures described here have broader significance. For example, The Gap, similar to Company A, sells its apparel and accessories through a number of different store concepts that include The Gap stores (including Gap Kids, Baby Gap, Gap Outlet and Gap Body), Old Navy, Banana Republic and Piper Lime. While The Gap focuses on casual and fashion apparel and accessories for men and women, Old Navy is positioned for the more value conscious consumer, and Banana Republic is positioned at price points that are higher than The Gap channel. Products are sold through retail stores and the Internet channel for each concept. Similarly, Target operates Target Stores, Mervyns and Dayton Hudson stores, which carry both private label brands and branded merchandise. Internet channels are also associated with each store concept at Target.

The objective of "presenting an attractive assortment" to the consumer is equally important to these retailers as well. For example, it is common practice to display complete apparel and accessory outfits from a given manufacturer, e.g., Ralph Lauren, both in stores and in the Internet channels. It is common knowledge across the retail industry that matching assortments displayed on the cover of catalogs, or displayed prominently in stores, generate a significantly larger level of sales than products stocked on shelves or racks. Thus the assortment selection and presentation design decisions are closely linked across many retail categories.

2. SUPPLY CHAIN DESCRIPTION

Company A's supply chain is illustrated in Figure 2-1. While the supply chain varies somewhat across brands, this figure illustrates the most general case. The overlap across supply chains for the various brands is minimal and

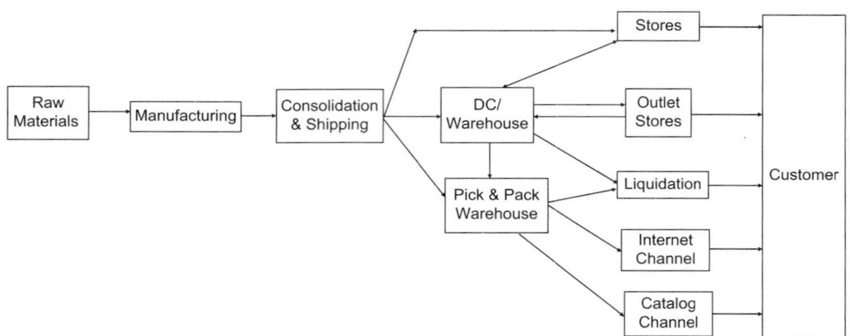

Figure 2-1. Retail Supply Chain for Company A

limited to sharing of warehouse space and merchandise handling capabilities at the distribution center (DC).

Since Company B is primarily a single channel retailer, its supply chain lacks the pick-&-pack warehouse, outlet stores, and the Internet and Catalog channels in the figure above.

Company A's products are sourced from both domestic and foreign suppliers. The foreign suppliers are located in 35 different countries, and are responsible for nearly two-thirds of the total merchandise purchased. A particular brand or concept that offers 60,000–70,000 SKUs may be sourced from as many as 1000 different vendors. Nearly 60% of the products are basics, which continue for at least two selling seasons. The planning calendar consists of four seasons, with the Fall season responsible for the majority of annual sales. Stores may carry both nationally known brands of products as well as private label products. Company B sources its products primarily from foreign vendors. It utilizes about 30 agents to obtain 36,000 SKUs from about 1000 active vendors. 65–70% of its furnishing products and almost 90% of its food products are basic (its core products can have a selling season that is 2–10 years long). It too plans for four separate seasons over the year.

Shipping from foreign sources is primarily by boat, in large metal shipping containers. Containers destined for multiple stores need to be sent to a DC and unpacked. Company A, with the more complex supply chain, operates three such DCs. The largest facility, with nearly 6 million square feet of space, is located in Memphis. It provides replenishments for all the stores, as well the sourcing for the direct-to-consumer shipments for the Internet and catalog channels for all products other than furniture. Furniture, given its physical size, is distributed through two separate distribution centers, one on the East coast and one on the West coast. Store-bound merchandise is then transferred to trucks for delivery. Direct-to-consumer shipments are handled by two independent shipping companies. Company B operates two DCs, one on each coast.

Demand fulfillment for their Internet business, when it is ready, will occur from a separate, outsourced DC on the east coast.

Merchandise can also follow a variety of paths during the selling process. Store customers usually pick up items at the store. But bulky items such as furniture are displayed in the store, while deliveries take place directly from a DC/ warehouse to the customer. In order to combine customer orders and reduce trucking costs, customer delivery time may require a lead time of several weeks. Items that are direct shipped are handled by third party logistics (TPLs) companies and delivered to the customer. Similarly, non-conveyable items that are purchased through the Internet or Catalog channels may ship directly to the customer from the DC/ warehouse. Thus, multiple items that the customer purchases at the same time may be delivered in different ways and at different times. The same customer may also shop in different channels at various times. Thus, the customers' level of satisfaction with their overall shopping experience in one channel will influence their future purchases in other channels. This cross channel interaction is not addressed in selecting inventory service levels.

Certain items in any channel may not sell as well as originally anticipated. Slow sellers or discontinued items in the stores are often sent to one of the retailer's outlet stores, and offered at a reduced price. The outlet channel may also be used for returned merchandise that the retailer does not wish to offer in the regular stores. Merchandise from the regular stores destined for the outlet stores is typically moved first to the DC, where it is consolidated and then allocated to the outlet stores based on their anticipated demands. In order to maintain an attractive presentation and selection in the outlets, about 30% to 40% of the outlet merchandise for Company A is sourced specifically for outlets, and consists of items that are not offered in regular stores. Some items that are no longer carried in stores may continue to be offered through the Internet or Catalog channels. Since customers can retain catalogs for some time, orders will sometimes be filled for items that are no longer carried in the most recent catalog.

3. SUPPLY CHAIN PLANNING PROCESSES

Let us now turn our attention to the various planning processes in these supply chains. We begin by describing a typical planning calendar (Figure 2-2), which can be 12–16 months long, and is implemented in a rolling horizon basis (our description of this calendar is primarily based on our discussions with Company B, although the process is very similar at Company A).

While the details of these steps are presented subsequently, we note that the first key interaction between the merchandising team and supply chain planning team occurs during the step of assortment selection process. As part of this step, not only do the teams formalize the assortment, they also perform financial

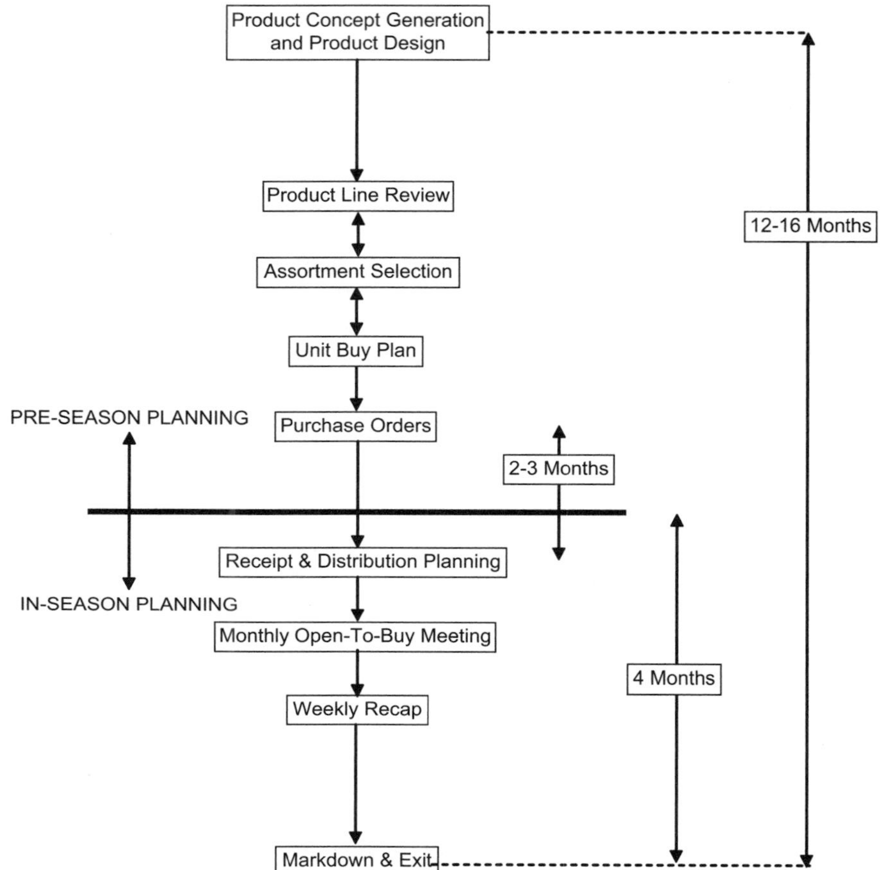

Figure 2-2. Retail Master Calendar

analyses to determine whether the sales targets, specified in the company's financial plans, will be met. This is based on a top-down analysis of the sales forecasts. Following this, when the unit buy plans are created, the teams forecast unit sales at the different stores for given pricing policies. This is a bottom-up analysis. A very important step at this point is the reconciliation between the top-down and bottom-up predictions. This step may lead to a revision of the company targets for sales and margins and/ or modifications in the assortment. These targets are further reviewed at the monthly review meetings, and may be revised, along with targets for initial markup, inventory turns and markdowns.

Decision making in this process tends to consist of a series of "what if" analyses, with little reliance on analytical optimization. Moreover, the process of revising company targets involves addressing a number of tradeoffs, which is often done in a subjective manner. Further, these decisions are greatly influenced by personal incentives. For instance, if the unit buy plans turn out to be higher than the financial targets, the teams would typically simply promise to meet the target, i.e, they would much rather perform better than predicted than to show a shortfall.

3.1 Product design and assortment planning

Retailer A has a highly "vertical structure" with respect to its planning processes. The planners assigned to the various processes tend to be specific to each brand, with minimal overlapping responsibilities across brands. The percentage of private label merchandise is small in the flagship brand, while it is quite high in other brands. Each brand has its own product design teams. As a specific example, in one brand, 40 *product designers* search the world for new product designs and material concepts. Merchandise is divided into a number of different categories, each with its own design team and buyers. The designers present their ideas to the *merchants* and *sourcing* specialists during a product line review process, where they evaluate sketches and samples of products, and consider pricing decisions. Upon approval, these specs are given to independent *sourcing agents*, spread across the world, who seek out the appropriate *vendors* for product prototypes.

Upon receipt of these prototypes, the merchants consider how the assortment as a whole will be presented to the consumer, and suggest appropriate modifications. This is a very important step in the process, since individual product design decisions must be made subject to the constraints and limitations imposed by the whole assortment. The assortment is also reviewed by the *visual and marketing* group, which specializes in creating store presentations. Finally, the products are adopted and handed over to the sourcing and *inventory teams*. The inventory team is responsible for producing high level forecasts, and determining if the product line can deliver its sales and revenue targets. Typically, the elapsed lead time from a new product's concept stage to delivery into the stores is about 12 months.

In this planning process, the central role in assortment decisions is played by merchants. The process architecture is illustrated in Figure 2-3 below, where the merchants are at the hub. Product design groups within a brand tend to work all year round, since about a third of the SKUs tend to be new at Company A each year. The in-store presentation changes frequently, giving consumers the impression of a rapidly changing assortment. Catalogs are also shipped to consumers frequently with different assortments of featured merchandise, corresponding to the season of the year. As noted previously, the total assortment of products in each of these channels turns over much less frequently than the presentations. Finally, the product lines in the three marketing channels overlap somewhat, but also contain many unique products.

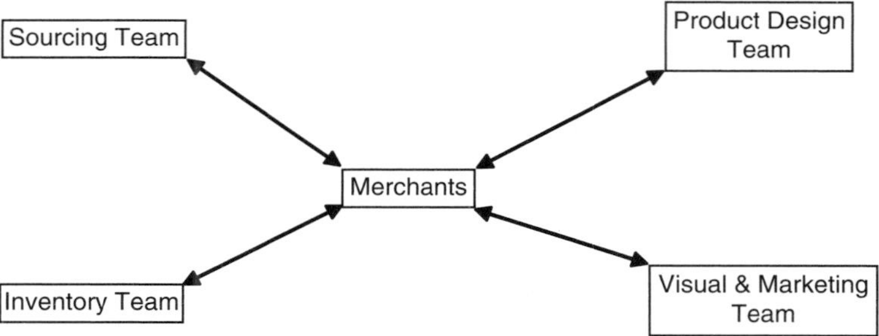

Figure 2-3. Product Design Process Architecture

3.2 Sourcing & vendor selection

As mentioned earlier, due to the retailer's vertical structure, each brand tends to have its own sourcing teams. This is a recognized weakness with regard to sourcing, since it does not exploit the potential synergies due to consolidation of buying across brands. As is typical of most retailers, Company A does not manufacture its own products. In fact, Company A manages most of its vendor interactions through independent agents, who are domain experts. These agents identify the vendors, ensure the ability of these vendors to execute purchase orders in a timely and financially sound manner, implement quality protocols in-line and for final goods, verify packaging, and determine the vendors' social compliance. Ensuring social compliance by vendors is becoming increasingly important for US based retailers, and continues to be a very difficult challenge.

The retailer evaluates the vendors primarily based on their past performance. Vendor evaluation score cards are selected for ongoing vendors, but no such metric is used to evaluate new vendors during the selection process. This retailer's sourcing organizations are generally not involved in the vendors' actual production planning process beyond sharing forecasts and placing purchase orders. This is in contrast to what we have observed at some other retailers who actively engage in vendor capacity planning (Agrawal *et al.* 2002) Also, because of capacity limitations, multiple vendors may sometimes be used for the same product.

The manufacturing process itself may take as long as three to five months to complete. But the manufacturing lead time can be as short as thirty days for products that consist primarily of upholstery or fabrics. The total order quantity for the merchandise is manufactured over a period of time and the goods are typically flowed to the retailer in multiple lots. For core products that are carried over multiple seasons, contracts often allow for modifications in order quantities within certain ranges, depending on the observed demand for the product.

3.3 Logistics planning

As mentioned earlier, shipping from foreign sources is primarily in large metal shipping containers. The shipping time for a container, including delivery to the DC or directly to stores, is from 30 to 40 days, depending on the final destination. Ideally, a shipment to the retailer fills one or more containers exactly. This objective may be used to influence lot size selection for both shipping and manufacturing. The allocation of merchandise across shipping containers can be quite complex. For example, it is highly desirable to have a dedicated shipping container that can be transported directly to a store. At the same time, stores have limited space and holding too much merchandise in the store at one time is not acceptable.

Planning the shipping needs for retailers is a complex but critical activity. For example, at Company B, logistics planning begins right after the merchandising plans are set for the following year. Unfortunately, merchandising plans do not specify how the percentage of imports relative to total purchases will change in the upcoming year. Nor do they specify how product inflows from particular countries may change. This information is important for logistics planning since securing shipping container capacity on specific freight lanes in a timely manner is critical to ensuring delivery reliability. This decision problem is dimensionally complex – Company B utilizes five different steam ship lines and uses about 7,000 40-ft containers annually. In the absence of the detailed capacity requirements, retailers use rudimentary forecasting methods for planning purposes.

Based on these rough forecasts, retailers negotiate rates for shipments with shipping companies. Rate negotiations typically happen in February and March for shipments starting in May through the following April. Contracts typically specify the total number of containers that will be used, with guaranteed minimums, but not the actual timing of the shipments. Rates have been hard to predict in the recent past due to significant uncertainty in the cost of fuel. The average cost of shipping a full container to the US is $3,200, and less-than-container shipments incur roughly a 33% price premium.

Containers destined for multiple stores need to be sent to the DCs to be unpacked. The merchandise is then transferred to trucks for delivery to the stores, which also adds to the shipping time. Retailers typically set aggressive targets for transfer time in the DC, e.g., less than 24 hours turnaround time. Depending on their country of origin and the quantity of items, some merchandise shipments do not fill a whole shipping container. In this case, the shipment is handled by local freight consolidators who pool shipments from multiple retailers. For these items, the retailer also needs to make arrangements for where the container will be unpacked and how the merchandise will be transported to its final destination. In order to facilitate shipping, the container requirements could thus potentially influence the retailer's choice of sourcing location or manufacturer. While the sourcing team at this retailer tries to deal with this problem subjectively, they do not consider the joint optimization of shipping and sourcing decisions in a systematic way.

The retailer also operates a “Pick and Pack” warehouse, where merchandise is “direct shipped” to customers from the Internet and Catalog channels. This requires special packaging that can be done at the manufacturing site. In some cases, the direct ship merchandise comes in larger packages that require additional set up for the automated pick and pack process at the warehouse. An important distinction is made between items that are “conveyable,” i.e., can be put on a conveyer belt and those that are not conveyable (items with very large dimensions or irregular shapes which cannot be handled by automated pick and pack equipment). Again, items shipped from the vendor to the pick and pack facility may not always fill a whole shipping container. In this case, they are combined with other retailers’ merchandise by a consolidator, and later separated and trucked to the pick and pack warehouse.

Shipments from the DCs to stores are primarily by truck. This shipping time was as high as 10 days, but has now shrunk to 2–3 days because of the use of TPLs like UPS. Oversized packages that are not handled by UPS are sent via other independent shippers.

Interestingly, we learned at Company B that domestic shipping can be more onerous than international shipping because the trucking industry in the US is in a state of disarray. We were told that from the retailers’ perspective, the performance of the trucking industry tends to be negatively correlated with the state of the construction industry, because the better the construction industry does, the fewer drivers are available for the trucking industry. Reliability of truck drivers and availability of equipment (trucks) capacity is a constant challenge. Finally, since shipments by trucks often require multiple handoffs due to the hub-and-spoke system used by shippers, numerous errors in shipping information and damages to products are introduced.

Appropriate packaging design is a very important issue for two reasons. First, it affects the probability of damage, which continues to present a significant challenge, especially for bulky items. For some items, the probability of damage was reported to be as high as 1/3 for each loading and unloading cycle. Packaging also affects the handling time and storage space required per item, and the need for repackaging at the DC. In order to minimize the complexity and cost associated with different packaging requirements across the channels, packaging tends to be designed for the most demanding channel (often the catalog/Internet channel). This can increase the product costs in other channels. Some retailers, such as Walmart, have achieved significant cost savings by redesigning their product packaging to facilitate shipping (Plambeck and Denend 2007.)

3.4 **Distribution planning and inventory management**

Company A operates in a centralized planning environment. Store managers do not place merchandise orders, but rely instead on decisions made by central planners. Nearly 50% of goods are on auto-replenishment programs, where replenishments come from the DC/ warehouse. Some branded merchandise can be replenished directly from the vendor. The systems in place for communication between stores and DCs are viewed as satisfactory, but they are still in the process of rolling out EDI linkages with vendors.

The frequency of shipping to stores presents an interesting challenge. Shipping less frequently reduces shipping costs, but increases the size of the shipments. Large shipments can generally be received by stores only before they open for business, which presents considerable staffing challenges. Consequently, smaller and more frequent shipments tend to be preferred, since they can be received by the store during normal working hours. Stores generally maintain only small back-rooms for stocking inventory, and may occasionally also rent off-site lockers for additional storage needs.

Scientific inventory management and demand forecasting is an acknowledged shortcoming of the present system at both of these retailers. Inventory management decisions are often made in an ad hoc manner, using rule of thumb weeks-of-supply (WOS) targets for merchandise at stores and in the DC/ warehouse, without a clear understanding the cost implications of over- or under-stocking. The result tends to be higher than optimal levels of inventory and an annual inventory turnover of less than 2.0 for Company A, which is well below that of some other home furnishings retailers. However, this retailer's strategy focuses on carrying the latest trends in home furnishings together with a fairly high markup. This has produced satisfactory results from a profitability standpoint, but the logistics planners believe that there are significant opportunities for cost reductions.

3.5 Clearance and markdown optimization

As mentioned earlier, unsold or slow-moving items are sent to one of the retailer's outlet stores, or sold through the Internet channel. It is important at some point to clear the discontinued items to make room for new merchandise. One option is to take markdowns at stores, but deeper price markdowns generally occur in outlet stores or on the Internet. A second liquidation option is to sell discontinued merchandise to a discounter, after removing labels that identify its origin. Some items may be donated to charitable organizations, which creates a tax deduction. Still others may simply be discarded.

The logistics planners that we spoke with felt that markdown planning and pricing decisions are not made in a scientific manner by this retailer. Often, the merchandise planners wait too long before implementing markdowns or liquidating products. This is also recognized as an opportunity for improving profits (see chapters 10, 11 and 12 for further discussion of pricing and markdown issues).

3.6 Cross-channel optimization

While Company A has done little to integrate many of the supply chain processes across the various brands, they do make use of cross-channel marketing. For instance, their advertising expense in the traditional print and mass media is minimal. In fact, their catalogs are used as the primary advertising mechanism, with about 400 million catalogs shipped annually. Many of their catalogs are shipped to areas where stores already exist, and this serves as an instrument to drive store traffic. To compensate the catalog channel for this service, which is significantly cheaper than actual advertising, they receive a fixed percentage of store revenue as a fee. Aggregate information about consumers and their buying behavior in the catalog channel is also used in making decisions about store location and for assessing the market potential of new products. This could likely produce additional benefits if cross channel supply chain interactions were included in the decision making process.

4. CONCLUSION

These discussions of the supply chain operations at two home furnishings retailers highlight a wide variety of unsolved analytical problems. One specific problem that is analytically challenging is the optimal use of containers to transport the flow of various quantities of merchandise from different supplier locations to the retailer's DC and stores, subject to delivery scheduling constraints. While some models exist in the literature for optimal container packing (Martello, *et al.*, 2000), the more general problem of optimally using an integer number of containers to deliver a flow of merchandise over time appears to be unsolved. For example, it may be advantageous, based on inventory versus shipping cost tradeoffs, to deliver some merchandise ahead of schedule and store it, in order to achieve the objective of exactly filling a container. A complete container that can be shipped to the retailer's DC avoids the additional expense of consolidation with another retailer's merchandise. A further objective is to ship a complete container directly to a store, if possible.

Chapter 6 in this volume discusses a number of papers that deal with the combined problems of assortment selection and inventory management. But modeling the life cycle costs associated with flowing the merchandise in the assortment through the retailer's complete supply chain is beyond the scope of the currently available methods. For example, how does the assortment selection affect the shipping container and inventory cost tradeoffs discussed above?

Additional aspects of assortment planning and inventory management are the presentation requirements for merchandise in stores and catalogs. Chapter 11 in this volume discusses several papers that have studied the impacts of inventory level on sales. But these models do not address the requirement to feature a combination of items that creates an attractive display. That is, assortment optimization models should somehow include these presentation effects.

The sequential nature of the retailer's decision making process is also an interesting variation on what existing supply chain models tend to assume. That is, assortment decisions are made first, followed by sourcing decisions, inventory ordering decisions, and finally shipping decisions. The timing for these retailer decisions is largely determined by the different lead times associated with each decision. That is, the two retailers described here have elected to postpone each separate decision as long as possible, rather than making them jointly. Conceptually, the overall problem could be modeled as one gigantic dynamic programming problem, but it would clearly be completely intractable. Models that capture the timing of these decisions in a way that includes sequentially updated states of information about demand could potentially be quite useful.

Finally, cross channel optimization clearly offers a number of opportunities for improving supply chain performance at both of these retailers. There are economies of scale across the channels in sourcing, in optimizing shipping containers, and in the use of trucks to deliver shipments to stores, which are currently not being exploited. In many cases, this is because retailers do not have methodologies that can capture these tradeoffs. Cross channel pricing tradeoffs are also important, in particular when a different channel is used to clear the excess merchandise from the original sales channel. There are also cross channel impacts of promotions, some of which are discussed in Kalyanam and Achabal (2005).

In summary, these two case studies illustrate the complexity of retailers' supply chain decisions in practice, and the gaps that exist between the currently available methodologies and the actual decision making environment. We hope that these discussions, as well as the methods and empirical studies presented in this volume, will provide the foundation for future research that will advance the state of the art in retail supply chain management and provide significant additional value for retailers' supply chain operations.

REFERENCES

Agrawal, N., S. A. Smith and A. A. Tsay. 2002. Multi-vendor sourcing in a retail supply chain. *Production and Operations Management*. 11(2) 157–182.

Kalyanam, K. and D. D. Achabal, (2005). Cross-channel optimization: A strategic roadmap for multichannel retailers. Working paper, Retail Management Institute, Santa Clara University, CA.

Martello, S., D. Pisinger and D. Vigo. 2000. The three dimensional bin packing problem. *Operations Research*. 48(2) 256–267.

Plambeck, E. and L. Denend. 2007. Walmart's sustainability strategy. Stanford Graduate School of Business Case, OIT-71.

Chapter 3
THE EFFECTS OF FIRM SIZE AND SALES GROWTH RATE ON INVENTORY TURNOVER PERFORMANCE IN THE U.S. RETAIL SECTOR

Vishal Gaur[1] and Saravanan Kesavan[2]
[1]*Johnson Graduate School of Management, Cornell University, Sage Hall, Ithaca, NY 14853, USA*
[2]*Kenan-Flagler Business School, University of North Carolina at Chapel Hill, Chapel Hill, NC 27599, USA*

1. INTRODUCTION

Inventory constitutes a significant fraction of the assets of a retail firm. Specifically, inventory is the largest asset on the balance sheet for 57% of publicly traded retailers in our dataset.[1] The ratio of inventory to total assets averages 35.1% with buildings, property, and equipment (net) constituting the next largest asset at 31%. Moreover, the ratio of inventory to current assets averages 58.4%. Inventory is not only large in dollar value but also critical to the performance of retailers. For example, according to Standard & Poor's industry survey on general retailing (Sack 2000), "Merchandise inventories are a retailer's most important asset, even though buildings, property and equipment usually exceed inventory value in dollar terms." Thus, the importance of improving inventory management in retail trade cannot be overemphasized.

The signals that managers and analysts use to determine how well a retailer is managing its inventory include inventory turnover (defined as the ratio of cost of goods sold to average inventory), inventory growth rate, and payables to inventory ratio. The statistics for these variables are publicly available from the financial statements of those retailers that are listed on the stock exchange (NYSE, AMEX or NASDAQ). Such data can be used to study a variety of research questions regarding the inventory productivity performance of each firm and of the retail sector as a whole.

---

[1] The data set consists of a large cross-section of US public listed retailers for the time-period 1985–2003. The data set is summarized in Section 3.

N. Agrawal, S.A. Smith (eds.), *Retail Supply Chain Management*,
DOI: 10.1007/978-0-387-78902-6_3,

For example, Gaur, Fisher and Raman (2005), henceforth referred to as GFR, conduct a descriptive investigation of inventory turnover performance of publicly listed U.S. retailers. They find that inventory turnover varies widely not only across firms but also within firms over time. They further show that a large fraction of the variation in inventory turnover can be explained by three performance variables obtained from public financial data: gross margin (the ratio of gross profit net of markdowns to net sales), capital intensity (the ratio of average fixed assets to average total assets), and sales surprise (the ratio of actual sales to expected sales for the year). They use the estimation results to propose a metric for benchmarking inventory productivity of retail firms.

In this paper, we extend the model of GFR to investigate the effects of firm size and sales growth rate on inventory turnover performance of U.S. retailers. The EOQ and newsvendor models, commonly used in theoretical operations management, show that inventory turnover should increase with the size of a firm due to economies of scale and scope. Several factors contributing to economies of scale and scope have been studied in the operations management literature, including statistical economies of scale (Eppen (1979), Eppen and Schrage (1981)), fixed costs in inventory and transportation models, and demand pooling effects in product variety. However, to our knowledge, there are no research papers using real data to estimate the effect of size on inventory turnover. Our results provide such estimates for retailers.

The relationship between sales growth rate and inventory turnover, while not directly studied in the academic literature, is commonly tracked by managers and analysts. For example, the aforementioned industry survey on general retailing by Standard & Poor's (Sack 2000) states that year-over-year growth in inventory should be in line with sales growth rate; if inventory growth exceeds sales growth rate, then it may be a warning that stores are over-stocked and vulnerable to markdowns. Raman et al. (2005) present a case study of a hedge fund investor who uses the ratio of sales growth rate to inventory growth rate as one of the metrics in making investment decisions on retail stock. The case presents several examples from financial performance of firms to illustrate this metric. It also makes a separate point that this relationship is ignored by financial investors. In this paper, we focus on examining evidence for the relationship of sales growth rate with inventory turnover, but do not assess its use by investors. We motivate this relationship using the operations management literature by using an instance of the newsboy model. For our analysis, we do not directly work with sales growth rate because we use a logarithmic regression model which precludes negative values of sales growth rate. Instead, we conduct our analysis using sales ratio, which we define as the ratio of sales in the current year to sales in the previous year.

The main results of our paper are as follows. First, we find that inventory turnover is positively correlated with firm size where size is defined as annual firm sales in the previous year. On average, in our data set, a 1% increase in firm size is associated with a 0.035% increase in inventory turnover (statistically significant at $p<0.0001$). We find evidence of diminishing returns to size: inventory turnover increases with size at a slower rate for large firms than for small firms. These results present evidence in support of the existence of economies of scale and scope in a retail setting.

Next, we find that inventory turnover is positively correlated with sales ratio. A 1% increase in this ratio is associated with a 0.38% increase in inventory turnover in our data set. We also find that inventory turnover is more sensitive to sales ratio when a firm is experiencing sales decline than when a firm is experiencing sales growth. A 1% increase in sales ratio is associated with 0.67% increase in inventory turnover when sales are declining and with 0.19% increase in inventory turnover when sales are increasing. Our results suggest that firms would find it harder to improve inventory turnover performance during periods of sales decline than during periods of sales growth. Thus, firms should use their forecast of future sales ratio to determine the amount of attention to give to inventory management.

The third main result of this paper is achieved through re-testing the hypotheses in GFR regarding gross margin, capital intensity and sales surprise on our data set. We test these hypotheses again because we use a larger and more recent data set than GFR. Our results for these tests are consistent with those obtained by GFR. We find that inventory turnover is negatively correlated with gross margin and positively correlated with capital intensity and sales surprise.

Our paper contributes to the academic literature by extending the methodology in GFR for empirical research on inventory productivity in retailing. We find that a significant fraction of the variation in inventory turnover for retailers can be explained by the selected performance variables. The models used in this paper and in GFR are useful to retail managers for comparing inventory turnover performance across firms and for a firm over time. They are also useful in helping retailers estimate inventory turnover as a function of their future growth, profit margin, and capital investment projections. With respect to the effects of firm size and sales ratio on inventory turnover, we describe several factors, based on the literature, which would imply either positive or negative correlations between size and inventory turns as well as between sales ratio and inventory turns. Thus, we set up competing hypotheses, and our tests enable us to state which of these effects will dominate. We believe that there is considerable scope for future research on these topics, and our results represent a first step.

The rest of this paper is organized as follows. Section 2 reviews the relevant literature; Section 3 describes our data set; and Section 4 summarizes the empirical model and findings from GFR that are useful in this paper. Section 5 presents our hypotheses, followed by the estimation model in Section 6, and the estimation results in Section 7. Finally, Section 8 discusses the limitations of our analysis and directions for future research.

2. LITERATURE REVIEW

In the recent years, several research papers in operations management have addressed questions on performance and drivers of performance through analysis of firm-level inventories, industry-level inventories and other financial data. The data typically used in these studies are provided by the U.S. Census Bureau, Standard & Poor's Compustat database, Center for Research in Security Prices (CRSP) database, and news archives such as ABI/Inform and LexisNexis. We review empirical research on operational performance relevant to our paper. This includes event-studies, panel data models, models relating inventory turnover to financial returns, and studies using case-based data.

Hendricks and Singhal (1996, 1997, 2001) conduct event-based studies to assess the impact of operational decisions on firm performance using public financial data. They find that firms that receive quality awards outperform a control sample on operating-income based measures such as increase in operating income, improvement in operating ratios, etc. Further they find that the stock market reacts positively when firms receive quality awards, and that there is a positive correlation between implementation of TQM principles and long term financial health of firms. Hendricks and Singhal (2005) investigate the effects of supply chain glitches on the financial performance of firms. Using a sample of 885 glitches announced by publicly traded firms, they compare changes in various operating performance metrics for the sample firms against a sample of control firms of similar size and from similar industries. They find that firms that experience glitches report lower sales growth, higher growth in cost, and higher growth in inventories relative to controls. Further, firms do not quickly recover from the negative economic consequences of glitches.

While the above papers study the impact of various operational events on firm performance, other researchers have sought to conduct time-series analysis of operational performance variables of firms. Rajagopalan and Malhotra (2001) investigate whether inventory turns for manufacturers have decreased with time due to the adoption of JIT principles. They study time trends in each of raw material inventory, work-in-process inventory and finished-goods inventory using aggregate industry-level time-series data from the U.S Census Bureau for 20 industrial sectors for the period 1961–1994. They find that raw material and work-in-process inventories decreased in a majority of industry sectors. However they do not find any overall trends in finished good inventories.

Chen et al. (2005) use firm-level inventory data from publicly traded manufacturing firms for the period 1981–2000 to study trends in inventory levels for each of raw material inventory, work-in-process inventory and finished-good inventory. They find that raw-material and work-in-process inventories have declined significantly while finished-goods inventory remained steady during this period. These results are consistent with Rajagopalan and Malhotra (2001) although, notably, the two papers use data with different granularity. Chen et al. also investigate whether abnormal inventory predicts future stock returns. Using the three-factor time-series regression model of stock returns (Fama and French 1993), they find that abnormally high and abnormally low inventories are associated with abnormally poor long-term stock returns.

Gaur et al. (1999) relate inventory turnover performance with stock returns of US retailers using a long-term contemporaneous analysis. They show that for time periods varying in length from 5 to 20 years, the cross-section of average stock returns is significantly positively correlated with average annual inventory turnover over the same period (controlling for gross margin). They also show that the cross-section of inventory turnover is negatively correlated with gross margin.

Gaur et al. (2005) use financial data for retail firms to investigate the correlation of inventory turnover with gross margin, capital intensity and sales surprise in a longitudinal study. They state that changes in inventory turnover cannot be directly interpreted as performance improvement or deterioration because they may be caused by changes in product portfolio, pricing, demand uncertainty, and many other firm-specific and environmental characteristics. They propose a benchmarking methodology that combines inventory turnover, gross margin, capital intensity and sales surprise to provide a metric of inventory productivity, which they term as adjusted inventory turnover.

Rumyantsev and Netessine (2007) use quarterly data from over 700 public US companies to test some of the theoretical insights derived from classical inventory models developed at the SKU level. They use proxies for demand uncertainty and lead time, and use a longitudinal study to show that inventory levels are positively correlated with demand uncertainty, lead times, and gross margins. The authors also find evidence for economies of scale as larger firms carry relatively lower levels of inventory compared to smaller firms.

Several researchers have studied the effects of specific operational decisions on firm performance. For example, Balakrishnan et al. (1996) study the effect of adoption of just-in-time (JIT) processes on return on assets (ROA). They compare a sample of 46 firms that adopted JIT processes against a matched sample of 46 control firms that did not. They do not find any significant ROA response to JIT adoption. Billesbach and Hayen (1994), Chang and Lee (1995), and Huson and Nanda (1995) study the impact of adopting JIT processes on inventory turns. Lieberman and Demeester (1999) study the impact of JIT processes on manufacturing productivity in the Japanese automotive industry. Their study suggests that reduction in inventory brought about by JIT practices enabled the firms to improve their productivity.

Our paper contributes to this research stream by extending Gaur et al. (2005) and Rumyantsev and Netessine (2007). We discuss various factors that could cause positive or negative correlations of size and sales growth rate with inventory turnover, and provide evidence regarding the existence of economies of scale and scope in retailing as well as the effect of growth rate of firms on their inventory turnover performance. Our results are useful to retailers to assess their performance changes over time.

3. DATA DESCRIPTION

We use financial data for all publicly listed U.S. retailers for the 19-year period 1985–2003 drawn from their annual income statements and quarterly and annual balance sheets. These data are obtained from Standard & Poor's Compustat database using the Wharton Research Data Services (WRDS).

The U.S. Department of Commerce assigns a four-digit Standard Industry Classification (SIC) code to each firm according to its primary industry segment. For example, the SIC code 5611 is assigned to the category "Men's and Boys' Clothing and Accessory Stores", 5621 is assigned to "Women's Clothing Stores", 5632 to "Women's Accessory and Specialty Stores", etc. We group together firms in similar product groups to form ten segments in the retailing industry. For example, all firms with SIC codes between 5600 and 5699 are collected in a single segment called apparel and accessories. Table 3-1 lists all the segments, the corresponding SIC codes, and examples of firms in each segment.

Table 3-1. Classification of Data into Retail Segments Using SIC Codes

Retail Industry Segment	SIC Codes	Number of firms	Number of observations	Examples of firms
Apparel And Accessory Stores	5600–5699	75	944	Ann Taylor, Filenes Basement, Gap, Limited
Catalog, Mail-Order Houses	5961	51	540	Amazon.com, Lands End, QVC, Spiegel
Department Stores	5311	26	374	Dillard's, Federated, J. C. Penney, Macy's, Sears
Drug & Proprietary Stores	5912	23	254	CVS, Eckerd, Rite Aid, Walgreen
Food Stores	5400, 5411	62	756	Albertsons, Hannaford Brothers, Kroger, Safeway
Hobby, Toy, And Game Shops	5945	11	118	Toys R Us
Home Furniture & Equip Stores	5700, 5712	24	260	Bed Bath & Beyond, Linens N' Things
Jewelry Stores	5944	17	210	Tiffany, Zale
Radio, TV, Consumer Electronics Stores	5731, 5734	20	276	Best Buy, Circuit City, Radio Shack, CompUSA
Variety Stores	5331, 5399	44	514	K-Mart, Target, Wal-Mart, Warehouse Club
Aggregate statistics		353	4246	

(a) Income Statement

	Notation	*Amount ($)*
Sales (net of markdowns)	S	100
Cost of Goods Sold (includes Occupancy and Distribution Costs)	CGS	(60)
Gross Profit		40
Selling, General & Administrative Expenses	SGA	(20)
Operating Profit	EBITDA	20
Depreciation & Amortization Expenses		(5)
Interest Costs		(6)
Profit Before Tax	PBT	9
Taxes		(4)
Net Profit	PAT	5

(b) Balance Sheet

Assets			*Liabilities*		
Fixed Assets (includes Owned Property and Capitalized Leases)	FA	30	Owner's Equity (includes Retained Earnings)	OE	40
Cash		15			
Inventory	Inv	45	Long-term Debt	LTD	20
Accounts Receivable		10	Accounts Payable		40
Total Assets	TA	100	Total Liabilities		100

Figure 3-1. Simplified View of Income Statement and Balance Sheet of a Retail Firm

Figure 3-1 presents a simplified view of an income statement and balance sheet that emphasizes the principal variables of interest in this paper. From Compustat annual data for firm i in segment s in year t, let S_{sit} denote the sales net of markdowns in dollars (Compustat annual field Data12), CGS_{sit} denote the corresponding cost of goods sold (Data41), and $LIFO_{sit}$ be the LIFO reserve (Data240). From Compustat quarterly data for firm i in segment s at the end of quarter q in year t, let GFA_{sitq} denote the gross fixed assets, comprised of buildings, property, and equipment (Compustat quarterly field Data118), and Inv_{sitq} denote the inventory valued at cost (Data38). From these data, we compute the following performance variables:

$$\text{Inventory turnover (also called inventory turns)}, \mathrm{IT_{sit}} = \frac{\mathrm{CGS_{sit}}}{\left(\frac{1}{4}\sum_{=1}^{4}\mathrm{Inv_{sitq}}\right) + \mathrm{LIFO_{sit}}},$$

$$\text{Gross margin}, \mathrm{GM_{sit}} = \frac{\mathrm{S_{sit}} - \mathrm{CGS_{sit}}}{\mathrm{S_{sit}}},$$

$$\text{Capital intensity}, \mathrm{CI_{sit}} = \frac{\sum_{q=1}^{4}\mathrm{GFA_{sitq}}}{\sum_{q=1}^{4}\mathrm{Inv_{sitq}} + 4\cdot\mathrm{LIFO_{sit}} + \sum_{q=1}^{4}\mathrm{GFA_{sitq}}}, \text{ and}$$

$$\text{Sales ratio}, \mathrm{g_{sit}} = \frac{\mathrm{S_{sit}}}{\mathrm{S_{si,t-1}}}$$

It is useful to note the following aspects of the measurement of these variables.

1. The Compustat database identifies ten methods for inventory valuation. Four of these are commonly used by retailers: FIFO (first in first out), LIFO (last in first out), average cost method, and retail method. The LIFO reserves of a firm vary depending on the method of valuation used, and adding back the LIFO reserves provides us a FIFO valuation of inventory.
2. The cost of goods sold line on the income statement comprises a number of expenses other than the purchase cost of merchandise. Costs of warehousing, distribution, freight, occupancy, and insurance can all be included in $\mathrm{CGS_{sit}}$. Further, the components of $\mathrm{CGS_{sit}}$ may vary from company to company. Most commonly, occupancy costs may be a separate line item on the income statement rather than being included in $\mathrm{CGS_{sit}}$. This lack of uniformity in reporting reduces the comparability of results among retailers. Thus, we restrict our analysis to comparisons within firm. Compustat indicates whether a firm changed its accounting policies with respect to a particular variable during a year; it provides footnotes to variables containing this information. We use these footnotes to identify firms that underwent accounting policy changes, and exclude them from our sample.
3. In the computation of inventory turns and capital intensity, we calculate average inventory and average gross fixed assets using quarterly closing values in order to control for systematic seasonal changes in these variables during the year. LIFO reserves are reported annually. We add the annual LIFO reserves to the average quarterly inventory to compute average inventory.

After computing all the variables, we omit from our data set those firms that have less than five consecutive years of data available for any sub-period during 1985–2003; there are too few observations for these firms to conduct time-series analysis. These missing data are caused by new firms entering the industry

Table 3-2. Summary Statistics of the Variables for each Retailing Segment

Retail Industry Segment (1)	Number of firms (2)	Number of annual observations (3)	Average Annual Sales($ million) (4)	Average Inventory Turnover (5)	Average Gross Margin (6)	Average Capital Intensity (7)	Average Sales Ratio (8)	Median Inventory Turnover (9)	Median Gross Margin (10)	Median Capital Intensity (11)	Median Sales Ratio (12)
Apparel And	75	944	1201.8	4.60	0.36	0.60	1.14	4.20	0.35	0.62	1.10
Accessory				2.15	0.08	0.14	0.38				
Stores											
Catalog, Mail- Order	51	540	489.3	8.63	0.38	0.50	1.62	5.39	0.38	0.50	1.18
Houses				9.10	0.17	0.18	2.83				
Department Stores	26	374	7068.6	4.61	0.34	0.65	1.08	3.55	0.35	0.66	1.05
				3.93	0.07	0.12	0.20				
Drug & Proprietary	23	254	2327.6	5.26	0.29	0.48	1.17	4.37	0.29	0.50	1.11
Stores				2.91	0.07	0.12	0.28				
Food Stores	62	756	5518.4	10.81	0.26	0.76	1.08	9.98	0.26	0.77	1.05
				4.41	0.06	0.08	0.20				
Hobby, Toy, And	11	118	1638.1	3.16	0.34	0.47	1.18	2.75	0.35	0.45	1.14
Game Shops				1.31	0.07	0.14	0.30				
Home Furniture &	24	260	391.1	4.76	0.42	0.58	1.23	3.11	0.43	0.57	1.13
Equip Stores				7.40	0.09	0.15	0.61				
Jewelry Stores	17	210	485.0	3.18	0.41	0.39	1.12	1.54	0.46	0.37	1.10
				7.04	0.15	0.14	0.36				
Radio,TV, Cons	20	276	1779.1	4.27	0.32	0.45	1.17	3.97	0.30	0.46	1.14
Electr Stores				1.71	0.12	0.10	0.24				
Variety Stores	44	514	7763.8	4.63	0.29	0.52	1.12	3.72	0.28	0.52	1.10

Table 3-2. (continued)

Retail Industry Segment (1)	Number of firms (2)	Number of annual observations (3)	Average Annual Sales($ million) (4)	Average Inventory Turnover (5)	Average Gross Margin (6)	Average Capital Intensity (7)	Average Sales Ratio (8)	Median Inventory Turnover (9)	Median Gross Margin (10)	Median Capital Intensity (11)	Median Sales Ratio (12)
				3.00	0.09	0.15	0.22				
Aggregate statistics	353	4246	3222.8	6.14	0.33	0.58	1.19	4.36	0.32	0.59	1.09
				5.53	0.11	0.17	1.05				

Note: The values given in columns (5)–(8) are the mean and standard deviation of each variable for the respective segment. The "Aggregate statistics" row refers to the complete data set.

during the period of the data set, and by existing firms getting de-listed due to mergers, acquisitions, liquidations, etc. Further, we omit firms that had missing data or accounting changes other than at the beginning or the end of the measurement period. These missing data are caused by bankruptcy filings and subsequent emergence from bankruptcy, leading to fresh-start accounting.

Our final data set contains 4,246 observations across 353 firms, an average of 12.03 years of data per firm. Table 3-2 presents summary statistics by retailing segment for the performance variables used in our study. It lists the mean, median and standard deviation by segment for each variable. Observe that food retailers have the highest median inventory turns of 10.0 and the lowest median gross margin of 0.26. On the other hand, jewelry retailers have the lowest median inventory turns of 1.54 and the highest median gross margin of 0.46. Also note that the coefficient of variation of inventory turnover (the ratio of standard deviation of IT_{sit} to mean IT_{sit}) is quite high: it is larger than 50% for six out of ten retail segments and its average value across all segments is 74%. This statistic shows that inventory turnover has a large variation even within each retail segment. Table 3-3 shows the Pearson correlation coefficients for ($\log IT_{sit} - \log IT_{si}$), ($\log GM_{sit} - \log GM_{si}$), ($\log CI_{sit} - \log CI_{si}$), ($\log S_{si,t-1} - \log S_{si}$) and ($\log g_{sit} - \log g_{si}$) for our data set. Here, we use log-values of all variables because we shall construct a multiplicative regression model in the rest of this paper. We compute the correlation coefficients for mean-centered log-values of variables because our model seeks to explain intra-firm variation in inventory turns. Mean centering is done by subtracting out the mean for each variable for each firm from the data columns; for example, $\log IT_{si}$ denotes the average of $\log IT_{sit}$ for firm i in segment s. Notice that ($\log IT_{sit} - \log IT_{si}$) is negatively correlated with ($\log GM_{sit} - \log GM_{si}$) and ($\log S_{si,t-1} - \log S_{si}$), and positively correlated with ($\log CI_{sit} - \log CI_{si}$) and ($\log g_{sit} - \log g_{si}$). Testing hypotheses on

Table 3-3. Pearson Correlation Coefficients for all Mean-Centered Variables

	$\log GM_{sit} - \log GM_{si}$	$\log CI_{sit} - \log CI_{si}$	$\log S_{si,t-1} - \log S_{si}$	$\log g_{sit} - \log g_{si}$
$\log IT_{sit} - \log IT_{si}$	–0.2747	0.1762	–0.04269	0.2651
	<.0001	<.0001	0.0081	<.0001
$\log GM_{sit} - \log GM_{si}$		0.0514	–0.0102	0.0509
		0.0014	0.5265	0.0016
$\log CI_{sit} - \log CI_{si}$			0.2501	–0.1830
			<.0001	<.0001
$\log S_{si,t-1} - \log S_{si}$				–0.4838
				<.0001

Note: For every pair of variables, the Table provides the Pearson's correlation coefficient and its p-value for the hypothesis H_1: $|\rho| \neq 0$.

these correlations will require a multivariate model which is discussed in subsequent sections.

4. ADJUSTED INVENTORY TURNOVER

GFR study the correlation of inventory turnover with gross margin, capital intensity and sales surprise using data for 311 publicly listed U.S. retailers for the period 1985–2000. In their paper, gross margin, and capital intensity are defined as shown in §3. Sales surprise, denoted SS_{sit}, is defined as the ratio of current year sales to the forecast of current year sales, where the forecast is computed by GFR using a time-series forecasting method. GFR hypothesize that inventory turnover is negatively correlated with gross margin, and positively correlated with capital intensity and sales surprise.

GFR use the following empirical model to test their hypotheses:

$$\log IT_{sit} = F_i + c_t + b_s^1 \log GM_{sit} + b_s^2 \log CI_{sit} + b_s^3 \log SS_{sit} + \varepsilon_{sit}. \quad (1)$$

Here, F_i is the time-invariant firm-specific fixed effect for firm i, c_t is the year-specific fixed effect for year t, b^1_s, b^2_s, b^3_s are the coefficients of log GM_{sit}, log CI_{sit}, and log SS_{sit}, respectively, for segment s, and ε_{sit} denotes the error term for the observation for year t for firm i in segment s. The hypotheses of GFR imply that, for each segment s, b^1_s must be less than zero, and b^2_s and b^3_s must be greater than zero. The main features of this model are as follows:

1. The model has a log-linear specification. Thus, it is assumed that a multiplicative model is suitable to represent the relationship between inventory turns, gross margin, capital intensity and sales surprise. This assumption is supported in GFR with simulation analysis.
2. The model includes an intercept for each firm in order to control for differences across firms. Note from the discussion in §3 that inventory turnover may not be comparable across firms due to differences in accounting policies for cost of goods sold. Other factors that can confound comparisons across firms include differences in managerial efficiency, marketing, real estate strategy, etc. Since data on these factors are omitted in GFR, attention is focused on year-to-year variations within a firm only. We call such a model an intra-firm model.

GFR find strong support for all three hypotheses in their data set. Based on these results, they propose a tradeoff curve that computes the expected inventory turnover of a firm for given values of gross margin, capital intensity, and sales surprise. They term the distance of the firm from its tradeoff curve as its *Adjusted Inventory Turnover*, denoted AIT, and use it as a metric for benchmarking inventory productivity of retailers by controlling for differences in gross margin, capital intensity, and sales surprise. The value of AIT for firm *i* in segment *s* in year *t* is computed as

$$\log \mathrm{AIT}_{sit} = \log \mathrm{IT}_{sit} - b^1 \log \mathrm{GM}_{sit} - b^2 \log \mathrm{CI}_{sit} - b^3 \log \mathrm{SS}_{sit} \quad (2)$$

or, equivalently, as

$$\mathrm{AIT}_{sit} = \mathrm{IT}_{sit}(\mathrm{GM}_{sit})^{-b^1}(\mathrm{CI}_{sit})^{-b^2}(\mathrm{SS}_{sit})^{-b^3} \quad (3)$$

Note that log AIT_{sit} is equal to the sum of the fixed effects terms, F_i and c_t, and the residual error, ε_{sit}, in (1). Thus, it captures the amount of variation in log IT_{sit} that is not explained by the regressors in (1). According to these results, managers of firms with low AIT should investigate whether their firms are less efficient than their peers, and identify steps they might take in order to improve their inventory productivity.

We employ the methodology from GFR in this paper. In particular, we use an intra-firm model with a log-linear specification. We use log GM_{sit} and log CI_{sit} as control variables for testing our hypotheses because GFR found them to be correlated with log IT_{sit} and they may further be correlated with firm size and sales ratio. We, however, do not use sales surprise in our model because data on managements' forecasts of sales are not available to us. If we were to estimate sales forecasts using our own time-series forecasting methods, then log SS_{sit} and log g_{sit} would be highly correlated and cause collinearity in the model. Hence, in the model in this paper, we replace log SS_{sit} by log g_{sit}.

5. HYPOTHESES

In this section, we discuss various reasons why inventory turnover can be correlated with firm size and sales ratio. We find that there are arguments in favor of both positive and negative correlation between inventory turns and size as well as between inventory turns and sales ratio. We also find that the effects of size and sales ratio on inventory turnover can vary across firms depending on their supply chain characteristics, business environment and growth strategy. Thus, we identify the mediating variables that are expected to cause size and sales ratio to be correlated with inventory turnover. Since we do not have data on the mediating variables, our hypotheses are limited to testing which effects dominate, positive or negative. We set up competing hypotheses to test these effects. The task of identifying the causes of these correlations is deferred to future research.

5.1 Effect of firm size on inventory turnover

We explain arguments for inventory turnover to be positively correlated with size using the effects of economies of scale and scope. We also discuss hindrances to economies of scale and scope that may reduce their effect or cause a negative correlation. Subsequently, we frame competing hypotheses to test the sign of correlation between inventory turnover and size. We measure size by the mean annual sales of the retailer lagged by one year, i.e., $S_{si,t-1}$ is the measure of size for year t for firm i in segment s.

Economies of scale and scope can manifest themselves for each item, or in a growth of number of stores, or in a growth of number of items at each retail location. In all three cases, we would expect inventory to increase less than linearly in sales, so that size and inventory turnover would be positively correlated. In the first case, if the mean demand for items at a retail location increases and the retailer maintains a fixed service level, then its safety stock requirement at the location increases less than proportionately because standard deviation of demand typically increases in the square root of mean demand. This relationship is precise when demand follows a Poisson distribution. For other distributions, this relationship has been tested by estimating the first two moments of the distribution. For example, Silver et al. (1998: p.126, 342) estimate the standard deviation of demand as $\sigma = a \cdot (\text{mean})^b$. They state that $0.5<b<1$ is typical and "this relationship has been observed to give a reasonable fit for many organizations."[2] As another example, Gaur et al. (2007) estimate the relationship among analysts' forecasts of total sales of firms, actual sales realizations and standard deviation of total sales. Their results are consistent with Silver et al. (1998), with the average estimated value of b across several data sets being 0.71. Therefore, if safety stock increases less rapidly than cycle stock as sales increase, then inventory turnover should increase with the size of each location due to economies of scale.

Second, inventory turnover should increase with sales when a retailer expands its geographical market by opening new retail locations which are served by existing warehouses or distribution centers. Eppen (1979) and Eppen and Schrage (1981) showed how pooling inventory in a centralized location can lead to a reduction in safety stock due to risk pooling. In their models, safety stock grows as $\sqrt{n}$ in the number of locations n if inventory is pooled at a central location rather than distributed across the n locations. Thus, as a firm adds new retail locations, it can achieve a more than proportionate reduction in its inventory level, and a corresponding increase in inventory turnover due to economies of scale in its distribution network.

Third, as a retailer grows in size, it is able to provide more frequent shipments to its stores due to economies of scale and/or economies of scope in fixed

[2] This section of Silver et al. (1998) focuses on estimation of demand uncertainty. It does not refer to this relationship as economies of scale.

replenishment costs as explained by the EOQ model. For example, such economies of scale and scope can be realized in transportation costs through better utilization of labor and transportation capacity. They would result in an increase in inventory turnover with the size of the firm.

The above three contributing factors may exist for different firms in different years in varying measures depending on the actions taken by the firms. For example, suppose that a firm increases size in a particular year by adding more products to its assortment without affecting the demand for existing products. For this action, the third argument would contribute to economies of scope, but the first and second arguments would not apply. Our hypotheses do not specify the above three effects separately, but instead specify the average tendency across the cross-section of retail firms for the years included in our data set. This implies that any differences in economies of scale and scope across firms or over time will contribute to the residuals in our model.

Apart from differences across firms, there could be hindrances to economies of scale and scope that may result in a negative correlation between size and inventory turns. First, economies of scale and scope require that a retailer's supply chain infrastructure have excess capacity. For example, distribution centers should be able to meet the requirements of new stores being added, and transportation logistics should be able to handle increase in volume of shipments. If a retailer does not have excess capacity in its supply chain infrastructure, it may need to add new capacity in order to grow. Such hindrances may create diseconomies of scale, implying that size and inventory turnover may be negatively correlated with each other. Second, it is often harder to manage a large firm than a small firm because their operations are more complex. Thus, firms may be unable to exploit operational synergies as they grow in size.[3]

Thus, the above discussion shows that a number of hypotheses can be formulated to estimate different drivers of economies of scale and scope effects among retailers. As a first step, we test the following hypotheses.

Hypothesis 1(a). *Inventory turnover of a firm is positively correlated with changes in its size.*

Hypothesis 1(b). *Inventory turnover of a firm is negatively correlated with changes in its size.*

Here, we use the retailer's sales lagged by one year as a measure of size. Our hypotheses may also be set up using relative sales, i.e., the ratio of sales lagged

[3] A counter argument is that as a retailer increases in size, it might have better forecasting tools and thus, might be better able to get the right product to the right place (and therefore, increase turns). Retailers' ability to forecast may even vary non-linearly in size: they may be really good at forecasting when they are very small (not listed publicly, and hence, omitted from our data set), have difficulty as they grow and until they have reached a size such that they have good systems in place and are incorporating sophisticated decision support tools. We incorporate such differences in systems in our model by using capital intensity as a control variable.

by one year to sales at the beginning of the time horizon for the firm. Since we use an intra-firm model, these two measures of size are equivalent.

5.2 Effect of sales ratio on inventory turnover

We identify reasons why sales ratio can be either positively or negatively correlated with inventory turnover. We construct both arguments using the newsboy model.

First consider the arguments for a positive correlation between sales ratio and inventory turnover. Consider a given retailer with known sales in period t-1 making inventory decisions for the next period, t. The retailer first determines the inventory level, q, for an item and then fulfills random demand over one period. Given the value of q, as realized demand increases, sales increase, and thus, sales ratio increases. Further, as realized demand increases, the retailer's average inventory over the period declines. Thus, its inventory turns increase. This implies a positive correlation between sales ratio and inventory turnover. We call this reasoning the *positive effect* of sales ratio on inventory turnover.

Now suppose that the retailer increases q in order to target a higher sales growth rate. As q increases, expected sales increase, and thus, expected sales ratio increases. However, it can also be shown that as q increases, average inventory increases more than proportionately than sales, and expected inventory turnover declines. Alternatively, a retailer may reduce q in order to improve its cash flows. In such a case, the retailer would find its expected inventory turns increasing, but expected sales and expected sales ratio decreasing. This implies a negative correlation between sales ratio and inventory turnover. We call this reasoning the *negative effect* of sales ratio on inventory turnover.

We now try to characterize the situations in which one or the other of these two effects will dominate. Changes in the inventory level or the service level of a retailer can be driven by a number of factors. There is extensive literature on how firms forecast sales growth. Makridakis and Wheelwright (1998) state that organizations need to consider several factors such as overall economy, their customers, distributors, competitors, etc. Further from an operations standpoint the firm needs to take into account its inventory levels, capacity constraints, ability to procure inventory from its suppliers, etc. before forecasting sales growth. Once a sales growth rate has been forecasted for the firm it plans to meet this target. The firm has competing objectives in setting its sales growth rate. Some of the common goals are profits, return on investment, market share, product leadership, etc. Hence, it is possible that the overall strategy of the firm may dictate growth while maintaining or improving inventory turnover or it may require the firm to pursue growth at the cost of excess inventory in the short-term.

For example, suppose that a retailer has a large untapped market potential. This is not an uncommon situation because a retailer cannot realize its full market potential overnight. Instead, its growth rate is limited by its capacity to hire and train employees, add new stores, and expand various functions of its organization such as distribution logistics, merchandising, accounting, information systems, etc. Thus, the growth rate of such a retailer can be restricted by its capacity and budget constraints. We expect that for such a retailer, sales could exceed inventory hence the *positive effect* will dominate so that there will be a positive correlation between sales ratio and inventory turnover.

Alternatively, consider a retailer that is close to saturating its market and has a small untapped market potential. Such a retailer may try to increase its sales growth rate by pushing more inventory to its stores. For example, it may increase service levels of existing products in order to stimulate demand. Or it may open new stores or expand its product line. As the retailer saturates its market, it realizes diminishing sales growth from each new store, store expansion, or new product line. However, all these activities require a fixed inventory outlay to stock the shelves. Therefore, we expect that for such a retailer, the *negative effect* will dominate so that there will be a negative correlation between sales ratio and inventory turnover.

In practice, it is difficult to estimate the market potentials of retailers and classify them into one type or the other. Therefore, we shall estimate the relationship between sales ratio and inventory turnover pooled across all retailers. We set up Hypotheses 2(a)–(b) to test whether positive correlation dominates of negative correlation dominates in our data set.

We also expect that retailers who experience sales decline will find it harder to manage inventory than retailers who experience sales growth because retailers who experience sales decline have to additionally find ways to dispose off excess inventory. Thus, we divide sales ratio into two regions: the *sales expansion region* where $g_{sit} \geq 1$, and the *sales contraction region* where $0 < g_{sit} \leq 1$. We set up Hypothesis 3 comparing these two regions in order to test whether inventory turnover is more sensitive to decline in sales or to increase in sales. Figure 3-2 depicts the relationship proposed in Hypothesis 3.

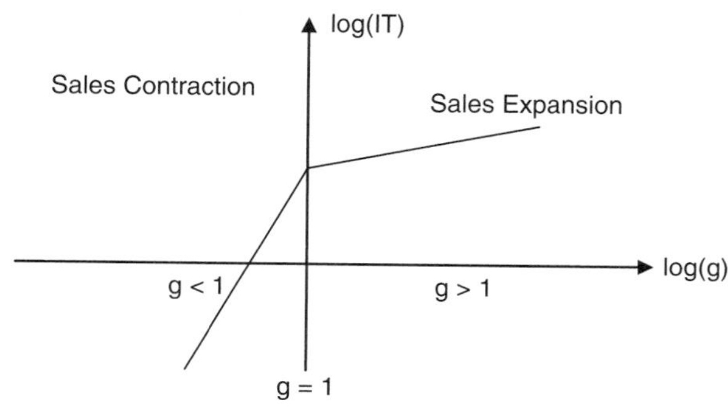

Figure 3-2. Illustration of Hypothesis 3
Note: This figure depicts a piecewise linear fit between the logarithm of inventory turnover, log(IT), and the logarithm of sales ratio, log(g), because we use a log-log model to test our hypotheses.

Hypothesis 2(a). *Inventory turnover of a firm is positively correlated with changes in its sales ratio in the sales expansion region as well as the sales contraction region.*

Hypothesis 2(b). *Inventory turnover of a firm is negatively correlated with changes in its sales ratio in the sales expansion region as well as the sales contraction region.*

Hypothesis 3. *Inventory turnover of a firm is more sensitive to sales ratio in the sales contraction region than in the sales expansion region.*

6. MODEL

We first estimate model (1) to re-test the hypotheses in GFR with our data set. Then, we modify the model in GFR to test our hypotheses. The model is specified as follows:

$$\log \mathrm{IT}_{sit} = F_i + c_t + b^1 \log \mathrm{GM}_{sit} + b^2 \log \mathrm{CI}_{sit} + b^4 \log S_{si,t-1} + b^5 \log g_{sit} + b^6 \max[0, \log g_{sit}] + \varepsilon_{sit}. \tag{4}$$

Here, F_i is the time-invariant firm-specific fixed effect for firm i; c_t is the year-specific fixed effect for year t; b^1, b^2, b^4, b^5, and b^6 are the coefficients of log GM_{sit}, log CI_{sit}, log $S_{si,t-1}$, log g_{sit}, and log[max(0, g_{sit})], respectively; and ε_{sit} denotes the error term for the observation for year t for firm i in segment s. Hypothesis 1(a) implies that $b^4 > 0$, Hypothesis 2(a) implies that $b^5 > 0$ and $b^5 + b^6 > 0$, and Hypothesis 3 implies that $b^6 < 0$. The main features of this model are as discussed in §4.

We estimate several variations of (4) to test our hypotheses. For example, we add the quadratic term, $[\log S_{si,t-1}]^2$, to test whether the effect of firm size on inventory turnover shows decreasing or increasing economies of scale. We also partition our data by firm size in order to study whether sales ratio has different effects on inventory turns for large and small firms. In another modification, we estimate the coefficients of the explanatory variables separately for each segment to test if the results are consistent across all segments or are driven by only a few of the segments in the data set. We use ordinary least squares estimation for simplicity. The estimators thus obtained are consistent in the presence of heteroscedasticity.

Table 3-4. Re-test of the Hypotheses in Gaur, Fisher, Raman (2005)

	Estimate	Std. Error
R^2 (%)	93.86	
log GM_{sit}	-0.287^{***}	0.024
log CI_{sit}	0.633^{***}	0.037
log SS_{sit}	0.034^{***}	0.008

***: statistically significant at p<0.0001.

Table 3-5. OLS Regression Estimates for Model (4)

	Model (1) without quadratic size term			Model (1) with quadratic size term		
	Estimate	Std. error	Std. Coeff. estimate	Estimate	Std. error	Std. Coeff. estimate
(1)	(2)	(3)	(4)	(5)	(6)	(7)
R^2 (%)	94.06			94.09		
log GM_{sit}	–0.364***	0.047	–0.302***	–0.347***	0.023	–0.302***
log CI_{sit}	0.687***	0.036	0.271***	0.712***	0.037	0.279***
log $S_{si,t-1}$	0.035***	0.011	0.078***	0.105***	0.023	0.165***
$[\log S_{si,t-1}]^2$				–0.006***	0.001	–0.092***
log g_{sit}	0.670***	0.050	0.691***	0.669***	0.048	0.694***
max{0, log g_{sit} }	–0.480** >	0.061	–0.388**	–0.454**	0.061	–0.375**

***,**: statistically significant at $p<0.0001$, $p=0.05$.

7. RESULTS

Table 3-4 shows the results for model (1). The three hypotheses in GFR are supported for our larger and more recent data set. The coefficient of gross margin is –0.287, the coefficient of capital intensity is 0.633, and the coefficient of sales surprise if 0.034. All three coefficients are statistically significant at $p<0.0001$.

Table 3-5 shows the fit statistics and coefficients' estimates for model (4) in columns (2)–(4). The F-statistic for the model is significant at $p<0.0001$, and the R^2 value is 92.5%. The rest of this section discusses the support for hypotheses regarding size and sales ratio.

First, consider the test of Hypotheses 1(a)–(b). We find that inventory turns are positively correlated with size, supporting Hypothesis 1(a). A 1% increase in the size of a firm leads to a 0.035% increase in inventory turns ($p<0.0001$).[4] Note that the effect of size on inventory turns appears to be small compared to other explanatory variables. This may be so because log $S_{si,t-1}$ has a higher standard deviation than the other explanatory variables. In order to control for this difference, we compute the standardized coefficient estimates as shown in column (4) of the Table 3-5 (see Schroeder et al. (1986, p. 31–32) for a description of standardized coefficients). The standardized coefficient of log $S_{si,t-1}$ is 0.078; thus, size still has a smaller effect on inventory turns compared to other variables in our model.

We now investigate whether the coefficient of log $S_{si,t-1}$ differs across firms and across model specifications. The object of this analysis is to characterize how the effects of economies of scale and scope vary across our data set. We first investigate the presence of diminishing economies. Since we have so far shown a linear relationship between log IT_{sit} and log $S_{si,t-1}$, the coefficient of log

[4] Relative size, Sales(i,t–1)/Sales(i,0), yields identical results in an intra-firm model.

$S_{si,t-1}$ in this model can be biased downwards if there are diminishing economies of scale and scope. To address this possibility, we add a quadratic term, $[\log S_{si,t-1}]^2$, to model (4). Columns (5)–(7) in Table 3-5 show the estimation results for this model. We find that the coefficient of $\log S_{si,t-1}$ increases from 0.035 to 0.105, and the coefficient of $[\log S_{si,t-1}]^2$ is –0.006 ($p<0.01$). Thus, we see that the quadratic model supports the hypothesis that there are diminishing returns to scale as firm size increases.

Another way to identify diminishing economies of scale is to perform the regression separately for small and large firms. We classify firms as small or large using the following approach. We compute the median of $\log S_{si,t-1}$ for every firm, and then use these values to compute the 25th percentile and the median of $\log S_{si,t-1}$ for each segment. In the first regression, firms whose median value of $\log S_{sit}$ falls below the 25th percentile are classified as small firms and the remaining as large firms. In the second regression, the cut-off point is set at the median. Table 3-6 shows the results for the first regression in columns (2)–(5) and for the second regression in columns (6)–(9). We see that in the first regression, the coefficient of $\log S_{si,t-1}$ is 0.11 ($p<0.0001$) for small firms, and is not statistically significant for large firms. In the second regression, the coefficient of $\log S_{si,t-1}$ is 0.06 ($p<0.0001$) for small firms, and is again not significant for large firms. The comparison of estimates between small and large firms is consistent with the results from the quadratic model, and provides strong support for the hypothesis that there are diminishing economies of scale as firm size increases. Note that the decrease in the coefficient estimate for small firms from 0.11 to 0.06 when we increase the set of small firms from the first quartile to the first two quartiles of size distribution is also consistent with the diminishing economies to scale argument.

The coefficient of $\log S_{si,t-1}$ may also differ across retail segments. To investigate this possibility, we estimate the coefficients of the model separately for each retail segment. Table 3-7 shows the results obtained. We find that four of the ten segments have positive and statistically significant ($p<0.01$) coefficient estimates, one segment has negative and statistically significant ($p<0.01$) coefficient estimate, and the remaining five segments do not show any statistical relationship. Where positive, the coefficient estimate ranges between 0.06 and 0.16. Jewelry stores have a negative and statistically coefficient estimate of –0.223. We find that the result for jewelry stores is not caused by the presence of any outliers, rather it holds consistently across firms. This suggests that the arguments for economies of scale and scope may not apply to jewelry products because the costs of distribution and logistics that these arguments are based on may not be critical to jewelry retailers.

In summary, we have shown two important relationships between firm size and inventory turnover. The first relationship supports the hypothesis that inventory turnover increases with size. The second relationship relates to diminishing returns to scale.

We now consider the tests of Hypotheses 2(a)–(b) and 3. The results in columns (2)–(4) of Table 3-5 show that inventory turnover is positively

Table 3-6. Regression Estimates for Model (4) Obtained after Partitioning Firms Based on Size

	Small firms (1st quartile)		Large firms (2nd – 4th quartiles)		Small firms (below median)		Large firms (above median)	
	Estimate	Std. error	Estimate	Std. error	Estimate	Std. error	Estimate	Std. error
(1)	(2)	(3)	(4)	(5)	(6)	(7)	(8)	(9)
R-square (%)	89.31		94.24		90.49		94.84	
log GM_{sit}	-0.349^{***}	0.039	-0.332^{***}	0.019	-0.371^{***}	0.029	-0.317^{***}	0.019
log CI_{sit}	0.343^{***}	0.048	0.432^{***}	0.023	0.391^{***}	0.033	0.387^{***}	0.026
log $S_{si,t-1}$	0.108^{***}	0.021	0.001	0.008	0.063^{***}	0.013	0.004	0.011
log g_{sit}	0.773^{***}	0.057	0.502^{***}	0.042	0.712^{***}	0.042	0.497^{***}	0.055
max{0, log g_{sit} }	-0.593^{***}	0.084	-0.283^{***}	0.052	-0.533^{***}	0.056	-0.232^{***}	0.069

***: statistically significant at $p<0.01$.

Table 3-7. Segment-Wise Coefficients' Estimates for Model (4)

Retail Segment	log GM_{sit}	log CI_{sit}	Log $S_{si,t-1}$	log g_{sit}
Apparel And Accessory Stores	-0.166^{***}	0.848^{***}	0.016	0.243^{***}
Catalog, Mail-Order Houses	-0.319^{***}	0.195^{***}	0.148^{***}	0.429^{***}
Department Stores	-0.334^{***}	1.049^{***}	–0.008	0.414^{***}
Drug & Proprietary Stores	-0.212^{***}	0.321^{***}	0.158^{***}	0.562^{***}
Food Stores	-0.393^{***}	1.287^{***}	–0.029	0.492^{***}
Hobby, Toy, And Game Shops	-0.894^{***}	0.307	–0.024	0.408^{***}
Home Furnishings & Equip Stores	–0.024	0.680^{***}	0.129^{***}	0.508^{***}
Jewelry Stores	-0.683^{***}	0.439^{***}	-0.223^{***}	0.308^{***}
Radio,TV,Cons Electr Stores	-0.330^{***}	0.389^{***}	0.062^{***}	0.307^{***}
Variety Stores	-0.187^{***}	0.122^{***}	0.009	0.223^{***}

***,**,*: statistically significant at $p<0.01$, $p=0.05$, and $p=0.1$.

correlated with sales ratio in model (4). The coefficient of log g_{sit} is 0.67 and the coefficient of max{0, log g_{sit}} is –0.48. This implies that a 1% increase in g_{sit} is associated with a 0.67% increase in inventory turns in the sales contraction region and with a 0.19% (= 0.67 – 0.48) increase in inventory turns in the sales expansion region. Both these coefficients are statistically significant at $p<0.0001$. Thus, we find that inventory turnover is positively correlated with sales ratio in both the regions, providing support for Hypotheses 2(a). Moreover, the coefficient of max{0, log g_{sit}} is negative and statistically significant, providing strong support for Hypothesis 3. The average value of the coefficient of log g_{sit} obtained by doing a regression omitting the variable max{0, log g_{sit}} is 0.38.

Columns (5)–(7) in Table 3-5 show the coefficient estimates for sales ratio when the model is quadratic in firm size. We find that the estimates and standard errors of these coefficients are similar to those obtained when the model is linear in size. Therefore, they also support Hypotheses 2 and 3. The results from the separate regressions for small and large firms in Table 3-6 also support our hypotheses.

The coefficients of log g_{sit} and max{0, log g_{sit}} in Tables 3-5 and 3-6 show that the effect of a change in sales ratio on inventory turnover is significantly lower when $g_{sit} > 1$ than when $g_{sit} \leq 1$. In Table 3-5, the coefficient of log g_{sit} is lower in the sales expansion region than in the sales contraction region by 0.48 in the linear model and by 0.454 in the quadratic model. This result confirms our intuition that firms would find it harder to improve inventory turnover during periods of sales decline than during periods of sales growth. Further, Table 3-6 shows that the coefficient estimates of log g_{sit} differ significantly across small and large firms in the sales contraction region, but are statistically similar in the sales expansion region. For example, when the smallest 25% of firms are classified as small, the coefficient estimates for small and large firms are 0.773 and 0.502, respectively, in the sales contraction region, and 0.180 (= 0.773 –0.593) and 0.219 (= 0.502 – 0.283), respectively, in the sales expansion region. Thus, we observe that during

Table 3-8. Example Showing the Effect of Volatility in Sales Ratio on Expected Inventory Turnover

	Probability distribution of g_{sit}	Expected multiplicative effect on inventory turnover due to variation in sales ratio*	
		Firm classified as small	Firm classified as large
Scenario A	$g_{sit} = 1.2$ with probability 0.5 $g_{sit} = 0.8$ w. p. 0.5.	$[(1.2)^{0.18} + (0.8)^{0.77}]/2 = 0.938$	$[(1.2)^{0.22} + (0.8)^{0.50}]/2 = 0.968$
Scenario B	$g_{sit} = 1.1$ with probability 0.5 $g_{sit} = 0.9$ w. p. 0.5.	$[(1.1)^{0.18} + (0.9)^{0.77}]/2 = 0.970$	$[(1.1)^{0.22} + (0.8)^{0.50}]/2 = 0.985$

*For the purpose of this table, we classify a firm as small if its size belongs to the first quartile of its retail segment and as large otherwise. Thus, we use the coefficients' estimates in Columns 2 and 4 of Table 3-4 for our computations. All computations are done assuming that (i) the effects of GM_{sit} and CI_{sit} are normalized to zero, (ii) the effect of diminishing returns to scale is negligible for small changes in size, and (iii) the firm size and sales ratio are normalized to 1.0 in the base case.

periods of sales decline, inventory turns for small firms are more sensitive to sales ratio than for large firms. But during periods of sales expansion, there is no significant difference in the coefficient of sales ratio between small and large firms. The coefficients' estimates for the case in which small and large firms are defined by the median tell the same story.

We explain this result with an example. Consider the effect of volatility in sales growth on the inventory turnover of a firm over a period of one year. Table 3-8 shows two growth scenarios for the firm and their effects on inventory turnover. In both scenarios, the firm's expected sales ratio is zero (i.e., $E[g_{sit}] = 1$). The scenarios differ in the standard deviation of sales ratio. We examine each scenario using the coefficients' estimates for a small firm and for a large firm obtained from Table 3-6. For example, in scenario A, we find that the expected inventory turnover of the firm is 93.8% of what it would have been if g_{sit} were a constant equal to 1. We make the following observations by comparing all the cases in this example:

(i) The firm's expected inventory turnover declines in each case even though its total expected sales are equal to the sales in the previous year. Thus, volatility in sales has a negative effect on inventory turnover.

(ii) The decline in expected inventory turnover is higher if the firm experiences more variation in g_{sit} (i.e., Scenario A) than if the firm experiences less variation in g_{sit} (i.e., Scenario B). For example, for a small firm, expected inventory turns decline by 6.2% in Scenario A and by 3.0% in Scenario B.

(iii) The decline in inventory turnover is higher if the firm is small than if the firm is large. Further, the difference between large and small firms increases as the standard deviation of g_{sit} increases.

Thus, this example shows the effect of volatility in sales ratio on inventory turnover using our results. Interestingly, the inferences from the example are analogous to those from the newsboy model in inventory theory. Further, it shows that a firm with more volatile sales has two ways to improve its inventory turnover: either it should target a sufficiently high growth rate that compensates for the effect of volatility in sales ratio on inventory turnover, or it should reduce its inventory and offer a lower service level.

As with firm size, we analyze whether the coefficient of log g_{sit} is consistent across segments. Table 3-7 shows the coefficients' estimates obtained for each segment. We find that the coefficient of log g_{sit} varies significantly across segments ($p<0.0001$). However, sales growth consistently has a large positive coefficient for each segment. Its value ranges between 0.22 for variety stores to 0.56 for Drug and Proprietary stores.

In summary, we find strong support for the hypotheses that inventory turnover is positively correlated with sales ratio and that inventory turnover is more sensitive to sales ratio in the sales contraction region than in the sales expansion region. We also find that the latter effect is stronger for small firms than for large firms.

8. CONCLUSIONS AND DIRECTIONS FOR FUTURE RESEARCH

Our paper highlights the importance of understanding inventory turnover performance of retailers. Like GFR, we find that inventory is a significant proportion of the assets of a retailer. However, inventory turnover varies widely across retailers and for a retailer over time. We have shown that a significant proportion of the within-firm variation in inventory turnover is explained by changes in firm size, sales ratio and variables identified by GFR. In particular, inventory turnover of a firm is positively correlated with both size and sales ratio. Our results support the arguments of economies of scale and scope studied in the operations management literature. We use a data set of 353 publicly listed U.S. retailers for the period 1985–2003 in our analysis. This data set is larger and more recent than that used by GFR. Thus, we also examine the hypotheses formulated in GFR regarding the correlations of inventory turnover with gross margin, capital intensity and sales surprise. We find that inventory turnover is strongly negatively correlated with gross margin and positively correlated with capital intensity in our data set. These results are consistent with those obtained in GFR.

Our results are useful to retailers for benchmarking their inventory turnover performance against their peers. Since the correlations estimated by us are based on a large set of firms, they provide estimates of the average change in inventory turnover associated with given changes in gross margin, capital intensity, size and sales ratio. A positive residual for a firm in our model indicates that the firm achieved higher inventory turnover than its peer group after controlling for differences in the explanatory variables, while a negative residual indicates otherwise. Thus, managers may use these residuals to investigate reasons for differences in inventory turnover performance across firms or for a firm over time. The fixed effects in our model may be used similarly by managers for benchmarking. Another application of our results is related to the difference between the coefficients of sales ratio during periods of sales growth and sales decline. This result shows that aggregate retail inventory changes with sales in a manner that is consistent with the newsboy model in inventory theory. This result also implies that managers should pay more attention to managing

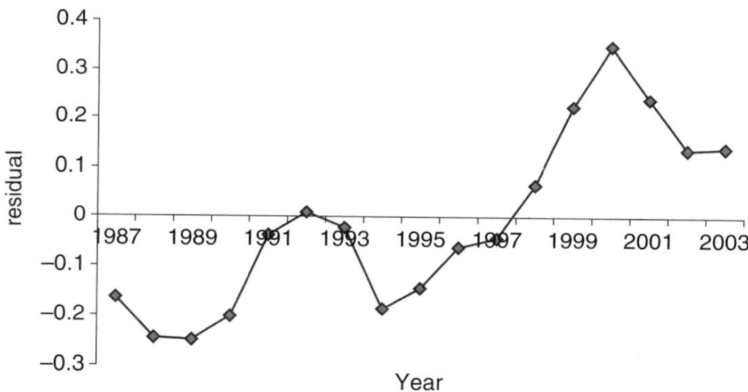

Figure 3-3. Time-Series Plot of Residuals from Model (4) for Best Buy Stores, Inc

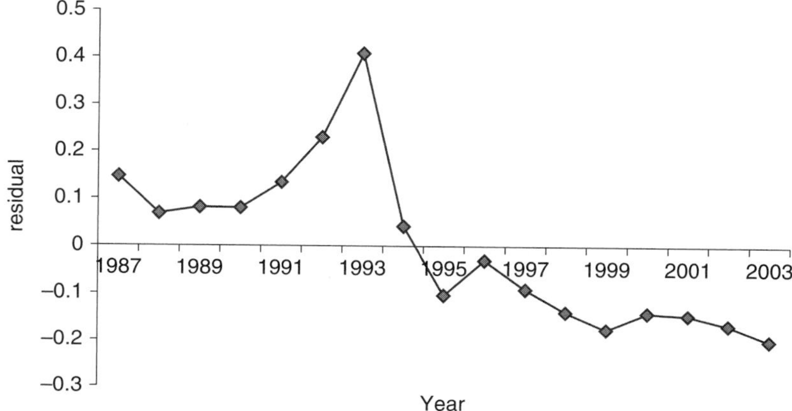

Figure 3-4. Time-Series Plot of Residuals from Model (4) for Jennifer Convertibles, Inc

inventory when a firm is small, or when a firm is going through a period of sales decline, or when a firm faces more volatility in sales.

Our paper suggests three possible directions for future research on aggregate-level inventory management in retailing.

1. Modeling of aggregate-level inventory decisions: Even though the variables in our model are statistically significant, there is still a considerable amount of variation in inventory turnover that remains unexplained. For example, we find that the residuals from our model show differing patterns across firms. There are firms whose residuals have consistently improved over time after controlling for changes in all the explanatory variables, and other firms whose residuals show a consistently declining trend. To illustrate this, Figures 3-3 and 3-4 show time-series plots of residuals from our model for Best Buy Stores, Inc. and Jennifer Convertibles, Inc., respectively. Notice that the residuals for Best Buy trend upwards with time while those for Jennifer Convertibles trend downwards. These unexplained but systematic differences suggest that there is scope for future research to better understand retailers' inventory turnover performance. Future research may also investigate models by which retailers make aggregate-level inventory decisions.

2. Explaining the drivers of inventory productivity using augmented data sets: Several operational factors can be said to contribute to the relationships of gross margin, capital intensity, firm size and sales ratio with inventory turnover. Since public financial data do not capture these operational factors, it is not possible to identify the drivers of inventory turnover using these data. A richer data set may be used in future research to examine the aforementioned relationships more closely. For example, the discussion in §5 identifies many variables that may be included in such a data set, for example, number of store locations, their store formats and square footage, number of warehouses and their square footage, same stores sales growth rates, etc. In a recent paper, Kesavan et al. (2007) construct such a data set by incorporating number of store locations, accounts payables, and several other variables. They apply a simultaneous equations model to estimate causal effects of sales, inventory and gross margin on each other. They further show that their model provides more accurate forecasts of sales than standard time-series models as well as equity analysts.

3. Examining the effects of firm lifecycle and bankruptcies on model estimation: Our data set consists of only publicly listed firms that have at least five consecutive years of data available. Since these firms would be above a certain size, our coefficient estimate for size could be subjected to selection bias. Also our coefficient estimates could be subjected to survival bias since slow growing firms could exit from our data set. Future research may examine how these factors affect the relationship of inventory management with other performance variables.

Acknowledgment The authors are thankful to Professor Ananth Raman for many helpful comments on this manuscript. The questions of the effects of firm size and sales growth rate on inventory turnover were suggested to the first author by Professors Marshall Fisher and

Ananth Raman. The authors are also thankful to seminar participants at Boston University, Cornell University, University of Michigan, and University of North Carolina for numerous suggestions that were helpful in this research.

REFERENCES

Balakrishnan, R., T. J. Linsmeier, M. Venkatachalam. 1996. Financial Benefits from JIT Adoption: Effects of Consumer Concentration and Cost Structure. *Accounting Review*, **71**, 183–205.

Billesbach, T. J., R. Hayen. 1994. Long-term Impact of JIT on Inventory Performance Measures. *Production and Inventory Management Journal,* First Quarter, 62–67.

Chang, D., S. M. Lee. 1995. Impact of JIT on Organizational Performance of U.S. Firms. *International Journal of Production Research*, **33**, 3053–3068.

Chen, H., M. Z. Frank., O. Q. Wu. 2005. What Actually Happened to the Inventories of American Companies Between 1981 and 2000? *Management Science,* **51**, 1015–1031.

Eppen, G. 1979. Effect of Centralization on Expected Costs in a Multi-location Newsboy Problem. *Management Science*, 25(5) 498–501.

Eppen, G., L. Schrage. 1981. Centralized Ordering Policies in a Multi-Warehouse System with Lead Times and Random Demand, in *Multi-Level Production/Inventory Control Systems: theory and Practice*, L. Schwarz (ed.), TIMS Studies in the Management Sciences, Vol. 16, North Holland, Amsterdam.

Fama E. F., K. R. French. 1993. Common Risk Factors in the Returns on Stocks and Bonds. *Journal of Financial Economics*, **33**(1) 3–56.

Gaur, V., M. L. Fisher, A. Raman. 1999. What explains superior retail performance? *Working paper, Cornell University.*

Gaur, V., M. L. Fisher, A. Raman. 2005. An Econometric Analysis of Inventory Turnover Performance in Retail Services. *Management Science,* **51**, 181–194.

Gaur, V., S. Kesavan, A. Raman, M. L. Fisher. 2007. Estimating Uncertainty Using Judgmental Forecasts, *M&SOM*, **9**(4) 480–491.

Hendricks, K. B., V. R. Singhal. 1996. Quality awards and the market value of firm: An empirical investigation. *Management Science,* **42**, 415–436.

Hendricks, K. B., V. R. Singhal. 1997. Does Implementing an Effective TQM Program Actually Improve Operating Performance: Empirical Evidence From Firms That Have Won Quality Awards. *Management Science*, **43**, 1258–1274.

Hendricks, K. B., V. R. Singhal. 2001. The Long-run Stock Price Performance of Firms with Effective TQM Programs as Proxied by Quality Award Winners. *Management Science*, **47**, 359–368.

Hendricks, K. B., V. R. Singhal. 2005. Association Between Supply Chain Glitches and Operating Performance. *Management Science*, **51**(5) 695–711.

Huson, M., D. Nanda. 1995. The Impact of Just-in-time Manufacturing on Firm Performance. *J. Operations Management*, **12**, 297–310.

Kesavan, S., V. Gaur, A. Raman. 2007. Incorporating price and inventory endogeneity in firm-level sales forecasting. Working Paper, Harvard Business School and University of North Carolina.

Lieberman, M. B., L. Demeester. 1999. Inventory Reduction and Productivity Growth: Linkages in the Japanese Automotive Industry. *Management Science*, **45**, 466–485.

Makridakis, S., S. C. Wheelwright., R. J. Hyndman. 1998. *Forecasting: Methods and Applications.*Third Edition, John Wiley & Sons, Inc.

Rajagopalan, S., A. Malhotra. 2001. Have U.S. Manufacturing Inventories Really Decreased? An Empirical Study. *M&SOM*, **3** 14–24.

Raman, A., V. Gaur., S. Kesavan. 2005. David Berman. Harvard Business School Case 605–081.

Rumyantsev, S., S. Netessine. 2007. What can be learnt from classical inventory models? A cross-industry exploratory investigation. *M&SOM,* **9**(4) 409–429.

Sack, K. 2000. *Retailing: General Industry Survey*, Standard & Poor's, New York.

Schroeder, L., D. Sqoquist, P. Stephan. 1986. *Understanding Regression Analysis.* Sage Publications.

Silver E. A., D. F. Pyke., R. Peterson. 1998. *Inventory Management and Production Planning and Scheduling.*Third Edition, John Wiley & Sons, Inc., New York.

Chapter 4
THE ROLE OF EXECUTION IN MANAGING PRODUCT AVAILABILITY

Nicole DeHoratius[1] **and Zeynep Ton**[2]
[1]*University of Chicago, Graduate School of Business, 5807 S. Woodlawn Ave., Chicago, IL 60637, USA*
[2]*Harvard Business School, Morgan Hall 425, Boston, MA 02163, USA*

1. INTRODUCTION

Several surveys show that a significant number of customers leave retail stores because they cannot find the products for which they are looking (e.g. Emmelhainz et al., 1991; Andersen Consulting, 1996; Gruen et al., 2002; Kurt Salmon Associates, 2002). Most research in operations management focuses on two factors to explain suboptimal product availability—poor assortment and poor inventory planning. Our research with several retailers during the last few years highlights a third factor, poor execution, or the failure to carry-out an operational plan. We find that even with the application of algorithms to select the appropriate stocking quantity and appropriate store assortment, retail customers still may face unnecessary stockouts. For example, after auditing 50 products at ten different stores, management at a specialty retailer found that only 16% of the stockouts could be attributed to statistical stockouts (cited in Ton, 2002). Instead, 24% of the stockouts were due to inventory record inaccuracy, discrepancies between the recorded and actual on-hand inventory quantity, and 60% were due to misplaced products, products that were physically present at the store but in locations where customers could not find them.

Inventory record inaccuracy and misplaced products are two examples of poor store execution. These problems affect product availability in two ways. First, they lead to stockouts and hence compromise retailers' service levels. When the actual level of inventory for a particular product is lower than the planned level due to either inventory record inaccuracy or product misplacement, the actual service level will be lower than the planned service level. At Borders Group Inc., a large retailer of entertainment products such as books, CDs, and DVDs, lost sales due to misplaced products reduced profits by 25% (Raman et al., 2001). Andersen Consulting (1996) estimates that sales lost due to products that are present in storage areas but not on the selling floor amount to \$560–\$960 million per year in the US supermarket industry.

N. Agrawal, S.A. Smith (eds.), *Retail Supply Chain Management*,
DOI: 10.1007/978-0-387-78902-6_4,

Second, for retailers that rely on automated replenishment systems to manage store inventory, execution problems affect future product availability through the distortion of historical sales and inventory data stored in these systems. Distortion of inventory data may prevent the triggering of a replenishment order when the system inventory is greater than the actual inventory or may unnecessarily trigger an order when the system inventory is less than actual inventory. Moreover, when a product that is actually out of stock is reported as in stock, the automated replenishment system may wrongly conclude that there is no demand. The system observes no sales for that item because it is not available to the customer. Thus, even when multiple customers are willing to purchase that item, the system may automatically reduce the forecast of future demand which in turn causes the retailer to stock less of it or even to drop the item from the assortment entirely.

Despite their prevalence and impact, research on execution problems is limited. Much of the work in the retailing context focuses on the drivers of these problems and only recently have researchers attempted to incorporate these problems into existing planning models. In this chapter we summarize the existing research on store execution and identify future research opportunities in this area. The chapter is organized as follows. In Section 2, we describe the magnitude and root causes of the two execution problems, based on specific, well-researched case studies. In Section 3, we describe the findings of the empirical studies that have identified factors that exacerbate the occurrence of execution problems. In Section 4, we describe the effect of execution problems on inventory planning and summarize how researchers have incorporated these problems into existing inventory models. Finally, in Section 5, we conclude with a discussion of future research opportunities.

2. RETAIL EXECUTION PROBLEMS

Evidence of execution problems exists in a number of different contexts. Distribution centers,[1] manufacturing firms,[2] financial services,[3] utility companies,[4] hospitals,[5] and government agencies[6] have all faced problems with misplaced products and/or record inaccuracy. The costs of poor execution in these

---

[1] See, for example, Bayers (2002), Millet (1994), and Rout (1976).

[2] See, for example, Hart (1998), Sheppard and Brown (1993), Tallman (1976), Brooks and Wilson (1993), Bergman (1988), Krajewski et al., (1987), Flores and Whybark (1987 & 1986), and Woolsey (1977).

[3] See, for example, Cassidy and Mierswinski (2004) and Capital Market Report (2000).

[4] See, for example, Woellert (2004) and Redman (1995).

[5] See, for example, McClain et al. (1992) and Young and Nie (1992).

[6] By the Numbers (2005), McCutcheon (1999), Galway and Hanks (1996), Laudon (1986), Schrady (1970), and Rinehart (1960).

contexts, as in retailing, have been shown to be substantial. We describe below the extent of such problems in retailing and identify how they arise.

2.1 Inventory record inaccuracy

At Gamma Corporation,[7] a leading retailer with hundreds of stores and over $10 billion in sales, physical audits revealed that inventory record inaccuracy was pervasive throughout the chain (DeHoratius and Raman, 2008). Discrepancies were found in 65% of the nearly 370,000 audited inventory records with the absolute difference between the recorded quantity and the on-hand quantity per item per store ranging from zero to 6,988 units (Figure 4-1). The average absolute discrepancy between the recorded quantity and the actual on-hand quantity was nearly five units, or 36% of the average target quantity. Of those records that were inaccurate, approximately 59% of them had positive discrepancies where the recorded quantity exceeded the on-hand quantity and nearly 41% of them had negative discrepancies where the on-hand quantity exceeded the recorded quantity. Interestingly, nearly each product that was stocked out in the store at the time of the audit showed a positive on-hand amount recorded in the inventory management system. In other words, these stockouts were invisible to corporate merchandise and inventory planners.

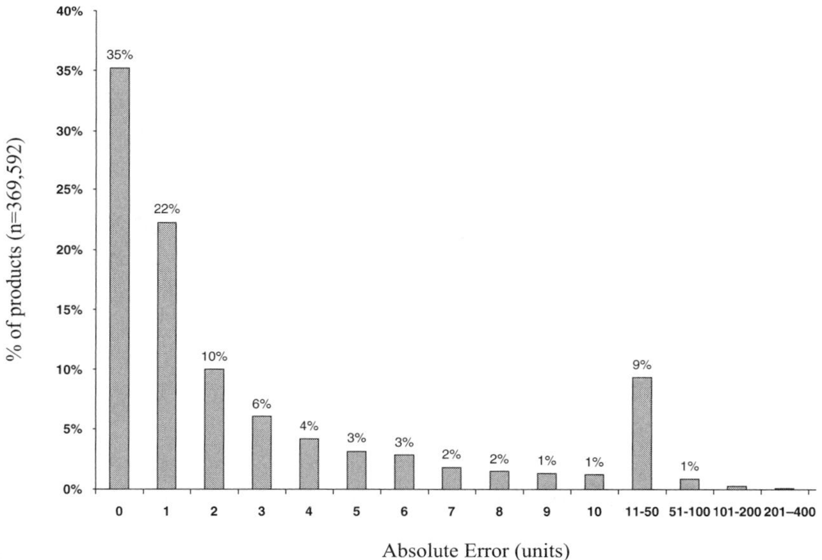

Figure 4-1. Histogram of the Absolute Difference Between System and Actual Inventory Measured in Units. (Source: Raman et al., 2001)

---

[7] Name disguised to preserve confidentiality.

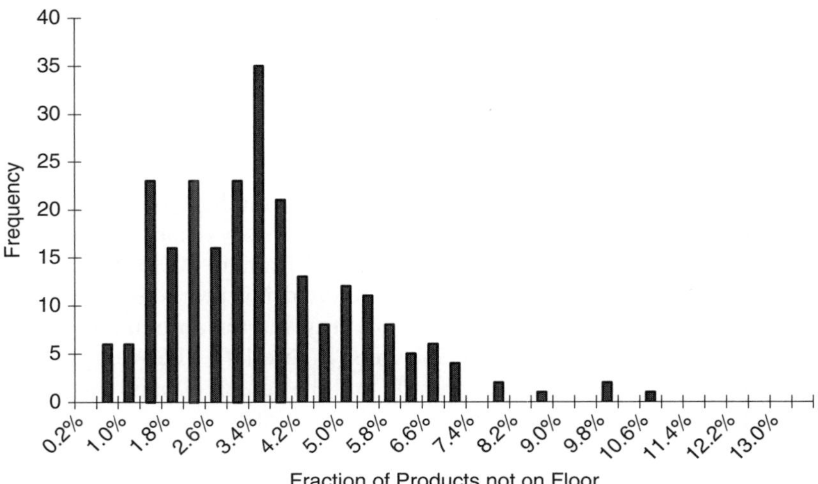

Figure 4-2. Histogram for the Fraction of Products That Are Not Available on the Sales Floor. (Source: Raman et al., 2001)

2.2 Misplaced products

Misplaced products, whether they are mis-shelved or left in storage areas, lead to stockouts if customers are unable to locate the inventory they seek. At Borders, two surveys showed that approximately 18% of customers who approached a salesperson for help experienced a phantom stockout (Ton and Raman, 2003). That is, the product was physically present at the store but could not be found even with the help of a salesperson. Physical audits at 242 Borders stores showed that, on average, 3.3% of a store's assortment (over 6,000 products per store) was placed in storage areas and had no presence on the selling floor (Figure 4-2). At some stores, nearly 10% of the assortment was missing from the selling floor. Note that these estimates of misplaced products are conservative because they do not include those products that have been mis-shelved either by customers or employees.

2.3 Root causes of execution problems

We identify three sources of poor execution: (1) poor process design, (2) an operating environment that makes it challenging for employees to conform to prescribed processes, and (3) employee errors. Poor process design may result from a number of factors including poorly specified work content, poorly specified sequence of activities, inadequate time given to perform work, and an absence of feedback on process quality. At most retail chains for example, extra units of inventory that do not fit into display shelves are kept in storage locations. When the level of inventory of a particular product on the shelf approaches zero, employees are supposed to replenish units of that product from the storage locations. Since existing systems do not track the location of stored products, store employees have to rely on their memory to determine where products are stored in order to replenish the shelves. Not surprisingly, this poor process design often leads to product misplacement. Furthermore, at many retail chains, employees have to manually enter the price lookup (PLU) into the registers. This process requires employees to remember the PLU codes of hundreds of different products, making errors inevitable. Executives at one supermarket chain, for example, told us that, on average, they sold 25% more "medium tomatoes" than the total amount shipped to their stores because store employees often entered the PLU code for "medium tomatoes" even when customers were buying other types of tomatoes, such as "organic tomatoes" or "vine ripe tomatoes". It is reasonable to expect inventory records to be more accurate at retail stores where electronic point-of-sale scanning is used.

Even when processes are well-designed, employees may deliberately choose not to follow them. In an operating environment where conformance to designed processes is not monitored or rewarded, employees may choose not to carry-out activities requiring substantial effort. In some cases, the operating environment may make it challenging for employees to follow designed processes. In numerous retail chains we have observed that instead of placing merchandise that does not fit into the display shelves into storage locations where extra merchandise is supposed to be stored, store employees hide it in places within the selling floor to avoid travelling to specified storage areas. This nonconformance causes misplaced products. Similarly, during the checkout process in a supermarket store employees sometimes choose not to scan two products that are identical in price but different in flavor (e.g., two liter bottle of Diet Coke and two liter bottle of Coca-Cola) separately, scanning one product twice instead. While the employees do not create a discrepancy between the value of the inventory sold and the amount due the store from the customer since the products are identically priced, this action does create a discrepancy in two inventory records. The recorded on-hand quantity for one product will be unnecessarily depleted by two units while the other product's record will remain at its current level despite the product leaving the store. Similar discrepancies arise when store employees do not properly record a product returned or exchanged by a customer.

Even when processes are well designed and employees intend to carry-out these processes, they may commit errors which lead to execution problems. Many retail activities are prone to employee error. For example, at some stores, standard operating procedures dictate that all products have presence on the selling floor. In shelving new merchandise, employees may fail to place some products on the selling floor and instead mistakenly take them to storage areas, leading to misplaced products. At some retail chains, distribution center employees may pick and ship the wrong product to the store leading to discrepancies between recorded and actual inventory quantities in that store. It is possible to observe numerous other examples of employee errors when examining retail store processes. Cognitive psychologists have studied human error for a long time and have identified the mechanisms by which errors are generated and how they can be reduced (for more information on human error, see Reason 2002).

Finally, execution problems within the context of retail stores can also be caused by customers. At many stores, when customers remove products from the shelves and subsequently decide not to purchase them, they may not return the products to their appropriate location but rather place them in the wrong location in the store. These products remain misplaced until store employees find them and place them in their proper locations. Customer or employee theft is another contributor to inventory record inaccuracy. Hollinger and Langton (2003) estimate that inventory theft costs US retailers close to 1.3% of annual sales or more than 26 billion dollars. Products that are removed from the store illegally are not removed from the inventory record until an audit is performed and the missing items identified.

3. FACTORS THAT EXACERBATE EXECUTION PROBLEMS

Two empirical studies examine specific drivers of misplaced products and inventory record inaccuracy. DeHoratius and Raman (2008) and Ton and Raman (2006) compare performance across retail stores within a chain and show large variation in execution performance across stores that are owned and operated by the same parent company, have the same incentives for store employees, use the same information technology systems, and are instructed to use the same standard operating procedures for shelving and replenishing inventory within the stores. As a result, these factors cannot explain the variation in performance. Instead, DeHoratius and Raman (2008) and Ton and Raman (2006) identify several alternative drivers of poor execution, namely inventory levels, product variety, employee turnover, lack of training, employee workload, and employee effort.

Note that the factors identified by DeHoratius and Raman (2008) and Ton and Raman (2006) contribute to execution problems by creating an operating environment that makes process conformance challenging or by making it more likely for store employees to make errors. How each of these factors contributes to execution problems is the subject of this section. Appendix 1 describes the research methodology used in these studies including a precise identification of the independent variables used, a list of the factors for which the authors controlled, and brief descriptions of their model estimation.

3.1 Inventory levels

Proponents of Just-in-Time (JIT) manufacturing have argued repeatedly that inventory hides process problems and thus inhibits process improvements (Schonberger, 1982; Hall, 1983; Krafcik, 1988). Production systems with high inventory levels have fewer learning opportunities and hence achieve lower quality over time. A similar effect is observed at retail stores. Stockouts at retail stores that result from poor inventory planning or from poor execution are similar to production problems. Although these stockouts are not desirable as they often lead to lost sales, they present opportunities for improvement analogous to production problems. Since the likelihood of a stockout is higher at lower inventory levels, all else being equal, stores with lower inventory levels are likely to have more learning opportunities.

In retail stores where each product is given a specific space on the selling floor, a visual inspection of the shelves allows store employees to identify the products that are stocked out. When a product on the selling floor is stocked out, store employees could check whether the recorded inventory level for the product matches the actual quantity observed in the store. If there is a discrepancy between the recorded inventory level and the actual inventory on the selling floor, employees could investigate whether the discrepancy is due to product misplacement or record inaccuracy. If the former, employees could attempt to locate the extra units and bring them back to the appropriate location. If the latter, the retailer can create a formal quality process that lets employees manually adjust the system inventory while also investigating the reason for the mismatch.

Retailers can learn from observing companies in other industries that maintain high levels of record accuracy. Arrow Electronics, a distributor of electronic parts and equipment, has close to 100% inventory record accuracy. Arrow takes advantage of periods when inventory levels are low. Specifically, Arrow has a mechanism that triggers counts when either system or physical inventory reaches zero. If a part is physically stocked out in a location, the picking operators are instructed to verify that the system inventory for that part is also zero. Similarly, if the system inventory is zero, the picking operators are instructed to verify that the physical inventory for the part is also zero. When there is a discrepancy between the system inventory levels and physical inventory levels, warehouse operators investigate the source of the problem and, when necessary, make inventory adjustments to the system (Raman and Ton, 2003).

Maintaining high inventory levels at retail stores causes execution problems not only by reducing opportunities to easily identify discrepancies but also by increasing the complexity in the operating environment. All else being constant (e.g. the size of the selling area), stores with higher levels of inventory often have more units stored in storage areas. Since the replenishment process from storage areas, like most operational processes, is prone to employee errors, there are more opportunities to make errors in replenishing merchandise to the shelf.

Thus, we expect more product misplacements in operating environments with high inventory levels.

Both DeHoratius and Raman (2008) and Ton and Raman (2006) provide empirical evidence to support the relationship between inventory levels and store execution. DeHoratius and Raman (2008) show that stores with higher inventory levels in a given selling area also have greater inventory record inaccuracy. Similarly, Ton and Raman (2006) show that stores with higher inventory levels per product also have a greater percentage of phantom products, defined as the products in storage areas but not on the selling floor. Ton and Raman (2007) confirm this finding using four years of data from the same research site.[8]

3.2 Product variety

As with earlier claims that higher product variety increases the complexity in manufacturing settings (e.g. Skinner, 1974; Anderson, 1995; Fisher et al., 1995; Macduffie et al., 1996; Fisher and Ittner, 1999), more variety at a retail store increases the confusion and complexity in the operating environment and hence causes more process nonconformance or employee errors that lead to the two execution problems. Increasing product variety, for example, increases the difficulty of differentiating products during the checkout process. Consequently, store employees may scan one product multiple times without recognizing that the customer is purchasing multiple different products, causing inventory record inaccuracy. Increasing product variety at a store also increases the number of steps performed in inventory replenishment at the stores. Given that stores have limited shelf space, store employees are required to move more units of products to storage areas at stores that have higher product variety. Since each step in replenishment is prone to errors, higher product variety is associated with more products that are in storage areas and not on the selling floor.

Both DeHoratius and Raman (2008) and Ton and Raman (2006) provide empirical evidence to support the relationship between product variety and store execution. DeHoratius and Raman (2008) show that stores with higher product variety also have greater inventory record inaccuracy. Similarly, Ton and Raman (2006) show that stores with more products in a given area also have a greater percentage of phantom products. Ton and Raman (2007) confirm this finding in their longitudinal study.

3.3 Employee turnover and training

The average employee turnover for US businesses in general is about 10%–15% (White, 2005). Retail stores, however, experience much higher rates of employee

---

[8] See appendix for details of this study.

turnover. According to the National Retail Federation, the average part-time and full-time employee turnover in the retail industry is 124% and 74% respectively. Ton and Raman (2006) report an average employee turnover of 112% for part-time employees and 65% for full-time employees at Borders stores. The authors show that stores with higher employee turnover also have a greater percentage of phantom products, suggesting these problems may be linked.

High levels of employee turnover affect store execution in numerous ways. First, employee turnover disrupts existing operations (Dalton and Todor, 1979; Bluedorn, 1982). When a store employee quits the store, there is often a period of finding and training a replacement. During this period, workload for existing employees is generally higher. Higher workload may lead to more errors and consequently more execution problems. Moreover, the departure of employees often causes demoralization of existing employees (Staw 1980; Steers and Mowday, 1981; Mobley, 1982). Demoralization may cause existing employees to make more errors in performing their jobs.

Second, employee turnover leads to a loss of accumulated experience (Argote and Epple, 1990; Nelson and Winter, 1982). As employees spend more time at the stores, they become better at performing their jobs and consequently make fewer errors. Ton and Raman (2006), for example, state that as employees spend more time at Borders stores, they become more familiar with the products in their sections and as a result become better at noticing those that are missing from the selling floor.

Third, because store employees typically leave their job within a year, retailers often choose not to invest in their training. In fact, the average training provided to new employees in the retail industry is merely seven hours (Managing Customer Service, 2001). As a result of limited training, new employees often start performing their jobs without a full understanding of the existing processes and their impact on store operations. Hence, they regularly commit process nonconformance (e.g., the checkout scanning example in Section 2.3). Ton and Raman (2006) provides empirical evidence for the positive effect of training on store execution. The authors find a negative association between the percentage of phantom products and the amount of training offered at the stores.

3.4 Employee workload

For most retailers, store labor represents the largest controllable expense at retail stores. In 2003, for example, selling, general, administrative expenses, which consist largely of store employee payroll expenses, represented approximately 20% percent of retail sales.[9] Consequently, many store managers are evaluated based on how well they manage payroll expenses at their stores.

[9] Source: Standard & Poor's Compustat, 427 public firms with SIC Codes between 5200 and 5999.

When store managers reduce payroll expenses—either by reducing the number of employees at the stores or reducing the number of hours worked—the amount of workload per employee increases. With increased workload, store employees are less likely to conform to designed processes. They are also likely to make more errors in performing their tasks. For example, a salesperson is more likely to scan two similar products that have the same price together instead of separately if he or she sees a long line of customers waiting to be checked out. It is often more difficult for customers and store managers to observe the accuracy of scanning than to observe the speed of scanning. Ton and Raman (2006) show that stores with higher employee workload, measured as payroll expenses as a percentage of sales, also have higher percentage of phantom products.

3.5 **Employee effort**

DeHoratius and Raman (2008) argue that employee effort affects inventory record inaccuracy. When store employees exert more effort into monitoring select products, the inventory records for these products are expected to be more accurate. Employee effort, however, is unobservable. Thus, the authors use two proxies, item cost and shipping method, for employee effort. They posit that employees exert more effort into monitoring expensive than inexpensive products and thus expensive items should be more accurate than inexpensive ones. Similarly, they argue that store employees monitor items shipped directly to the retail store from the vendor more closely than those items shipped to the store from the retailer's own distribution center. We discuss each of these proxies and their findings in turn.

Inventory shrinkage is a common problem at retail stores and store employees often spend considerable effort in shrink prevention activities. Inventory shrinkage has a direct impact on store operating profits and shrinkage of expensive products affects store profitability more than shrinkage of less expensive products. Given that store managers are often evaluated on their financial performance, controlling the shrinkage of expensive products is often a key priority for store personnel. Consequently, it is not unusual for store employees to monitor expensive and inexpensive items differentially. DeHoratius and Raman (2008) show that this differential treatment, in turn, leads to lower levels of record inaccuracy for expensive items relative to inexpensive ones.

DeHoratius and Raman (2008) also show that the magnitude and likelihood of inventory record inaccuracy is lower for those products shipped directly to the retail store from the vendor compared to those products shipped to the store from the retailer's own distribution center. They posit that store employees pay more attention to checking shipments that arrive from vendors. They do so because when the value ordered by the store exceeds the value shipped from the vendors, stores receive a credit from the vendor. Stores do not, however, receive a credit from the distribution centers unless the discrepancy between what was shipped and what was ordered exceeds a threshold more than thirty times the average cost of a single product. Consequently, store employees pay more attention to checking shipments that arrive from vendors to ensure invoice accuracy. Moreover, shipments from the vendors tend to contain fewer products and hence easier for store employees to inspect.

4. HOW EXECUTION PROBLEMS AFFECT INVENTORY PLANNING

Inventory planning at retail stores requires two main decisions, how much inventory to stock and when to replenish. The policies retailers establish with respect to these decisions are critical determinants of store performance (see Tayur et al, 1999; Graves and de Kok, 2003). We use two examples to demonstrate how inventory record inaccuracy and misplaced products affect each of these decisions. These examples are described in detail in DeHoratius (2002) and Ton (2002). We then summarize research that incorporates execution problems into existing inventory planning models.

4.1 Effect of inventory record inaccuracy on inventory planning

Management at Gamma received a letter of complaint from a regular customer noting that a specific product he sought was persistently out of stock (DeHoratius, 2002). He stated that the product failed to be replenished even after bringing the stockout to the attention of the store manager. After researching the problem, Gamma management discovered that, although the product was out of stock, inventory records showed 42 units on-hand in that store. Because the inventory record showed that there was sufficient on-hand inventory to meet demand, the automatic replenishment system failed to release additional inventory to the store even though, in reality, the shelf was empty.

Sales records also revealed that this store had not sold a single unit of this product, a product that typically sold one unit per week per store, during the past seven weeks. The demand forecast was then automatically updated to reflect the recent low levels of sales, namely zero sold in seven weeks. Therefore, not only were customers unable to find the product on the shelf during the time when the product was out of stock but their demand was less likely to be met in the future. Moreover, it is important to note that the product might have remained out of stock until the next physical audit or cycle count had this customer not written to Gamma. Without inventory to sell, the recorded quantity would remain at 42 units and never fall below the targeted reorder point for this product.

DeHoratius and Raman (2008) found the lost revenue due to stockouts caused by record inaccuracy problems at Gamma amounted to 1.09% of Gamma's retail sales and 3.34% of its gross profit. They derived this estimate from examining those items similar to the one above – items that were out of stock at the store but with a positive on-hand quantity sufficiently large so as to prevent the automated replenishment system from triggering an order.

4.2 Effect of misplaced products on inventory planning

Figure 4-3 shows the cumulative number of customers who entered a store and the cumulative sales for a particular product, a type of bread, between 8:00 am and 8:00 pm. As shown in the figure, the cumulative number of customers entering the store steadily increased from 8:00 am to 8:00 pm. The particular product, on the other hand, was selling well until about 12:30 pm, did not sell at all from 12:30 to 4:00 pm, started selling again after 4:00 pm, and stopped selling after 6:00 pm.

During both of these periods when there were no sales for this particular product, the system inventory level for this product was positive. As a result, a simple interpretation of these sales and inventory data would be that the in-stock for this product was 100%, and that there was no demand for this product between 12:30 pm and 4 pm, and after 6 pm. The reality, however, was quite different. Between 12:30 pm and 4 pm, the inventory was located in the backroom, and was not available to the customers. At 6 pm the product stocked out.

Although one could argue that even if the product was available for sale no customer would have chosen to purchase it during 12:30 and 4 pm, we believe this to be highly unlikely. Given that there was no change in the rate at which customers entered the store during this period, it is likely that the store lost sales as a result of this product misplacement.

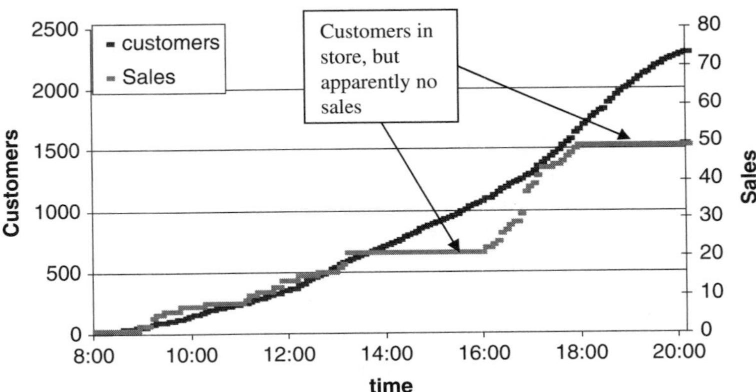

Figure 4-3. Store Sales versus Customer Entrances (Source: Ton, 2002)

4.3 Incorporating execution problems into existing research streams

The two examples above demonstrate the challenge faced by retailers when deciding how much and when to replenish. Because retailers are unable to observe all customer actions, they rely on data to infer the actions of customers and plan accordingly. Yet, as the two examples reveal, these data can be misleading. In reality, execution errors lead to additional uncertainty and recently several researchers have begun to incorporate such uncertainty into both tactical and strategic planning models. This research shows the impact execution problems have on tactical decisions such as safety stock calculations, ordering policies, and the timing of inventory counts as well as strategic decisions such as channel coordination and technology choice.

Among the papers addressing tactical decisions in the presence of execution problems, Iglehart and Morey (1972) were the first to determine the optimal buffer stock that protects against shortages caused by record inaccuracy. Moreover, they determine the optimal frequency of inventory counts by taking into account the cost of holding buffer stock and the cost of conducting inventory counts to correct record inaccuracies. More recently, Kök and Shang (2007) derive the optimal joint inspection and replenishment policy by minimizing the total inventory and inspection cost. They create an inspection adjusted base stock policy which adjusts the replenishment order according to the level of inaccuracy and the chosen inspection policy. They argue that the order quantity needs to be increased to accommodate the additional uncertainty caused by record inaccuracy, that inaccuracy accumulates over time, and that discrepancies can be corrected through inspection. Thus, if inspection cost is high, their model suggests auditing less frequently and carrying additional inventory to buffer against record inaccuracy.

Iglehart and Morey (1972) and Kök and Shang (2007) assume error in inventory records are random with a mean of zero. Thus, the discrepancy between the inventory record and actual inventory can take either sign. Moreover, both papers allow for a correlation between an item's demand and the magnitude of its inventory discrepancy. Emma (1966) and DeHoratius and Raman (2008) show empirically that the more frequently an item sells (i.e., the greater the demand) the greater the record inaccuracy. By taking into account these empirical findings, Iglehart and Morey (1972) and Kök and Shang (2007) offer practical solutions to record inaccuracy. However, one factor that limits the applicability of Kök and Shang's (2007) findings to the retail context is their backlogging assumption. In most retail settings, unfilled demand is lost rather than backlogged.

Offering an alternative solution to record inaccuracy, DeHoratius et al. (2008) propose the maintenance of a probabilistic inventory record to account for the presence of inventory record inaccuracy in retail systems. This probabilistic inventory record would take the place of the point estimate commonly used in retail to track inventory holdings. They model inventory inaccuracy through an "invisible" demand process that can either deplete or replenish physical but not recorded inventory. Using a periodic review process with unobserved lost sales, they demonstrate that the impact of inventory record inaccuracy can be mitigated through this probabilistic approach to inventory planning. Furthermore, they do so while taking into account the product characteristics that have been shown to impact record inaccuracy such as the cost of a product and its annual selling quantity (DeHoratius and Raman, 2008).

Kang and Gershwin (2005) focus on one source of inventory record inaccuracy, namely theft, whereby the quantity of units recorded ends up being greater than that actually found on the retail shelf for a given item. Fleisch and Tellkamp (2005) also analyze inventory shortages caused by either record inaccuracy or misplaced products. Through simulation, these studies demonstrate that even small rates of inventory record inaccuracy and misplaced products can result in substantial lost sales and suboptimal retail performance.

Camdereli and Swaminathan (2005), Rekik et al. (2008), Atali et al. (2005), and Gaukler et al. (2007) focus primarily on strategic planning in the face of execution problems. Camdereli and Swaminathan (2005) not only derive the optimal inventory policy for a retailer that knows the proportion of its inventory that is misplaced but also show that decreasing the proportion of misplaced products impacts channel parties differently. They identify conditions for channel coordination in the face of reduced product availability due to misplaced products. Rekik et al. (2008) also examine the impact of reduced product availability due to misplaced products. Unlike Camdereli and Swaminathan (2005), the objective of Rekik et al. (2008) is to explore how the use of RFID technology can mitigate the cost of product misplacement. Gaukler et al. (2007) also examine the role of RFID by evaluating whether the use of this technology can improve in-store, shelf replenishment processes and hence product availability. They also discuss the impact of execution problems on channel coordination and the differential benefit RFID technology has among channel members. Similar to Rekik et al. (2008) and Gaukler et al. (2007), Atali et al. (2005) examine the value of RFID technology in reducing execution errors by comparing an inventory system with and without the visibility such technology provides. However, unlike the previously cited papers, Atali et al. (2005) evaluate not only product misplacement, one-sided errors that deplete inventory levels and reduce product availability, but also execution problems than can result in the actual inventory level exceeding the recorded level.

5. FUTURE RESEARCH OPPORTUNITIES

There are several research opportunities for those interested in the impact of store execution on product availability, in particular, and on retail supply chains, in general. For example, the widely accepted theoretical relationship between inventory levels and product availability is that increasing inventory levels is associated with increased service levels and thus increased store sales. The empirical findings summarized in this chapter suggest that increasing inventory levels also increases the occurrence of misplaced products and inventory record inaccuracy. Thus, through its effect on store execution, increasing inventory levels may also compromise service levels. There is an opportunity for management scholars to develop models where both the direct and indirect effects of increasing inventory levels on product availability are considered and to examine empirically the direct and indirect effects of inventory levels on stores sales. In Ton and Raman's (2007) longitudinal study, the direct positive effect of increasing inventory levels on sales is larger than the indirect negative effect through store execution. There may, however, be settings where the indirect negative effects outweigh the direct positive effects.

There is also opportunity to incorporate execution problems into models that estimate demand and assess forecast accuracy in the presence of stockouts (Wecker 1978; Agrawal and Smith 1996; Nahmias 1994; Raman and Zotteri 2000; Smith and Agrawal 2000). Raman and Zotteri, for example, argue that sales data could be used along with inventory data to incorporate lost sales estimates into the estimation of demand. More specifically, the authors generate an estimate of the lost sales by using inventory data to identify when a stockout occurred. Once it is known when the stockout occurred, the historical sales rate can be appropriately extrapolated to determine lost sales during the stocked out period. Thus, both observed demand when a product is in stock (e.g., sales history) and an estimate of unobserved demand when product is out of stock (lost sales estimation) can be used to estimate the demand more accurately. Consider a common situation, however, where there are no units of inventory for the product (either due to inventory record inaccuracy or misplacement) while the inventory records show a positive value for inventory. In this situation, sales for the product will be zero, while the inventory for the product will be shown to be positive. Consequently, the above analysis will conclude that there was no demand for the product during the period where the inventory record was inaccurate or when the product was misplaced and the demand estimation will be inaccurate. The demand estimation could be improved by assigning a probability that the inventory record is not a true reflection of reality when the system inventory level for the product is positive but there are no observed sales.

Note that efforts to compensate for execution problems with robust decision support tools and efforts to prevent execution problems through, among others, improved process design, improved conformance to designed processes and error prevention are not mutually exclusive. There is much to gain from better understanding the ways to eliminate or reduce the prevalence of execution problems while simultaneously designing decision tools robust enough to account for the existence of execution problems. As we reviewed in this chapter, existing research has identified numerous factors that exacerbate the occurrence of execution problems. These factors are largely under the control of retail managers. For example, retail managers can choose to invest more in training or spend more on payroll expenses to reduce the occurrence of execution problems. These actions, however, result in increased costs. Given that higher profitability is the overreaching goal, researchers and retail managers can develop models that optimize store profitability given the relevant costs and benefits of changes to training or payroll expenses. Moreover, retailers could institute a process improvement effort that identifies and corrects those employee actions leading to errors in product location or record inaccuracy.

Additional empirical research opportunities exist. For example, the findings of DeHoratius and Raman (2008) and Ton and Raman (2006) need to be tested using data from other retailers in order to determine their generalizability. The effect of human resource variables such as employee turnover, training, and workload on inventory record inaccuracy also needs to be examined. Moreover, opportunities exist for researchers to examine the impact of process design, technology, or employee incentives on execution. Given that most retailers do not alter process design, technology usage, or employee incentives across their own stores, researchers wishing to examine these factors would need to compare execution across several different retailers.

More research is also needed to examine the relationship between storage area size and store execution. Proponents of lean production systems have argued that smaller repair areas force employees to introduce procedures to reduce defects and hence production systems with smaller repair areas are associated with higher quality. Most retail stores include areas for storing extra merchandise that do not fit into the display shelves. These storage areas are similar to the repair areas in manufacturing plants. Consistent with the lean production system argument, Ton and Raman (2006) hypothesize, but are unable to support with their data, a relationship between smaller storage areas and product misplacement. Nevertheless, the authors cite anecdotal evidence suggesting that smaller storage areas force store employees to better monitor products in the storage areas and to quickly replenish the selling floor with units from storage locations, lowering the level of product misplacements. Specifically, store execution suffered tremendously at one particular store when the store's storage capacity increased from one year to next.

One potential reason for the lack of statistical significance may be the imperfect measure the authors used for storage capacity. Although stores typically have multiple storage areas, Ton and Raman (2006) used size of the backroom as a surrogate for total storage in a store. In addition, it might be *how* the storage area is managed rather than its size that affects the percentage of products that are in storage but not on the selling floor. Ton (2002) observed a great deal of variation in the utilization of the storage areas. Some backrooms were very well organized, with products clearly categorized, and each shelf well displayed with labels that indicated what merchandise was stored on that shelf. In other backrooms there were no labels on the shelves and multiple products were stacked on top of each other. Some backrooms were so messy that there were boxes and carts in between the shelves that prevented employees from gaining access to large areas of the storage space.

There is also opportunity to conduct empirical research on the consequences of poor store execution. DeHoratius and Raman (2008) and Ton and Raman (2007) show the negative effect of poor store execution on store sales. Similar research can be conducted to examine the effect of store execution on other financial or non-financial measures of store performance. In addition, researchers may examine the effect of store execution on the performance of retail supply chain initiatives such as vendor managed inventory or collaborative planning, forecasting and replenishment (CPFR) programs.

Note that in this chapter we focused solely on the effect of execution on managing product availability. The execution problems described in this chapter have implications beyond managing product availability. Accurate inventory data may allow a company to make same-day delivery promises or to integrate online and physical store operations (see Raman and Ton, 2003 and Ton and Raman, 2003 for teaching case examples). Thus, researchers can identify the impact of execution problems on business strategy as well as performance.

APPENDIX 1

DeHoratius and Raman (2008)

Research Site: The authors examine the drivers of inventory record inaccuracy using data from Gamma Corporation, a large specialty retailer with over ten billion dollars in annual sales. Gamma uses electronic point-of-sale scanning for all its sales and an automated replenishment system for inventory replenishment.

Data: The authors collected data from physical audits of 37 Gamma stores in 1999. These data included detailed information about each stock-keeping-unit (SKU) contained in each store, amounting to a total of 369,567 observations, or SKU-Store combinations. Physical audits revealed the recorded quantity (the recorded number of inventory units for each SKU at a specific store) as well as the on-hand quantity (the actual number of inventory units present at the store for each SKU). In addition to SKU level data, the authors collected both store and product category data and complemented their quantitative analysis with extensive fieldwork.

Dependent Variable: The dependent variable is the inventory record inaccuracy of each SKU in each store. Inventory record inaccuracy (IRI) is measured as the absolute difference between the recorded and actual quantity for each SKU-store combination.

Independent variables: SKU level variables are: the cost of the item, its annual selling quantity, and whether the item had been shipped to the store from one of Gamma's distribution centers or directly from the vendor. Store level variables are: total number of units in a given selling area, product variety, and the number of days between the current and previous physical audit.

Empirical Model: Because these data have a multi-level structure (SKUs are contained within stores and product categories), the authors fit a series of hierarchical linear models to examine the drivers of IRI. In addition to the independent variables noted above, the empirical model includes the following control variable, a dummy for each region in which a store is located (REGION_ONE$_k$,REGION_TWO$_k$). Equation (1) below summarizes their model.

$$\begin{aligned} IRI_{ijk} = &\Theta_0 + b_{00j} + c_{00k} + e_{ijk} + \pi_1 * (QUANTITY\_SOLD_{ijk}) \\ &+ \pi_2 * (ITEM\_COST_{ijk}) + \pi_3 * (DOLLAR\_VOLUME_{ijk}) \\ &+ \pi_4 * (VENDOR_i) + \pi_5 * (VENDOR\_COST_{ijk}) \\ &+ \gamma_{001} * (REGION\_ONE_k) + \gamma_{002} * (REGION\_TWO_k) \\ &+ \gamma_{003} * (DENSITY_k) + \gamma_{004} * (VARIETY_k) + \gamma_{005} * (DAYS_k) \end{aligned} \tag{1}$$

where

IRI_{ijk} is the record inaccuracy of item i ($i = 1. \ldots, n_{jk}$) in product category j ($j = 1. \ldots \ldots, 68$) and store k ($k = 1, \ldots \ldots, 37$).
Θ_0 is a fixed intercept parameter.
The random main effect of product category j is $b_{00j} \sim N(0, \tau_{boo})$.
The random main effect of store k is $c_{00\,k} \sim N(0, \tau_{coo})$.
The random item effect is $e_{ijk} \sim N(0, \sigma^2)$.
τ_{boo}, τ_{coo}, and σ^2 define the variance in IRI between product categories, stores, and items, respectively.
π_1–π_5 are the fixed item level coefficients and γ_{001}-γ_{005} are the fixed store level coefficients.

Each of the variables is defined below:

QUANTITY_SOLD$_{ijk}$ is the annual selling quantity of item i in product category j and store k
ITEM_COST$_{ijk}$ is the cost of item i in product category j and store k
DOLLAR_VOLUME$_{ijk}$ is the interaction between the cost of the item and its annual selling quantity
VENDOR$_i$ is a dichotomous variable that takes the value of one if the item is shipped direct to the store from the vendor and takes the value of zero if the item is shipped to the store from the retail-owned distribution center.
VENDOR_COST$_{ijk}$ is an interaction term between the way in which an item was shipped to the store and its cost
DENSITY$_k$ is the total number of units in a store divided by that store's selling area (units per square foot)
VARIETY$_k$ is the number of different merchandise categories within a store
DAYS$_k$ measures the number of days between audits for a given store.

Findings: The authors find significant positive relationships between IRI and an item's annual selling quantity, store inventory density, store product variety, and the number of days since the last store audit. A significant negative relationship exists between IRI and an item's cost as well as its dollar volume. The way in which an item is shipped to the store is a significant predictor of IRI such that items shipped direct to the store from the vendor are more accurate than items shipped from the retail distribution center. This relationship, however, depends on the cost of an item. Specially, the difference between vendor-shipped and DC-shipped items is greater for inexpensive items than for expensive ones.

Ton and Raman (2006)

Research Site: The authors examine the drivers of misplaced products using data from Borders Group, a large retailer of entertainment products such as books, CDs, and DVDs. To ensure product availability, the retailer has invested heavily in information technology and merchandise planning to make sure that the right product is sent to the right store at the right time.

Data: The authors collected data from physical audits of 242 Borders stores in 1999. Physical audits provide data on the total units of inventory at the store, total number of products at the store, and the number and dollar value of the products that were present in storage areas but not on the selling floor. In addition to physical audit data, the authors collected data on store attributes and human resource characteristics. The authors complemented their empirical data with extensive fieldwork.

Dependent Variable: The dependent variable, % phantom products, is the percentage of products that are in storage areas but not on the selling floor. The authors call these products "phantom" because they are physically present in the store and often shown as available in retailers' merchandising systems, but in fact are unavailable to customers.

Independent variables: The authors use the following independent variables: inventory level per product, total number of products in a given area, size of the storage area, employee workload, employee turnover, store manager turnover, and the number of trainers at the store.

Empirical Model: The authors estimate the parameters of equation (2) using ordinary least square estimator to examine the drivers of % phantom products. In addition to all independent variables, the empirical model includes the following control variables: store sales, store age, seasonality, unemployment rate, and a dummy variable for each region. Note that, one variable, store sales, is an endogenous variable and hence the authors employ instrumental variable estimation to cope with endogeneity. The authors use corporate sales as an instrument for store sales.

$$\begin{aligned}
\%PHANTOM_PRODUCTS_i = {} & \beta_0 + \beta_1\, SEASONALITY_i + \beta_2\, UNEMPLOYMENT_RATE_i + \beta_3\, LN(AGE)_i \\
& + \beta_4\, SALES_i + \beta_5\, WAGE_i + \beta_{6j}\, REGION_i + \beta_7\, INVENTORY_DEPTH_i \\
& + \beta_8\, PRODUCT_DENSITY_i + \beta_9\, STORAGE_SIZE_i \qquad (2) \\
& + \beta_{10}\, LABOR_INTENSITY_i + \beta_{11}\, SMTURNOVER_i + \beta_{12}\, FTTURNOVER_i \\
& + \beta_{13}\, PTTURNOVER_i + \beta_{14}\, TRAINING_i + \varepsilon_i
\end{aligned}$$

Each of the variables is defined below:

%PHANTOM_PRODUCTS$_i$ is the number of products in storage but not on floor in store *j* divided by the total number of products in store *i*.

SEASONALITY$_{ij}$ is the seasonality index for month *j* in which the audit is conducted at store *i*. The seasonality index for month *j* is calculated as:

$$\theta_j = \frac{\sum_{i=1}^{242} S_{ij}}{\left(\sum_{j=1}^{12} \sum_{i=1}^{242} S_{ij} \middle/ 12 \right)}$$

UNEMPLOYMENT_RATE$_i$ is the unemployment rate of the metropolitan statistical area in which the store is located in 1999.

LN (AGE)$_i$ is the natural log of the age of store *i* (in months) during the time of the audit.

SALES$_i$ is the total sales at store *i* in 1999.

WAGE$_i$ is the average hourly wage at store *i* in 1999.

REGION$_j$ are 17 dummy variables indicating region in which store *i* is located.

INVENTORY_DEPTH$_i$ is the total number of units in store *i* divided by the number of products in store *i*.

PRODUCT_DENSITY$_i$ is the number of products in store *i* divided by the total selling area of store *i*.

STORAGE_SIZE$_i$ is the backroom area of store *i* divided by the total selling area of store *i*.

LABOR_INTENSITY$_i$ is the payroll expenses at store *i* in 1999 divided by sales at store *i* in 1999.

SM TURNOVER$_i$ is a dummy variable indicating the departure of store manager at store *i* in 1999.

FT TURNOVER$_i$ is the total number of full-time employees in store *i* that departed in 1999 divided by the average number of full-time employees in store *i*.

PT TURNOVER$_i$ is the total number of part-time employees in store *i* that departed in 1999 divided by the average number of part-time employees in store *i*.

TRAINING$_i$ is the total number of "trainer months" at store *i* in 1999.

Findings: The authors find significant positive relationships between % phantom products and inventory level per product, total number of products in a given area, employee workload, and store manager turnover. The authors find partial support for the positive relationship between employee turnover and % phantom products. The authors also find a significant negative relationship between % phantom products and the amount of training at the store.

Ton and Raman (2007)

Research Site: The authors examine the effect of product variety and inventory levels on store sales using data from Borders Group.

Data: The authors collected data from physical audits of all Borders stores from 1999 to 2002. The dataset includes 356 stores, some of which opened between 1999 and 2002. As a result the authors do not have four years of data for all 356 stores.

Dependent Variables: The authors use two dependent variables. First is the percentage of phantom products, products that are in storage areas but not on the selling floor. The second dependent variable is store sales.

Independent variables: The authors use the following independent variables: inventory level per product, total number of products at a store.

Empirical Model: The authors estimate the parameters of equation (3) to examine the effect of product variety and inventory levels on % phantom products and estimate the parameters of equation (4) to examine the effect of % phantom products on store sales. In both equations, the authors control for *factors that vary over time for stores and are different across stores* (seasonality, unemployment rate in the store's metropolitan statistical area, amount of labor used in a month, employee turnover, full-time employees as a percentage of total employees, store manager turnover, and the number of competitors in the local market), *factors that vary over time but are invariant across stores* (year fixed effects), and *factors that are time-invariant for a store but vary across stores* (store fixed effects).

The authors use ordinary least squares (OLS) estimators in estimating both equations (3) and (4) and report the heteroskedasticity robust standard errors for OLS. In addition to OLS estimators, the authors also treat equations (3) and (4) as seemingly unrelated regressions (SUR) allowing for correlation in the error terms across two equations. In addition, because of autocorrelation in the error terms of equation (4), the authors consider a flexible structure of the variance covariance matrix of the errors with first-order autocorrelation and estimate the parameters of (4) using maximum likelihood estimation.

$$\begin{aligned}\%PHANTOM\_PRODUCTS_{it} = \alpha_i + \lambda_t + \beta_1\, PRODUCT\_VARIETY_{it} \\ + \beta_2 INVENTORY\_LEVEL_{it} + X_{it}\beta + \varepsilon_{it}\end{aligned} \tag{3}$$

$$\begin{aligned}SALES_{it} = \eta_i + \theta_t + \gamma_1\, \%PHANTOM\_PRODUCTS_{it} + \gamma_2\, PRODUCT\_VARIETY_{it} \\ + \gamma_3\, INVENTORY\_LEVEL_{it} + X_{it}\gamma + \varepsilon_{it}\end{aligned} \tag{4}$$

Each of the variables is defined below:

$\%PHANTOM\_PRODUCTS_{it}$ is products that are in storage areas but not on floor at store i in year t at the time of the physical audit divided by the # of products at store i in year t at the time of the physical audit
$SALES_{it}$ is sales during the month preceding the audit at store i in year t
$PRODUCT\_VARIETY_{it}$ is the # of products at the store at the time of the physical audit at store i in year t
$INVENTORY\_LEVEL_{it}$ is the # of units at the store at the time of the physical audit at store i in year t divided by the # of products at the store at the time of the physical audit at store i in year t
The vector X_{it} represents the following control variables:
$SEASONALITY_j$ is the seasonality index for month j in which the audit is conducted at store. Let S_{ijt} = sales at store i in month j in year t. Then the seasonality index for month j is.

$$\frac{\sum_{t=1}^{4}\sum_{i=1}^{267} S_{ijt}}{\left(\sum_{t=1}^{4}\sum_{j=1}^{12}\sum_{i=1}^{267} S_{ijt} \Big/ 48\right)}$$

$UNEMPLOYMENT_{it}$ is the unemployment rate of the metropolitan statistical area in which the store is located during the month preceding the audit at store i in year t.
$LABOR_{it}$ is the payroll expenses during the month preceding the audit at store i in year t.
$EMPLOYEE\_TURNOVER_{it}$ is the fraction of employees that are charged with managing inventory that had left during the month preceding the audit at store i in year t.
$PROPORTION\_FULL_{it}$ is the fraction of full-time employees during the month preceding the audit at store i in year t.
$SM\_TURNOVER_{it}$ is a dummy variable that has a value of 1 if the store manager had left the company voluntarily since the last physical audit at store i in year t.
$COMPETITION_{it}$ is the total number of Barnes & Noble and Borders stores in the area during the month preceding the audit at store i in year t.
$PLAN\_SALES_{it}$ is the forecasted sales during the month preceding the audit at store i in year t.

Findings: The authors find that increasing both product variety and inventory level per product at a store is associated with an increase in % phantom products. The authors also find that an increase in % phantom products is associated with a decrease in store sales. Consequently, increasing product variety and inventory levels has an indirect effect on store sales. This indirect negative effect, however, is smaller than the direct positive effect of increasing inventory levels and product variety on store sales.

REFERENCES

Agrawal, N., and Smith, S. A., 1996, Estimating Negative Binomial Demand for Retail Inventory Management with Unobservable Lost Sales, *Naval Res. Log.* **43**: 839–861.

Andersen Consulting, 1996, *Where to Look for Incremental Sales Gains: The Retail Problem of Out-of-Stock Merchandise.*

Anderson, P., 1995, *Technology. The Blackwell Encyclopedic Dictionary of Organizational Behavior*, Blackwell Publishers, Inc., Cambridge, pp. 557–560.

Argote, L., and Epple, D., 1990, Learning Curves in Manufacturing, *Science* **247**: 920–924.

Atali, A., Lee, H., and Özer, Ö., 2005, If the Inventory Manager Knew: Value of RFID under Imperfect Inventory Information, *M&SOM Conference Proceedings.*

Bayers, C., 2002, The Last Laugh, *Business 2.0* **3**(9): 86–93.

Bergman, R. P., 1988, A B Count Frequency Selection For Cycle Counting Supporting MRP II, *Prod. and Inv. Mgt. Rev.* 35–36.

Bluedorn, A., 1982, The Theories of Turnover: Causes, Effects, and Meaning, *Res. in the Soc. of Org.* **1**: 75–128.

Brooks, R. B., and Wilson, L. W., 1993, *Inventory Record Accuracy. Unleashing the Power of Cycle Counting*, Oliver Wight Publications Inc., Essex Junction.

By the Numbers, 2005, *Government Executive* **37**(6): 12.

Camdereli, A. Z., and Swaminathan, J. M., 2005, Coordination of a Supply Chain Under Misplaced Inventory, University of North Carolina Working Paper OTIM-2005–02

Capital Markets Report (June 22, 2000); http://www.erisks.com.default.asp.

Cassidy, A., and Mierswinski, E., 2004, *Mistakes Do Happen: A Look at Errors in Consumer Credit Reports.* U.S. Public Interest Research Group Report.

Dalton, D., and Todor, W., 1979, Turnover Turned Over: An Expanded and Positive Perspective, *Acad. of Mgt. Rev.* **4**(2): 225–235.

DeHoratius, N., 2002, *Critical Determinants of Retail Execution*, Harvard Business School Unpublished Dissertation.

DeHoratius, N., Mersereau, A., and Schrage, L., 2008, Retail Inventory Management When Records are Inaccurate, *Manufacturing and Service Operations Management* **10**(2): 257–277.

DeHoratius, N., and Raman, A., 2008, Inventory Record Inaccuracy: An Empirical Analysis, *Mgt. Sci.* **54**(4): 627–641.

Emma, C. K., 1966, Observations on Physical Inventory and Stock Record Error, Interim Report 1, Department of Navy Supply Systems Command, Mechanicsburg, PA.

Emmelhainz, L. W., Emmelhainz, M. A., and Scott, J. R., 1991, Logistics Implications of Retail Stockouts, *J. of Bus. Log.* **12**(2): 129–142.

Fisher, M.L., Jain A., and MacDuffie, J.P., 1995, Strategies for Product Variety: Lessons from the Auto Industry, in: *Redesigning the Firm,* B. Kogut and E. Bowmanm eds., Oxford University Press, New York, pp. 116–154.

Fisher, M. L., and Ittner, C. D., 1999, The Impact of Product Variety on Automobile Assembly Operations: Empirical Evidence and Simulation Analysis, *Mgt. Sci.* **45**(6): 771–786.

Fleisch, E., and Tellkamp, C., 2005, Inventory Inaccuracy and Supply Chain Performance: A Simulation Study of Retail Supply Chains, *Intern. J. of Prod. Econ.* **95**(3): 373–85.

Flores, B. E., and Whybark, D. C., 1986, Multiple Criteria ABC analysis, *Inter. J. of Oper. and Prod. Mgt.* **6**(3): 38–45.

Flores, B. E., and Whybark, D. C., 1987, Implementing Multiple Criteria ABC Analysis, *J. of Oper. Mgt.* **7**: 79–85.

Galway, L. A., and Hanks, C. H., 1996, *Data Quality Problems in Army Logistics*, MR-721-A RAND.

Gaukler, G. M., Seifert, R. W., and Hausman, W. H., 2007, Item-Level RFID in the Retail Supply Chain, *Prod. and Oper. Mgt.* **16**(1): 65–76.

Graves, S. G., and de Kok, A. G., 2003, *Handbooks in Operations Research and Management Science, v. 11: Supply Chain Management: Design, Coordination, and Operations*, Elsevier Publishers, Amsterdam.

Gruen, T. W., Corsten, D. S., and Bharadwaj, S., 2002, *Retail Out-of-Stocks: A Worldwide Examination of Extent, Causes and Consumer Responses*. Grocery Manufacturers of America, The Food Marketing Institute, and CIES.

Hall, R.V., 1983, *Zero Inventories*, Dow Jones-Irvin Inc., Homewood.

Hart, M. K., 1998, Improving Inventory Accuracy Using Control Charts, *Prod. and Inv. Mgt* **11**: 44–48.

Hollinger, R. C., and Langton, L., 2003, *National Retail Security Survey Final Report*. www.soc.ufl.edu/srp.htm.

Iglehart, D. L., and Morey, R. C., 1972, Inventory Systems with Imperfect Asset Information, *Mgt. Sci.* **18**(8): B388- B394.

Kang, Y., and Gershwin, S. B., 2005, Information Inaccuracy in Inventory Systems: Stock Loss and Stockout, *IIE Transactions* **37**(9): 843–59.

Kök, G. A., and Shang, K. H., 2007, Inspection and Replenishment Policies for Systems with Inventory Record Inaccuracy, *Manufacturing and Service Operations Management* **9**(2): 185–205.

Krajewski, L. J., King, B. E., Ritzman, L. P., and Wong, D. S., 1987, Kanban, MRP, and Shaping the Manufacturing Environment, *Mgt. Sci.* **33**(1): 39–57.

Krafcik, J. F., 1988, Triumph of the Lean Production System, *Sloan Management Review* **30**(1): 41–51.

Kurt Salmon Associates, 2002, *Biennial Consumer Outlook Survey*.

Laudon, K. C., 1986, Data Quality and Due Process in Large Interorganizational Record Systems, *Comm. of the ACM* **29**: 4–11.

MacDuffie, J. P., Sethuraman, K, and Fisher, M. L., 1996, Product Variety and Manufacturing Performance: Evidence from the International Automotive Assembly Plant Study, *Mgt. Sci.* **42**(3): 350–369.

Managing Customer Service, 2001, Hiring and Training Tips From Fortune's Top-Rated Employer, August 1.

McClain, J. O., Thomas, L. J., and Mazzola, J. B., 1992, *Operations Management: Production of Goods and Services*, Prentice-Hall /Pearson Education, Upper Saddle River.

McCutcheon, C., 1999, Pentagon's Ongoing Record of Billions in Lost Inventory Leads Hill to Demand Change, *Cong. Quart. Week* **57**(18): 1041.

Millet, I., 1994, A Novena to Saint Anthony, or How to Find Inventory by Not Looking, *Interfaces* **24**: 69–75.

Mobley, W., 1982, *Employee Turnover: Causes, Consequences, and Control*, Addison-Wesley, Reading.

Nahmias, S., 1994, Demand Estimation in Lost Sales Inventory Systems, *Nav. Res. Log.* **41**: 739–757.

Nelson, R., and Winter, S., 1982, *An Evolutionary Theory of Economic Change*, Harvard University Press, Cambridge.

Raman, A., DeHoratius, N., and Ton, Z., 2001, Execution: The Missing Link in Retail Operations, *California Management Review* **43** (3): 136–152.

Raman, A., and Ton, Z., 2003, Operational Execution at Arrow Electronics, Harvard Business School Case.

Raman, A., and Zotteri, G., 2000, Estimating Retail Demand and Lost Sales, Harvard Business School Working Paper .

Reason, J. 2002, *Human Error,* Cambridge University Press, Cambridge.

Redman, T., 1995, Improve Data Quality for Competitive Advantage, *Sloan Mgt. Rev.* **36**: 99–107.

Rekik, Y, Sahin, E., and Dallery, Y., 2008, Analysis of the Impact of RFID Technology on Reducing Product Misplacement Errors at the Retail Stores, *International Journal of Production Economics*, **112**(1): 264–278.

Rinehart, R. F., 1960, Effects and Causes of Discrepancies in Supply Operations, *Oper. Res.* **8**(4): 543–564.

Rout, W., 1976, That Damn Storeroom, *Prod. and Inv. Mgt.* **17:** 22–29.

Schonberger, R. J., 1982, *Japanese Manufacturing Techniques: Nine Hidden Lessons in Simplicity,* The Free Press, New York.

Schrady, D. A., 1970, Operational Definitions of Record Accuracy, *Naval Res. Log. Quart.* **17**(1): 133–142.

Sheppard, G. M., and Brown, K. A., 1993, Predicting Inventory Record-Keeping Errors with Discriminant Analysis: A Field Experiment, *Inter. J. of Prod. Econ.* **32**: 39–51

Skinner, W., 1974, The Focused Factory, *Har. Bus. Rev.* **53**(May June): 113–121.

Smith, S. A., and Agrawal, N., 2000, Management of Multi-Item Retail Inventory Systems with Demand Substitution, *Oper. Res.* **48**(1): 50–64.

Staw, B., 1980, The Consequences of Turnover, *J. of Occup. Beh.* **1**(4): 253–273.

Steers, R., and Mowday, R., 1981, Employee Turnover and Post-Decision Accommodation Processes, *Res. in Org. Beh.* **3**: 235–281.

Tallman, J., 1976, A Practical Approach to installing a Cycle Inventory Program, *Prod. and Inv. Mgmt.* **17**(4): 1–16.

Tayur, S., Ganeshan, R., and Magazine, M., 1999, *Quantitative Models for Supply Chain Management*, Kluwer Academic Publishers, Boston.

Ton, Z., 2002, *The Role of Store Execution in Managing Product Availability*, Harvard Business School Unpublished Dissertation.

Ton, Z., and Raman, A., 2006, Cross Sectional Analysis of Phantom Products at Retail Stores, Harvard Business School Working Paper.

Ton, Z., and Raman, A., 2007, The Effect of Product Variety and Inventory Levels on Retail Store Sales: A Longitudinal Study, Harvard Business School Working Paper.

Ton, Z., and Raman, A., 2003, Borders Group, Inc., Harvard Business School Case.

Wecker, W. E., 1978, Predicting Demand From Sales Data in the Presence of Stockouts, *Mgt. Sci.* **24**: 1043–1054.

White, E., 2005, New Recipe: To Keep Employees, Domino's Decides It's Not All About Pay, *The Wall Street Journal*, February 17, A1.

Woolsey G., 1977, The Warehouse Model That Couldn't Be and The Inventory that Couldn't Be Zero, *Interfaces* **7**(3): 14–17.

Woellert, L., June 28, 2004, Shortchanged on Long Distance, *BusinessWeek* **3889**: 13.

Young, S. T. and Nie, W. D., 1992, A Cycle-Count Model Considering Inventory Policy and Record Variance, *Prod. and Inv. Mgmt.* **33**(1): 11–16.

Chapter 5
CATEGORY CAPTAINSHIP PRACTICES IN THE RETAIL INDUSTRY

Mümin Kurtuluş[1] and L. Beril Toktay[2]
[1]*Owen Graduate School of Management, Vanderbilt University, 401 21st Avenue South, Nashville, TN 37203, USA*
[2]*College of Management, Georgia Institute of Technology, 800 West Peachtree Street NW, Atlanta, GA 30308-0520, USA*

1. INTRODUCTION

A product category is defined as a group of products that consumers perceive to be interrelated and/or substitutable (Nielsen Marketing Research 1992). Soft drinks, oral care products, and frozen vegetables are some examples of retail categories. Categories can be viewed as the smallest strategic business unit within a retailer. Thus, retailers have refocused their efforts on managing entire product categories as a single business unit, a practice called category management. The goal is to improve business performance through focusing on delivering consumer value. In particular, retail category management involves decisions such as merchandizing product assortment, determining retail prices, and allocating shelfspace to each product on the basis of category goals. Unlike in the traditional approach where retailers managed their product portfolio on a brand-by-brand or SKU-by-SKU basis, category management emphasizes the management of product categories as a whole and allows the retailers to capture the synergies that may arise as a result of grouping the products together. Various synergies such as promotion coordination, store traffic driving strategies, and substitution patterns can be captured by grouping the products together. However, category management requires that a lot of resources be dedicated to understanding the consumer response to the assortment, pricing and shelf placement decisions of products within a category.

Recently, a new trend has emerged: Retailers have started to outsource retail category management to a chosen supplier on whom they rely for strategic recommendations and insights, a practice often referred to as "category captainship." Factors such as the increase in the number of product categories offered at the retailers, combined with the scarcity of the resources required to

N. Agrawal, S.A. Smith (eds.), *Retail Supply Chain Management*,
DOI: 10.1007/978-0-387-78902-6_5,

manage each category effectively have given rise to this new trend. In a typical category captain arrangement, the retailer shares all relevant information such as sales data, pricing, turnover, and shelf placement of the brands with the category captain. The category captain, in return, performs analysis about the category and provides the retailer with a detailed plan that includes recommendations about which brands to include in the category, where to locate each brand on the shelf, how to display the products, how much space to allocate to each brand, which new brands to include and which old brands to exclude from the category, and how to price the products in the category. The retailer is free to accept or reject any of the recommendations provided by the category captain.

Category captainship practices vary depending on the retailer, resulting in a continuum of practices. At one end of the spectrum, some retailers implement the category captain's recommendations as they are; at the other end, some retailers filter the recommendations provided by the category captain and verify their appropriateness before deciding on the implementation (Steiner 2001). Retailers usually design the category captainship contracts to be short term (one to two years at most) in order to keep the flexibility of being able to renegotiate the contracts or rotate the category captain. In addition, category captainship contracts usually include target levels such as target profit and target sales to be achieved by the category captains.

1.1 Category captainship implementations in practice

Many retailers and manufacturers in the consumer goods industry practice category captainship and report positive benefits. Retailers such as Wal-Mart, Metro, Safeway, and Kroger practice category captainship in some of their product categories and usually assign manufacturers such as Kraft Foods, P&G, Kellogg and Danone to serve as category captains because of their established brands in the market and their resource availability. Below are some specific examples of category captainship implementations from practice.

Carrefour, the second largest retailer in the world, recently asked Colgate to serve as category captain in the oral care category. Based on a number of consumer studies, Colgate suggested that Carrefour restructure the display in the oral care category so as to merchandise toothbrush products above toothpaste products, as opposed to merchandising them next to each other. As a result of the restructuring, Carrefour reported 6–16% sales increase in the oral care categories in its retail markets. Colgate also benefited from this sales increase (ECR Conference 2004). The sales increase in the oral care category came at a little cost to the entire channel because Colgate mostly utilized its already existing consumer studies and its expertise in the oral care category. If Carrefour was to conduct the research necessary for such a restructuring, it would have been much more expensive.

Similarly, Ross Products serves as category captain for Safeway in the infant formula category (Progressive Grocer 2004). Safeway asked Ross Products to examine the category and prescribe solutions to improve the profitability of the category. Ross' assessment of the category revealed that the category was under-merchandised: the infant formula subcategory was contributing 34% of the baby care category's dollar volume, but was receiving only 11% of the shelfspace. Ross recommended some changes in shelfspace positioning, and also reviewed and revised the pricing to boost profitability. After implementing the recommendations, the category boasted 9.2% sales growth benefiting both Safeway and Ross Products (Progressive Grocer 2004). One could argue that Safeway could have developed a similar prescription to improve the performance in the infant formula category without using Ross Products as a category captain, however, the cost of doing so would have been much higher as Safeway does not have the expertise that Ross Products does.

General Mills serves as category captain for some of its retail partners in the Baking Ingredients and Mixes category (Progressive Grocer 2004). General Mills' recommendations are focused around SKU rationalization and variety-vs-duplication analysis. SKU rationalization is aimed at reducing the number of SKUs to reduce consumer confusion at the shelf and thus create growth. Similarly, excessive duplication does not add much in incremental volume. Removing duplications allows for expanded product variety, which in turn can generate more sales in the category and help it grow. One of the retailers for which General Mills serves as category captain has seen a 10.2% increase in base dollar volume since General Mills' SKU rationalization efforts (Progressive Grocer 2004).

Although category captains are common in the grocery and consumer products industries, category captainship practices are making an appearance in apparel retailing as well. VF Corp., NC based manufacturer of brands such as Lee and Wrangler, serves as category captain for a number of its retail partners in the jeans category (Apparel Magazine 2005). VF Corp works with its retail partners to determine the product mix to be offered in each region, how products will be displayed on the sales floor, and how inventory levels will be managed in the category. Inspired by the success in the jeans category, VF Corp is looking forward to take on category captainship responsibility in other categories such as sports licensing and outdoor performance apparel categories.

These examples, and many other successful category captainship implementations, demonstrate that by working together, retailers can considerably benefit from their manufacturers' expertise in managing their categories and deliver consumer value through supply chain collaboration. However, conflict of interest between the retailer and the category captain or between competing manufacturers could be an issue. First, what is in the best interest of the category captain may not be the best for the retailer. Second, the category captain may take advantage of its position and disadvantage competitor manufacturers. It is not surprising that there is an emerging debate on whether or not category captainship poses some antitrust challenges.

While there are many cases under investigation due to claims of antitrust practices, one publicly known and well-documented example where some antitrust issues have been important is the United States Tobacco Co. vs. Conwood Co. case. United States Tobacco Co. (UST), the biggest company in the smokeless-tobacco category, was recently condemned to pay a $1.05 billion antitrust award to Conwood, the second biggest competitor in the category (Greenberger 2003). Conwood had sued UST, the category captain, and had claimed that UST used its position as category captain to exclude competition and provide an advantage to its own brands. The court ruled that UST's practices resulted in unlawful monopolization, harming competition, and consequently, the consumers. This example clearly illustrates that category captainship practices might have negative impact on both the non-captain manufacturers and end consumers. Monopolization in the category may result in lower variety and higher prices, which in turn may harm the consumers. Similarly, many other category captainship arrangements in the tortillas, cranberries, and carbonated soft drinks categories are before the court regarding category captainship misconduct (Desrochers et al. 2003).

To summarize, while many retailer-manufacturer dyads claim positive benefits from their category captainship implementations, there is also evidence concerning negative impacts of using category captains. Retailers planning to implement category captainship should develop an understanding of the pros and cons of such practices and should weigh potential advantages and disadvantages of using category captains for category management. The goal of this chapter is to provide an overview of the existing research on category captainship, and identify research directions that would improve our understanding of its impact.

The chapter is organized as follows. We start by reviewing the literature on category captainship in Section 2. In Section 3, we discuss the potential impact of category captainship practices on the retailing industry. Section 4 offers some future research directions.

2. REVIEW OF EXISTING RESEARCH ON CATEGORY CAPTAINSHIP

Despite a decade of implementation, there is limited academic research concerning category captainship. The existing research on category captainship can be grouped into four broad categories that aim to answer the following questions:

- Under what conditions will category captainship partnerships emerge in equilibrium?
- What is the impact of the retailer delegating the pricing decision to a category captain?

- What is the impact of the retailer delegating the assortment selection decision to a category captain? How should the retailer structure the manufacturer relationship for maximum benefit?
- What are the antitrust issues that may arise as a result of using category captains for category management? What can be done to avoid these antitrust issues?

The limited research in this field is due to some challenges such as the broad scope of category captainship implementations and continuum of category captainship implementations. In general, category captainship implementations include recommendations about retail category management decisions such as pricing, assortment, shelfspace management, promotions, etc. However, researchers usually focus on recommendations in only one of these areas, limiting their research and findings to a subset of category captainship implementations. In addition, while some retailers implement their category captain's recommendations as they are, others use them only after modifying the recommendations. Researchers usually focus on one end of this spectrum where the retailer implements the recommendations as they are and ignore all other possibilities. In section 4, we propose some avenues for future research that could potentially overcome these challenges and improve our understanding of category captainship practices.

In what follows, we will review the literature by describing their contributions to each of the questions in the above outline. Niraj and Narasimhan (2003), Wang et al. (2003), and Kurtuluş and Toktay (2005) all consider $N \geq 2$ competing manufacturers that sell their differentiated products to consumers through a common retailer. All of these papers utilize linear downward sloping demand functions that are commonly used in marketing and economics. The demand for products when $N = 2$ is given by

$$q_1 = a_1 - p_1 + \theta(p_2 - p_1) \qquad q_2 = a_2 - p_2 + \theta(p_1 - p_2) \tag{1}$$

where p_1 and p_2 are the retail prices of the two products and $\theta \in [0, 1]$ is the cross-price sensitivity.

The parameters in the demand system have the following interpretation: If the retail prices for both products are the same, the relative demand for each product is determined by the parameters a_1 and a_2. Therefore, the parameters a_1 and a_2 can be interpreted as the relative *brand strength* of each product. The parameter θ is the cross-price sensitivity parameter that shows by how much the demand for product j increases as a function of a unit price increase in product i. The assumption $\theta \in [0, 1]$ implies that the products are substitutable. As θ increases, the demand for product i, q_i, becomes more sensitive to price changes of product j, p_j. Therefore, we interpret the parameter θ as being the *degree of product differentiation*; the higher the parameter θ, the less differentiated the products are.

This type of linear demand system that is consistent with Shubik and Levitan (1980) is widely used in marketing (McGuire and Staelin 1983, Choi 1991,

Wang et al. 2003) and economics (Vives 2000, and references therein). The demand functions can be justified on the basis of an underlying consumer utility model: They are derived by assuming that consumers maximize the utility they obtain from consuming quantities q_1 and q_2 at prices p_1 and p_2, respectively. The utility representation is useful as it allows researchers to investigate how consumers are influenced by different pricing policies and different product assortments via a calculation of the consumer surplus.

2.1 Emergence of category captainship

Niraj and Narasimhan (2003) consider a model where two manufacturers ($N = 2$) sell their differentiated products through a common retailer and define category management as an information sharing alliance between the retailer and all manufacturers in the category. Category captainship, on the other hand, is defined as an exclusive information sharing alliance between the retailer and only one manufacturer. The paper investigates whether or not such exclusive information sharing alliances would emerge in equilibrium, and if so under what conditions.

The paper assumes that $a_1 = a_2 = a$ and demand uncertainty is captured by a two-point distribution for the brand strength parameter a. The demand strength parameter a is either high (H) or low (L) with probability λ and 1- λ, respectively. Both the retailer and the manufacturers observe private signals that carry information regarding the realization of the uncertain parameter. Information reduces uncertainty by reducing the conditional variance of the underlying random variable. The quality of information available to the partners in the supply chain is captured by parameter R, which the authors call reliability and which is characterized by the following conditional probabilities:

$$P(H \mid H) = P(L \mid L) = R \quad \text{and} \quad P(L \mid H) = P(H \mid L) = 1 - R$$

with $R \in (0.5, 1]$.

The authors assume the two manufacturers to be symmetric and have a baseline reliability of μ. The retailer's baseline reliability is denoted by ρ. The authors also introduce the parameter σ to capture the degree of complementarity of the information sources. There are three possible information sharing arrangements: (i) If both manufacturers decide to offer a partnership and the retailer accepts the partnership, then both manufacturers have reliabilities of $\mu + \sigma$ and the retailer's reliability is $\rho + 2\sigma$; (ii) If only one manufacturer offers a partnership and the retailer accepts the partnership, then one of the manufacturer's reliability remains at μ, the reliability of the manufacturer offering the partnership increases to $\mu + \sigma$ and the retailer's reliability becomes $\rho + \sigma$; (iii) If the retailer does not accept the partnership proposal, all firms remain at the baseline level of reliabilities.

Formation of an information sharing alliance increases (i) the total channel profit; and (ii) the reliability of information available to the party participating in the alliance. While participation in an information sharing alliance leads to more reliable information regarding the market (parameter a), it also increases the other firm's reliability which makes it more difficult for the information-sharing party to appropriate a bigger share of the total channel profit.

The sequence of events is as follows. First, the manufacturers, independently and simultaneously, offer an information sharing alliance. Neither, one or both manufacturers may propose an alliance and the retailer can accept or reject any of the proposed alliances. Second, the manufacturers and the retailer play a pricing game where the manufacturers act as a Stackelberg leader and set their wholesale prices simultaneously and then the retailer sets the retail prices.

The game is solved backwards. First, the authors solve the pricing game that takes place between the manufacturers and the retailer for given levels of reliabilities. The retailer sets retail prices for given reliability and wholesale prices by maximizing its total expected category profit

$$\Pi_R = E[(p_1 - w_1)q_1 + (p_2 - w_2)q_2]$$

where the expectation is over the true state of the world, the retailers signal, and the signals received by both manufacturers. Then, expecting the retailer's pricing responses, the manufacturers set wholesale prices by maximizing their respective profits. Manufacturer i's profit is

$$\Pi_i(w_i, w_j) = E\left[w_i q_i(w_i, w_j)\right] \quad \text{for } i = 1, 2 \text{ and } i \neq j,$$

where the expectation is over the true state of the world, the retailer's signal and manufacturers' signals. It is also assumed that the manufacturers' production costs are zero. Second, after determining the wholesale and retail prices for given levels of reliabilities, the supply chain partners decide on whether to form an information sharing alliance or not. The manufacturers' information sharing decision will be based on a comparison of their expected profits in all possible scenarios. Similarly, the retailer's acceptance or rejection decision is based on a comparison of its expected profit in the baseline scenario, and the non-exclusive and exclusive information sharing alliance scenarios.

It turns out that the main determinant of whether or not an information sharing alliance will be formed is the reliability of information available to each player in the supply chain. First, unless both the retailer's and the manufacturers' reliabilities are very high, at least one of the manufacturers will offer an information sharing alliance. Second, the retailer accepts any proposed alliance when the reliability of information available to the retailer is higher than the reliability of the information available to the manufacturers. Third, category captainship type of exclusive information sharing alliances will emerge when the reliability of information available to the retailer is high and the reliability of information available to the manufacturers is at intermediate levels.

The intuition behind this result is as follows: We already said that when the reliability of information available to the retailer and both manufacturers is very high, none of the manufacturers will offer an information sharing alliance. However, as the reliability of information available to the manufacturers decreases, it becomes feasible for one of the manufacturers to offer an information sharing alliance. As the reliability of the information available to the manufacturers decreases further, it becomes profitable for both manufacturers to offer an information sharing alliance. The retailer accepts a proposed alliance if and only if the retailer's potential gain from an increase in reliability outweighs the potential harm of increased reliability at the manufacturers. The authors argue that their results potentially explain why more sophisticated retailers such as Wal-Mart, who are likely to have higher reliability of information, entered into category management and category captainship type of information sharing alliances before other retailers.

2.2 Delegation of the pricing decisions

The idea of an upstream party in a supply chain (such as a manufacturer) interfering with the retailer's pricing decisions is not new. There is a large amount of research in economics on so-called Resale Price Maintenance (RPM) practices where a manufacturer imposes a minimum or a maximum resale price to the retailers (e.g., Gilligian 1986, Overstreet 1983 and references therein). While the RPM and category captainship practices are similar in the sense that the manufacturer interferes with retailer's pricing decisions, there are also significant differences between the two. While with resale price maintenance, the manufacturer imposes a retail price on its own products only, in category captainship, the manufacturer might recommend retail prices for all products in the category. Also, while RPM practices are manufacturer driven, category captainship practices are mostly retailer driven.

Wang et al. (2003) and Kurtuluş and Toktay (2005) take the category captainship decision as given and consider how each stakeholder in the supply chain is affected when the retailer delegates the pricing decisions to one of its leading manufacturers.

Wang et al. (2003) consider a game theoretic model with N manufacturers that sell their products through a common retailer and investigate whether it is profitable for the retailer to delegate pricing authority to a category captain. The demand model used in Wang et al. is a generalized and slightly modified version of the demand system described in (1). The demand for product i is given by

$$q_i = \frac{1}{N}\left[a - p_i + \frac{1}{N-1}\sum_{i \neq j}^{N} \theta(p_j - p_i)\right]$$

In the absence of a category captain, the manufacturers act as Stackelberg leaders and offer wholesale prices $(w_1, w_2, ..., w_N)$ to the retailer at stage one of the game. Then at stage two, given the wholesale prices, the retailer sets the retail prices to maximize total category profit

$$\max_{p_1,...,p_N} \sum_{i-1}^{N} (p_i - w_i)q_i.$$

The game is solved through backward induction. First, the retailer solves the above optimization problem for given wholesale prices and determines the quantity responses and then each manufacturer sets its own wholesale price in expectation of the quantity demanded of its own product, $\hat{q}_i(w_1, ..., w_N)$, to maximize profit. The production costs are assumed to be zero for all the products. At stage one of the game, each manufacturer solves

$$\max_{w_i} w_i\hat{q}_i(w_1, ..., w_N).$$

The paper assumes, without loss of generality, that the manufacturer with index one (the first manufacturer) is the category captain. Category captainship is modeled as being an alliance between the retailer and the manufacturer of the first brand. In other words, under category captainship, the retailer and the category captain act as an integrated firm. In this model, after the N-1 manufacturers offer their wholesale prices $(w_2, w_3, ..., w_N)$ the alliance (where the category captain and the retailer act as an integrated firm) sets the retail prices to maximize the alliance profit

$$\max_{p_1,...,p_N} p_1q_1 + \sum_{i=2}^{N} (p_i - w_i)q_i.$$

Then, given the quantity responses $\hat{q}_i(w_2, ..., w_N), i = 2$, the manufacturers set their wholesale prices.

The main result in Wang et al. is that using a category captain for category management is profitable for both the retailer and the category captain. The intuition is as follows: After the retailer and the category captain form an alliance, the alliance will gain from the category captain's brand (i.e., coordination between the retailer and the captain) and will lose from selling other brands in the category. It turns out that both the channel coordination effect and the competition effect have a positive impact on the joint profit gain, therefore benefiting both the retailer and the category captain. On the other hand, category captainship generally does not benefit the non-captain manufacturers due to increased pressure from the channel. Furthermore, the paper also identifies conditions under which category captainship can benefit all participating partners. Category captainship may benefit all parties in the supply chain if (i) the category captain has the authority to choose the retail price for its own brand only (i.e., partial delegation); and (ii) the non-captain manufacturer behaves strategically (i.e., adjusts its own wholesale price to the use of a category captain in the supply chain).

In addition, the paper identifies conditions under which category captainship is more beneficial for the alliance members. The paper finds that the profitability of using a category captain is higher if the product category (1) has fewer products (lower N); (2) has higher price competition among products (higher cross-price sensitivity θ) and (3) has no store brand as opposed to having a store brand. The inclusion of a store brand modifies the demand system slightly and therefore the alliance profit. When there is a store brand, the alliance sets the retail prices to maximize the alliance profit

$$\max_{p_1,\ldots,p_N} p_1 q_1 + \sum_{i-2}^{N} (p_i - w_i) q_i + p_s q_s,$$

where q_s and p_s are the demand and price for the store brand and q_i and q_s are given by

$$q_i = \frac{1}{N+1}\left[a - p_i + \frac{1}{N}\sum_{i \neq j}^{N} \left[\theta(p_j - p_i) + \delta(p_s - p_i)\right]\right] \quad \text{for } i = 1, \ldots, N$$

$$q_s = \frac{1}{N+1}\left[a - p_s + \frac{1}{N}\sum_{j}^{N} \delta(p_j - p_s)\right].$$

The parameter δ in the above equations is the cross-price sensitivity between the manufacturers' brands and the store brand.

The model also offers some insights as to which manufacturer should be selected as a category captain. The ideal category captain is the manufacturer who has a higher brand strength (i.e., higher a) and a higher cross-price sensitivity. This finding is in line with the current practice where retailers assign their leading manufacturers as category captains.

The proliferation of product variety in conjunction with the relative scarcity of retail shelfspace has resulted in intensified manufacturer competition. As reported in Quelch and Kenny (1994), the number of products in the consumer goods industry increased by 16% per year between 1985 and 1992 while shelfspace increased by only 1.5% per year during the same period. To capture this effect, Kurtuluş and Toktay (2005) consider two manufacturers who sell through a common shelfspace-constrained retailer. Category captainship is modeled as the delegation of pricing authority to a leading manufacturer, as in Wang et al.

The shelfspace constraint, S, imposes that $q_1 + q_2 \leq S$. In their model, q_1 and q_2 can be viewed as demand rates for each product per replenishment period; the retailer prices the products so that the total demand rate does not exceed the shelfspace availability. Another interpretation would be to view q_1 and q_2 as the long-term volumes to be purchased and sold subject to a total volume target for the category.

The paper considers two scenarios that represent traditional retail category management and category captainship. In the first scenario, the retailer is responsible for managing the category and determines retail prices (and shelf-space allocations). The manufacturers, on the other hand, compete for the limited shelfspace at the retailer. First, given the wholesale prices, the retailer solves the following optimization problem:

$$\max_{p_1, p_2} \ (p_1 - w_1)q_1 + (p_2 - w_2)q_2$$
$$\text{s.t.} \ \ q_1 + q_2 \leq S$$
$$q_1 \geq 0, \ q_2 \geq 0$$

The authors fully characterize the quantity responses $\hat{q}_1(w_1, w_2)$ and $\hat{q}_2(w_1, w_2)$, which are the optimal quantities determined in the above optimization problem for given wholesale prices (w_1, w_2).

Then, anticipating the retailer's response functions $\hat{q}_1(w_1, w_2)$ and $\hat{q}_2(w_1, w_2)$, the manufacturers play a simultaneous move wholesale price game. Each manufacturer maximizes

$$\Pi_i(w_i, w_j) = (w_i - c_i)\hat{q}_i(w_i, w_j) \quad \text{for } i, j = 1, 2 \text{ and } i \neq j,$$

where c_i is manufacturer i's production cost.

In the second scenario called "category captainship", the authors assume that the retailer delegates pricing decisions to one of its manufacturers and implements the recommendations as they are in return for a target category profit K. First, for given (w_2, S, K), the authors solve the following optimization problem faced by the category captain:

$$\max_{p_1, p_2} \ (w_1 - c_1)q_1$$
$$\text{s.t.} \ \ (p_1 - w_1)q_1 + (p_2 - w_2)q_2 \geq K$$
$$q_1 + q_2 \leq S$$
$$q_1 \geq 0, \ q_2 \geq 0$$

Note that under category captainship, the category captain sets prices to maximize its own profit but has to deliver a target profit to the retailer. The authors characterize the quantity responses $\hat{q}_1(w_2)$ and $\hat{q}_2(w_2)$ for all possible w_2. Then, the non-captain manufacturer sets its wholesale price w_2 in expectation of $\hat{q}_2(w_2)$ by maximizing its profit $(w_2 - c_2)\hat{q}_2(w_2)$.

The results in Kurtuluş and Toktay (2005) are based on a comparison of the two scenarios described above. The main insight of the paper is that given the limited shelfspace at the retailer, category captainship practices may result in competitive exclusion. Competitive exclusion refers to the phenomenon where the category captain takes advantage of its position to advantage its own brand and disadvantage competitors' products. Their results suggest that in some cases, the category captain would indeed prefer to go so far as to exclude the non-captain manufacturer's brand from the category. The UST vs. Conwood case is a good example of a high level of competitive exclusion. In practice, competitive exclusion may take many different forms, most of them less extreme than completely excluding competitors. For example, displaying the non-captain manufacturers' brands at the bottom of the shelf, or promoting the non-captain manufacturers' brands at a less desirable time are some less obvious forms of competitive exclusion.

According to this paper, competitive exclusion is more likely to occur when the difference, in terms of market share, between the manufacturers is large. For example, this result suggests that if Coca Cola is assigned as category captain in the soft drinks category, Coca Cola would prefer excluding a manufacturer with a small brand from the category rather than excluding a bigger manufacturer such as Pepsi. The intuition is as follows: The target profit level set by the retailer would allow Coca Cola to deliver the target profit level without a small brand in the category; however, Coca Cola cannot meet the target profit level if it excludes a bigger manufacturer such as Pepsi.

Setting a target profit level for the category partially prevents the category captain from excluding the non-captain manufacturers, but competitive exclusion cannot be prevented completely in this manner. A natural question to ask is: What measures can the retailer take to avoid competitive exclusion? One obvious solution would be for the retailer to mandate that the category captain not exclude any of the brands in the category. However, as we mentioned already, exclusion may take many different and non-obvious forms, which may make it difficult for the retailer to monitor the exclusion of the non-captain brands from the category. A second measure is for the retailer to filter the category captain's recommendations before implementing them. This would avoid the more blatant forms of exclusion. Of course, for the same reason as before, it may not be easy for the retailer to detect biased recommendations when they are subtle.

2.3 Delegation of the assortment selection decision

In both Wang et al. and Kurtuluş and Toktay (2005), the retailer delegates the pricing authority to a leading manufacturer. However, in practice, the scope of category captainship is broader than making price recommendations. Retailers usually rely on their category captains for assortment recommendations as well. Kurtuluş and Toktay (2006) consider a model where the retailer delegates the assortment selection decision in the category to a leading manufacturer. The goal of this research is to answer how the assortment offered to the consumers at the retailers will change if category captainship is implemented and how the retailer can benefit the most from its category captains for assortment recommendations.

There is an emerging literature on retail assortment planning where the main focus is on retailer's optimal assortment selection (Kök et al. 2006). Kurtuluş and Toktay (2006) contribute to this emerging stream by investigating how retail assortment under category captainship may differ from that under retail category management. The authors extend some of the results proposed in van Ryzin and Mahajan (1999) and show that these results could be maintained under some conditions even if the assortment decision is delegated to the category captain.

The paper considers a two-stage supply chain with multiple manufacturers that produce differentiated products and sell their products to the consumers through a common retailer. The paper uses the multinomial logit (MNL) consumer choice model (see van Ryzin and Mahajan 1999 and Cachon et al. 2005). Let $N=\{1,2,...,n\}$ denote the set of manufacturers. Let $S \subseteq N$ denote the subset of variants that retailer decides to include in the retail assortment. A customer either purchases one of the variants in S or does not purchase anything. Let variant 0 represent the no-purchase option of the customers.

Let $U_i=(u_i\text{-}r_i)+\xi_i$ denote the utility that a consumer gets from variant i. The parameter u_i can be interpreted as consumers reservation price of buying variant i and r_i is the retail price of variant i. Then, the term $u_i\text{-}r_i$ is consumers' expected net utility from variant i. The paper assumes that the variants are labeled in decreasing net utility order: $u_i\text{-}r_i \geq u_j\text{-}r_j$ for all $i<j$. The term ξ_i is a mean-zero Gumbel-distributed error term that creates consumer heterogeneity. It is also assumed that ξ_i's are independent across end products. (See Kök et al. 2006 for an excellent review and discussion of MNL models in the operations literature).

Given the choice set S and the no-purchase option 0, a consumer buys the option with the highest utility. Let $q_i(S) = P[U_i = \max\{U_j : j \in S \cup \{0\}\}]$ denote the probability that variant i has the maximum utility given that the customer is offered assortment S. Similarly, $q_0(S)$ is the probability that an incoming customer selects the no-purchase option. The probability $q_i(S)$ is given by

$$q_i(S) = \frac{v_i}{v_0 + \sum_{j \in S} v_j} \tag{2}$$

where $v_i = e^{(u_i - r_i)/\mu}$ is interpreted as the population's preference for item $i \in S$. Let also λ denote the rate of customers entering the store. The demand for variant i is given by $\lambda\, q_i(S)$.

First, the paper considers a model where the retailer is responsible for selecting the retail assortment. The manufacturers offer their wholesale prices and in response, the retailer decides which items to include in the retail assortment. The retailer's objective is to select S to maximize the expected profit

$$\sum_{S \subseteq N} [m_i \lambda q_i(S) - \sigma(q_i(S))],$$

where m_i is the retailer's margin for product i, and $\sigma(.)$ is the operational cost associated with including variant i in the retailer's assortment. The authors assume that $\sigma(.)$ is increasing and concave. Both Cachon et al. 2005 and van Ryzin and Mahajan 1999 assume similar cost structures.

Second, the paper considers a model where the retailer delegates the assortment selection decision to a leading manufacturer using various strategies such as target profit, target sales and target variety. Setting target profit and target sales levels in a category is quite common in practice. The retailer delegates the assortment decision to the category captain because the category captain can invest in activities such as promotion planning, traffic driving strategies or consumer education that would potentially increase the rate of customers purchasing from the category. This way, the retailer can benefit from the category captain's expertise. To capture the category captain's expertise and superior knowledge about consumers, the authors assume that the category captain increases the rate of customers who would potentially shop in the category and denote this increase by Λ.

The category captain's problem is

$$\max_{S \subseteq N} \ (\lambda + \Lambda)\frac{w_1 v_1}{v_0 + \sum_{i \in S} v_i}$$

$$\text{s.t.} \quad (\text{TP}) \quad (\lambda + \Lambda)\frac{\sum_{i \in S} m_i v_i}{v_0 + \sum_{i \in S} v_i} - \sum_{i \in S} \sigma\left(\frac{v_i}{v_0 + \sum_{i \in S} v_i}\right) \geq K$$

$$(\text{TS}) \quad (\lambda + \Lambda)\frac{\sum_{i \in S} v_i}{v_0 + \sum_{i \in S} v_i} \geq \theta$$

$$(\text{TV}) \quad b \geq \bar{b}$$

where K is the target profit (TP) level, θ is the target sales (TS) level, and $\bar{b}$ is the target variety (TV) level.

The paper compares the performance of the three different strategies in delegating the assortment selection decisions. The main insight from the paper is that when the retailer has the power to offer a take-it-or-leave-it contract to the category captain, with target profit and target sales level contracts, the structure of the recommended assortment may be the same as the structure of the optimal assortment under retail category management. In plain words, this result implies that when the retailer is more powerful, the assortment offered to the consumers under retail category management would not be that different than the recommended assortment under category captainship. On the other hand, with a target variety level contract, the structure of the recommended assortment differs from the optimal assortment under retail category management. Therefore, it is not surprising that many retailers rely on their manufacturers for recommendations on assortment planning by setting profitability and sales volume levels in the categories.

Conversely, if the category captain has the power to offer a take-it-or-leave-it contract, the structure of the recommended assortment, with all three contracts, is usually not the same as the structure of the optimal assortment under retail category management and the variety offered to the consumers is lower as a result of which consumer surplus may decrease. In plain words, this result implies that if the category captain is more powerful, the assortment offered to the consumers under retail category management would not be the same as the recommended assortment under category captainship.

To summarize, Kurtuluş and Toktay (2006) suggest that retailers should consider implementing category captainship in categories where they are more powerful than their category captains so that they can properly align the incentives of their category captains by either target profit or target sales level contracts and benefit from category captain's resources.

2.4 Antitrust concerns

Recently, some economists have voiced antitrust concerns related to category captainship (Steiner 2001, Desrochers et al. 2003). In the US, the Antitrust Institute has voiced reservations about category captainship. In Europe, ECR has taken the lead to ensure that category captainship is implemented in compliance with European Union competition rules.

Desrochers et al. (2003) states that antitrust concerns related to category captainship practices focus around two issues: (1) competitive exclusion and (2) competitive collusion. Competitive exclusion refers to situations where the category captain takes advantage of its position to disadvantage other manufacturers. Kurtuluş and Toktay (2005) contribute to the ongoing debate by offering theoretical support for the existence of competitive exclusion. They suggest some ways to avoid competitive exclusion such as assigning a non-leader manufacturer as a category captain.

Another exclusion-based concern is that smaller competitors are denied the right to compete for category captainship because they do not have the necessary resources (Desrochers et al. 2003). Retailers usually assign one of their leading manufacturers to serve as a category captain because only those manufacturers have the necessary resources that can benefit the retailer. Big manufacturers already invest a great deal in consumer research and can use these resources toward helping retailers manage their categories better. The concern is that category captain manufacturers' power will be further enhanced and smaller manufacturers will be put at a disadvantage.

Competitive collusion concerns include the possibility that a category captain can use its role to facilitate collusion and limit the competition among rivals in the category (Desrochers et al. 2003). First, the category captain may transfer sensitive information such as pricing, merchandising, and promotion plans from one manufacturer to another. When manufacturers in the category know about their rivals' pricing, they might price more or less aggressively, or if they know about their rivals' promotion plans, they may promote their brands more selectively. Second, the category captain can coordinate its recommendations across the retailers for which it serves as category captain. Desrochers et al. suggest that if retailers are more selective in sharing sensitive data with their category captains, some forms of competitive collusion scenarios can be avoided.

To summarize, while category captainship practices in the retailing sector present a very valuable opportunity for the retailers to benefit from their category captain manufacturers' expertise and resources, these practices also open up an opportunity for the category captain manufacturers to take advantage of their positions as category captains and exclude competitors and restrict competition in the categories. While research shows that category captainship may have significant positive impact on retailer's and category captain's performances, economists (Desrochers et al.) also point out to some of the controversial issues surrounding category captainship practices and claim that these practices might harm the consumers.

3. IMPACT OF CATEGORY CAPTAINSHIP PRACTICES ON THE RETAIL INDUSTRY

In this section, we consider how category captainship practices could potentially change the nature of the manufacturer-retailer relationships and the landscape in the retail industry.

Practices such as category captainship delegate considerable power to the category captain manufacturers because in most cases they can effectively control outcomes in the category. While some retailers continue to work with their category captains and verify their recommendations, other retailers prefer to implement their category captain's recommendations due to lack of resources. While private information on the category captain's part makes it easier for the category captain to provide biased recommendations and control the outcomes in the category, it also makes it more difficult for the retailers to detect category captain's biased recommendations. Category captain's influence over the retailer also depends on the size of the retailer. Small retailers are more likely to accept and implement the category captain's exact recommendations, whereas larger retailers have more control over the process and are more likely to implement their category captain's recommendations after verifying them.

In order to decrease the amount of control given to the category captains, some retailers assign a second manufacturer in the category to serve as co-captains and use them as consultants to verify the category captain's recommendations. In addition, the retailers keep the option to renegotiate the category captainship contracts quite frequently by offering short term contracts of only one to two years. The short term nature of the category captainship agreements aims to balance the power in the supply chain.

A potential adverse effect of category captainship on retailers is the loss of capability to manage the categories internally. Retailers should be aware that category management requires a thorough understanding of consumer preferences and purchase patterns, a knowledge base that is hard to build once that expertise is lost.

Traditionally, manufacturers such as Procter&Gamble and Unilever were the main players in the consumer goods industry and retailers were primarily a means of reaching consumers. The early nineties saw an increase in the number of high quality new product introductions and the emergence of other strong manufacturers, which led to higher competition for shelfspace. This, combined with the retailers' awareness of the importance to be in contact with end consumers, provided the basis for a shift in power from manufacturers to retailers. Many retailers such as Wal-Mart, Carrefour, and Metro owe their rapid growth to these developments (Corstjens and Corstjens 1995).

As Corstjens and Corstjens describe in their influential book Store Wars, "...the giant retailers, now, stand as an obstacle between the manufacturers and the end consumers, about as welcome as a row of high-rise hotels between the manufacturer's villa and the beach." Their book describes the contemporary national brand manufacturers over the past decade as being in a continuous battle for shelfspace and mindspace at the retailers. It is therefore no surprise that manufacturers would advocate any initiative that can increase their influence over retail decisions, and category captainship is such a practice. But by outsourcing retail category management to their leading manufacturers, retailers may in the long-run lose their capabilities in managing their product categories and their knowledge about consumers. This loss of capability may prepare the basis for a shift in power back from the retailers to the manufacturers.

Given this changing landscape in the consumer goods supply chains over the past few decades, an intriguing question is what will happen to the retailer-manufacturer relationships and power balance in the consumer goods supply chains in the near future. With the growing popularity of category captainship practices in the retail industry, the number of manufacturer-retailer partnerships (e.g., Wal-Mart and P&G, Carrefour and Colgate) is increasing. While such partnerships will positively influence the partner manufacturers, they will also place the non-partnering manufacturers at a great disadvantage, forcing them to become a partner to a leading retailer. Manufacturers' battle for shelf-space and mindspace over the past decade has started to transform into a battle for being a partner (category captain) for a major retailer.

4. FUTURE RESEARCH DIRECTIONS

Although category captainship practices became quite prevalent in the retail industry over the past decade, the consequences of using category captains for category management are not fully understood by either academics or practitioners. Therefore, we believe that there is room for more original research in this field. We have identified five directions for future research that would help both academics and practitioners to better understand the consequences of category captainship practices.

First, existing research on category captainship assumes that the retailers either delegate the pricing or the assortment decision to a leading manufacturer. However, in practice, the scope of category captainship implementations is broader: retailers rely on their leading manufacturers for pricing, assortment, shelfspace management, promotions etc. Therefore, exiting models cannot fully capture the category captainship phenomenon. The question of how different category captainship arrangements impact the retailer and the manufacturers needs to be answered when the retailer relies on its category captain for a combination of assortment, pricing, shelfspace management and promotion planning recommendations.

New research can take advantage of the existing literature on joint inventory and pricing decisions in operations (see Petruzzi and Dada 1999, Elmaghraby and Keskinocak 2003, and Yano and Gilbert 2003 for literature reviews on different aspects of the joint pricing and inventory decisions) that could be used as the basis for investigating the impact of jointly delegating the shelfspace allocation and pricing decisions to a leading manufacturer. In addition, there is a literature on trade promotions in marketing (e.g., Lal and Villas-Boas 1998 and Kim and Staelin 1999) and operations (e.g., Iyer and Ye 2000 and Huchzermeier et al. 2002) that could be used as the basis for research to understand the impact of recommendations made by category captains to their retailers about different aspects of promotion planning.

We believe that specific aspects of category captainship practices could be investigated through mathematical models, but answering broader questions needs empirical research. In particular, empirically testing the impact of category captainship practices on the financial performance of the retailers and understanding when such practices would benefit the retailers would be a good starting point. Empirical research is also needed to test the hypothesis that category captainship may result in competitive exclusion. Such empirical research would provide a basis for the antitrust cases that are under investigation regarding category captainship misconduct.

Second, existing research on category captainship exclusively focuses on categories where products are substitutes. However, a product category sometimes can consist of complementary products such as toothpaste and toothbrush products in the oral care category. Future research should be conducted to understand the differences in category captainship implementations where the products are substitutes versus complements, and whether categories where the retailer offers complementary products are more suitable for category captainship.

Third, all of the existing models assume that the information available to the retailer and the suppliers is the same. However, in practice, the basis for category captainship relationships is the fact that the category captain often has better knowledge about some aspect of the category than the retailer does. While retailers deal with as many as hundreds of categories, a typical manufacturer usually focuses on only a few. Therefore, it would be appropriate to assume that the category captain has private information about some parameters. For example, the category captain may have better information about the cross price sensitivities, which would allow him to make more accurate pricing decisions. Including asymmetric information in the models would also change the dynamics in that the retailer may be at a disadvantage to evaluate the recommendations provided by the category captain. Existing research concludes that category captainship benefits both the retailer and the category captain. However, this result may change when the category captain has private information. Characterizing the conditions under which the retailer benefits from category captainship under asymmetric information would therefore be another fruitful avenue for research.

Fourth, future research should explore the value of having an independent third party providing category management services for retailers. Companies such as ACNeilsen collect and sell syndicated data and software that can be used for category management; however, they do not provide category management recommendations. Research is needed to understand the advantages and disadvantages of using a third party for category captainship. On one hand, retailers could take advantage of the expertise and resources of the third party providers without worrying about bias in the information provided. On the other hand, the retailers should be concerned about losing their internal category management capabilities. Another source of concern for the retailers is that these third party providers would provide recommendations to many retailers that compete for the same consumers, potentially causing the retailer to lose its competitive edge.

Finally and related to the last point above, information leakages and competitive collusion are other areas that need further research. Category captainship requires that the retailer share a lot of strategic information with its category captain. In practice, a leading manufacturer serves as a category captain for many retailers that are competing for the same consumers. A potential danger that a retailer sharing strategic information faces is the leakage of strategic information to other competing retailers. The tradeoff that retailers face is the benefit from category captainship versus the potential problems and loss of competitiveness that could arise from information leakage. Research to identify under what market conditions, and retailer and manufacturer characteristics these concerns are overcome by the benefits of category captainship would be valuable.

REFERENCES

Apparel Magazine 2005. VF on 'Three Cs' of Category Captainship, *Apparel Magazine*, Nov. 10, 2005.

Cachon, G.P., C. Terwiesch, and Y. Xu. 2005. Retail Assortment Planning in the Presence of Consumer Search. *Manufacturing & Service Operations Management*, **7**(4): 330–346.

Choi, S.C. 1991. Price Competition in a Channel Structure with a Common Retailer, *Marketing Science*, **10**(4): 271–296.

Corstjens, J., M. Corstjens. 1995. *Store Wars: The Battle for Mindspace and Shelfspace*, Wiley.

Desrochers, D.M., G.T. Gundlach, and A.A. Foer. 2003. Analysis of Antitrust Challenges to Category Captain Arrangements, *Journal of Public Policy & Marketing*, **22**(2): 201–215.

ECR Conference. 2004. Category Management is Here to Stay, Brussels, 2004 (http://www.ecrnet.org/conference/files/24-05-04/04-category%20management.ppt).

Elmaghraby, W., P. Keskinocak. (2003). Dynamic pricing in the presence of inventory considerations: Research overview, current practices and future directions, *Management Science*, **49**, 1287–309.

Gilligian, T. 1986. The Competitive Effects of Resale Price Maintenance. *RAND Journal of Economics*, **17**, 544–556.

Greenberger, R.S. 2003. UST Must Pay $ 1.05 Billion To a Big Tobacco Competitor. *Asian Wall Street Journal*. New York, NY: Jan. 15, p. A.8.

Huchzermeier, A., A. Iyer, and J. Freheit. 2002. The Supply Chain Impact of Smart Customers in a Promotional Environment. *Manufacturing & Service Operations Management*, **4** (3): 228.

Iyer, A. V., J. Ye. 2000. Assessing the Value of Information Sharing in a Promotional Retail Environment. *Manufacturing & Service Operations Management*, **2**(2): 128–143.

Kim, S. Y., R. Staelin. 1999. Manufacturer allowances and retail pass-through rates in a competitive environment. *Marketing Science*, **18** (1): 59–76.

Kök, A.G., M.L. Fisher, and R. Vaidyanathan. 2006. Assortment Planning: Review of Literature and Industry Practice. *Retail Supply Chain Management*. Eds. N. Agrawal and S. Smith.

Kurtuluş, M., L. B. Toktay. 2005. Category Captainship: Outsourcing Retail Category Management, INSEAD Working Paper.

Kurtuluş M., L. B. Toktay. 2006. Retail Assortment Planning under Category Captainship: Outsourcing Retail Category Management, Working Paper.

Lal, R., M. Villas-Boas. 1998. Price promotions and trade deals with multiproduct retailers. *Management Science*, **44** (7): 935–949.

McGuire, T.W., R. Staelin. 1983. An Industry Equilibrium Analysis of Downstream Vertical Integration. *Marketing Science*, **2**(2): 161–191.

Nielsen Marketing Research. 1992. *Category Management: Positioning Your Organization to Win*. Lincolnwood. IL: NTC Business Books.

Niraj, R., C. Narasimhan. 2003. Vertical Information Sharing in Distribution Channels. Washington University Working Paper.

Overstreet, T. 1983. *Resale Price Maintenance: Economic Theories and Empirical Evidence*. Washington, D.C.: Federal Trade Commission.

Petruzzi, N. C., M. Dada. (1999). Pricing and the newsvendor problem: A review with extensions, *Operations Research*, **47**, 183–94.

Progressive Grocer. 2004. Category Captains 2004: Captains of the Industry, *Progressive Grocer*, Nov. 15, 2004.

Shubik M.J., R.E. Levitan. 1980. *Market Structure and Behavior*. Cambridge: Harvard University Press.

Steiner, R.L. 2001. Category Management – A Pervasive, New Vertical/Horizontal Format, *Antitrust*, **15** (Spring), 77–81.

Quelch, J., D. Kenny. 1994. Extend profits, not the product lines. *Harvard Business Review* **72**(5): 153–160.

van Ryzin, G., and S. Mahajan. 1999. On the Relationship Between Inventory Costs and Variety Benefits in Retail Assortment. *Management Science*, **45**(11): 1496–1509.

Vives, X. 2000. *Oligopoly Pricing: old ideas and new tools*. Cambridge, MA. The MIT Press.

Wang Y, J.S. Raju, S.K. and Dhar. 2003. The Choice and Consequences of Using a Category Captain for Category Management. The Wharton School, University of Pennsylvania.

Yano, C., S.M. Gilbert. (2003). Coordinated pricing and production/procurement decisions: A review., in A. Chakravart and J. Eliashberg, eds, *Managing Business Interfaces: Marketing, Engineering and Manufacturing Perspectives*, Kluwer Academic Publishing.

Chapter 6
ASSORTMENT PLANNING: REVIEW OF LITERATURE AND INDUSTRY PRACTICE

A. Gürhan Kök[1], Marshall L. Fisher[2], and Ramnath Vaidyanathan[3]
[1]*Fuqua School of Business, Duke University, Durhan, NC, USA*
[2]*The Wharton School, University of Pennsylvania, PA, USA*
[3]*The Wharton School, University of Pennsylvania, PA, USA*

1. INTRODUCTION

A retailer's assortment is defined by the set of products carried in each store at each point in time. The goal of assortment planning is to specify an assortment that maximizes sales or gross margin subject to various constraints, such as a limited budget for purchase of products, limited shelf space for displaying products, and a variety of miscellaneous constraints such as a desire to have at least two vendors for each type of product.

Clearly the assortment a retailer carries has an enormous impact on sales and gross margin, and hence assortment planning has received high priority from retailers, consultants and software providers. However, no dominant solution has yet emerged for assortment planning, so assortment planning represents a wonderful opportunity for academia to contribute to enhancing retail practice. Moreover, an academic literature on assortment planning is beginning to emerge. The purpose of this chapter is to review the academic literature on assortment planning, to overview the approaches to assortment planning used by several retailers so as to provide some examples of practice, and to suggest directions for future research.

Retailers engage in assortment planning because they need to periodically revise their assortment. Several factors require a retailer to change their assortment, including seasons (the fall assortment for an apparel retailer will be different from the spring assortment), the introduction of new products and changes in consumer tastes.

Most retailers segment the stock keeping units (SKU) they carry into groups called categories. For example, for a consumer electronics retailer, a category might be personal computers. Within categories, they will usually define subcategories, such as laptops and desktops within the computer category. (The terminology used varies across retailers e.g. department, class and subclass

N. Agrawal, S.A. Smith (eds.), *Retail Supply Chain Management*,
DOI: 10.1007/978-0-387-78902-6_6,

may be used instead of category and subcategory, but the practice of grouping SKUs with similar attributes for planning purposes is universal.) Retailers focus most of their energy on deciding what fraction of their shelf space and product purchase budget to devote to each category and subcategory. For example, a consumer electronics retailer would worry more about how to divide their resources between laptops and desktops than about which specific models of each to carry, a decision that is usually left to a more junior buyer. The resource allocation decisions are based on their own historical sales in each subcategory, especially whether sales in a subcategory have been trending up or down, together with external information from a variety of sources such as industry shows, vendors and competitor moves.

Given fixed store space and financial resources, assortment planning requires a tradeoff between three elements: how many different categories does the retailer carry (called a retailer's breadth), how many SKUs do they carry in each category (called depth), and how much inventory do they stock of each SKU, which obviously affects their in-stock rate. The breadth vs depth tradeoff is a fundamental strategic choice faced by all retailers. Some, like department stores, will elect to carry a large number of different categories. Others, such as category killers like Toys 'R Us and Best Buy, will specialize in a smaller number of categories, but have great depth in each category.

We have all had the experience of going into a store looking for a particular product, not finding it, and settling for another similar product instead. This is called substitution, and the willingness of customers to substitute within a particular category is an important parameter in assortment planning. If customers have a high propensity to substitute in a category, then providing great depth and a high in-stock rate is less critical. The reverse is also true.

We can delineate three patterns with respect to customer substitution: (1) the customer shops a store repeatedly for a daily consumable and one day she finds it stocked out so she buys another. This is called stock-out based substitution. (2) a customer identifies a favorite product based on ads or what she has seen in other stores, but when she tries to find it in a particular store, she can't because they don't carry it, so see buys another product. This is called assortment based substitution. (3) the consumer chooses her favorite product from the ones she sees on the shelf in a store when she is shopping and buys it if it has higher utility than her no purchase option. In this case, there may be other products she would have preferred, (but she didn't see them either because the retailer didn't carry them or because they were stocked out), and in this sense we can say she substituted, although she may not be aware that these other products exist and hence doesn't herself think of her purchase decision as involving substitution. The first two patterns are common with daily consumables like food and the later with consumer durables like apparel or consumer electronics.

Assortment planning is a relatively new but quickly growing field of academic study. The academic approach to the assortment planning problem rests on the formulation of an optimization problem with which to choose the optimal set of products to be carried and the inventory level of each product.

Decisions for each product are interdependent because products are linked in considerations such as shelf space availability, substitutability between products, common vendors (brands), joint replenishment policies and so forth. Most of the literature focuses on a single category or subcategory of products at a given point in time. While a retailer might have a different assortment at each store, the academic literature has focused on determining a single assortment for a retailer, which could be viewed as either a common assortment to be carried at all stores or the solution to the assortment planning problem for a single store.

This chapter begins in Section 2 by briefly reviewing four streams of literature that assortment planning models build on: product variety and product line design, shelf space allocation, multi-product inventory systems and a consumer's perception of variety.

In Section 3, we discuss empirical results on consumer substitution behavior and present three demand models used in assortment planning: the multinomial logit, exogenous demand and locational choice models.

In Section 4, we describe optimization based assortment planning studies. Sections 4.1 through 4.3 review optimization approaches for the basic assortment planning problem. The models and solution methodologies in these papers vary because of differences in the underlying demand model and the application context. We then review variations on the basic assortment planning problem, including assortment planning with supply chain considerations in Section 4.4, assortment planning with demand learning and assortment changes during the selling season in Section 4.5, and multi-category assortment planning that considers the interactions between different categories due to existence of basket shopping consumers in Section 4.6.

In Section 5, we discuss demand and substitution estimation methodologies. The methods depend on the demand model and the type of data that is available.

In Section 6, we present industry approaches to assortment planning. We describe the assortment planning process at four prominent retailers: Electronics retailer Best Buy, book and music retailer Borders, Indian jewelry retailer Tanishq, and Dutch supermarket chain Albert Heijn. As will be seen, these companies take significantly different approaches and emphasize different aspects of the assortment problem.

In Section 7, we provide a critical comparison of the academic and industry approaches and use this to identify research opportunities to bridge the gap between the two approaches.

For an earlier overview of the assortment planning literature, see Mahajan and van Ryzin (1999).

2. RELATED LITERATURE

In this section, we briefly review the literature on topics related to assortment planning.

2.1 Product variety and product line design

Product selection and the availability of products has a high impact on the retailer's sales, and as a result gross profits and assortment planning has been the focus of numerous industry studies, mostly concerned with whether assortments were too broad or narrow. Retailers have increased product selection in all merchandise categories for a number of reasons, including heterogeneous customer preferences, consumers seeking variety and competition between brands: Quelch and Kenny (1994) report that the number of products in the market place increased by 16% per year between 1985 and 1992 while shelf space expanded only by 1.5% per year during the same period. This has raised questions as to whether rapid growth in variety is excessive. For example, many retailers are adopting an "efficient assortment" strategy, which primarily seeks to find the profit maximizing level of variety by eliminating low-selling products (Kurt Salmon Associates 1993), and "category management," which attempts to maximize profits within a category (AC Nielsen 1998). There is empirical evidence that variety levels have become so excessive that reducing variety does not decrease sales (Dreze et al. 1994, Broniarcyzk et al. 1998, Boatwright and Nunes, 2001). And from the perspective of operations within the store and across the supply chain, it is clear that variety is costly: a broader assortment implies less demand and inventory per product, which can lead to slow selling inventory, poor product availability, higher handling costs and greater markdown costs.

The literature that studies the economics of product variety is vast. The main model in this field is the oligopoly competition between single product firms based on Hotelling (1929). In the Hotelling model, consumers are distributed uniformly on a line segment and firms choose their positions on the line segment and their prices to maximize profits. Consumers' utility from each firm is decreasing in the firm's price and their physical distance to the firm. Each consumer chooses the firm that provides her the maximum utility. The objective is to find the number of firms, their locations and their prices in equilibrium and the resulting consumer welfare. Extensions of this model are used to study product differentiation. There are two types of product differentiation. In a horizontally differentiated market, products are different in features that can't be ordered. In that case, each of the products is ranked first for some of the consumers. A typical example is shirts of different color. In a vertically differentiated market, products can be ordered according to their objective quality from the highest to the lowest. A higher quality product is more desirable than a lower quality product for any consumer. Anderson et al. (1992) and Lancaster (1990) provide excellent reviews of this literature.

One of the outgrowths of the literature on the economics of product variety is the product line design problem pioneered by Mussa and Rosen (1978) and Moorthy (1984). A monopolist chooses a subset of products from a continuum of vertically differentiated products and their prices to be sold in a market to a variegated set of customer classes in order to maximize total profit. Consider cars as a product with a single attribute, say engine size. The monopolist's problem is to choose what size engines to put in the cars and how to price the final product. These papers assume convex production costs and do not consider operational issues such as fixed costs, changeover costs, and inventories. Joint consideration of marketing and production decisions in product line design is reviewed by Eliashberg and Steinberg (1993). Dobson and Kalish (1993) propose a mathematical programming solution for this problem in the presence of fixed costs for each product included in the assortment. Desai et al. (2001) study the product line design problem with component commonality. Netessine and Taylor (2005) extend Moorthy's (1984) work by using the Economic Order Quantity (EOQ) model to incorporate economies of scale. de Groote (1994) also considers concave production costs and analyzes the product line design problem in a horizontally differentiated market. He shows that the firm chooses a product line to cover the whole market and the product locations are equally spaced. Alptekinoglu (2004) extends this work to two competing firms, one offering infinite variety through mass customization and the other limited variety under mass production. He shows that the mass producer needs to reduce variety in order to mitigate the price competition. Chen et al. (1998) is the only paper that considers product positioning and pricing with inventory considerations. They show that the optimal solution for this model under stochastic demand can be constructed using dynamic programming.

These models were early treatments of assortment planning from the manufacturer's view that were precursors of similar models developed for retailing. The manufacturer's problem is one of product positioning in an attribute space (quality or some other attribute) and pricing. The retailer's problem is to select products from the product lines of several manufacturers. A more careful consideration of inventories at product level is needed in retail assortment planning, since inventories have a direct impact on both sales and costs for the retailer.

2.2 Multi-item inventory models

Multi-item inventory problems are also highly relevant to the assortment planning problem. The inventory management of multiple products under a single a shelf space or budget constraint is studied extensively in the operations literature and solutions using Lagrangian multipliers is presented in various textbooks, e.g., Hadley and Whitin (1963). Downs et al. (2002) describe a heuristic approximation to the multi-period version of this problem with lost sales. In these models, the demand of products are not dependent on others' inventory levels (i.e., there is no substitution between products).

The other group of inventory models with multiple products consider stock-out based substitution, focusing on the stocking decisions given a selection, but not the selection of the products. These models are based on an exogenous model of demand which we shall describe in the next section. Briefly, the total demand of a product is the sum of its own initial demand and the substitution demand from other products. Substitution demand from product k to j is a fixed proportion α_{kj} of the unsatisfied demand of product j. McGillivray and Silver (1978) first introduced the problem with two products. Parlar and Goyal (1984) study the decentralized version of the problem. Noonan (1995) and Rajaram and Tang (2001) present heuristic algorithms for the solution of the case with n products. Netessine and Rudi (2003) investigate the case with n products under centralized and decentralized management regimes. The complexity of the problem is prohibitive and it is not possible to obtain an explicit solution to the problem. Netessine and Rudi (2003) find that a decentralized regime carries more inventory than the centralized regime because of the competition effects. Mahajan and van Ryzin (2001b) establish similar results under dynamic customer substitution with the multinomial logit choice model. Parlar (1985) and Avsar and Baykal-Gursoy (2002) study the infinite horizon version of this problem under centralized and competitive scenarios respectively. Lippman and McCardle (1997) consider a single period model under decentralized management, where aggregate demand is a random variable and demand for each firm is a result of different rules of initial allocation and reallocation of excess demand. Bassok et al. (1999) consider an alternative substitution model, in which the retailer observes the entire demand before allocating the inventory to products. In this retailer controlled substitution model, the retailer may upgrade a customer to a higher quality product. The reallocation solution is obtained by solving a transportation problem.

The literature on assemble-to-order systems is also related. The demand for individual components are linked through the demand for finished goods. See Song and Zipkin (2003) for a review. An online retailer's order fulfillment problem when customers can order multiple products can be viewed as an assemble-to-order systems. Song (1998) estimates the order fill rate in such systems and discusses other examples from retailing.

2.3 Shelf space allocation models

In some product segments such as grocery and pharmaceuticals, how much shelf space is allocated to a given product category is an important component of the assortment planning process. This view seems especially relevant for fast moving products whose demand is sufficiently high that a significant amount of inventory is carried on the shelf. This contrasts with other categories e.g., shoes, music, books where only one or two units are carried for most SKUs, hence amount of inventory and shelf space are not critical decisions at product level.

As one example, Transworld Entertainment carries 50,000 SKUs in an average store but stock more than one of only the 300 best sellers.

In an influential paper Corstjens and Doyle (1981) suggest a method for allocating shelf space to categories. They perform store experiments to estimate sales of product i as $\alpha_i s_i^{\beta_i} \prod_j s_j^{\delta_{ij}}$, where s_i is the space allocated to product i, β_i is own space elasticity, and δ_{ij}s are the cross-space elasticities. Cost functions of the form $\gamma_i s_i^{\tau_i}$, are also estimated from the experiments. The problem of profit maximization with a shelf space constraint is solved within a geometric programming framework. Their results are significantly better than commercial algorithms that allocate space proportional to sales or to gross profit by ignoring interdependencies between product groups. The estimation and optimization procedures can not be applied to large problems, hence they elect to work with product groups rather than SKUs. Bultez and Naert (1988) apply the Corstjens and Doyle (1981) model at the brand level assuming symmetric cross elasticities (i.e., $\delta_{ij} = \delta$ for all i,j) within product groups. Their model is tested at four different Belgian supermarket chains, leading to encouraging results.

An interesting paper by Borin and Farris (1995) reports the sensitivity of the shelf space allocation models to forecast accuracy. They compare the solution with correct parameters to that with incorrect parameter estimates. Even when the error in parameter estimates are 24%, the net loss in category return on inventory is just over 5% compared to the optimal allocation based on true estimates. This proves the robustness of these models to estimation errors. Similar to these shelf space allocation papers, but using an inventory theoretic perspective, Urban (1998) models the own and cross product effects of displayed inventory on demand rate in a mathematical program and solves for shelf space allocation and optimal order-up-to quantities. He reports that on average a greedy heuristic yields solutions that are within 1% of a solution obtained by genetic programming.

Irion et al. (2004) extend the Corstjens and Doyle model to study the shelf space allocation problem at the product level. Demand for each product is a function of its own and other products' shelf space through own and cross shelf space elasticities. The cost for each product consists of linear purchasing costs, inventory costs from an economic order quantity model, and a fixed cost of being included in the assortment. The objective is to allocate (integer) number of facings to each product in order to maximize profits under a total shelf space availability constraint and lower and upper bounds on the number of facings for each product. The problem is transformed into a mixed integer program (MIP) with linear constraints and objective function through a series of linearization steps. The linearization framework is general enough to accommodate several extensions. However, there is no empirical evidence that product level demand can be modeled as a function of the shelf space allocated to the product itself and competing products via own and cross space elasticities.

Shelf space allocation papers do not explicitly address assortment selection and inventory decisions and ignore the stochastic nature of demand.

2.4 Perception of variety

Consumer choice models often assume that customers are perfectly knowledgeable about their preferences and the product offerings. Therefore, consumers are always better off when they choose from a broader set of products. However, empirical studies show that consumer choice is affected by their perception of the variety level rather than the real variety level. This perception can be influenced by the space devoted to a category, the presence or absence of a favorite item (Broniarczyk et al. 1998), or the arrangement of the assortment (Simonson 1999). Hoch et al. (1999) define a measure of the dissimilarity between product pairs as the count of attributes on which a product pair differs. They show that this measure is critical to the perception of variety of an assortment and that consumers are more satisfied with stores carrying those assortments perceived as offering high variety. van Herpen and Pieters (2002) find the impact of two attribute-based measures that significantly impact the perception of variety. These measures are entropy (whether all products have the same color or different colors) and dissociation between attributes (whether color and fabric choice across products are uncorrelated). The perception of variety at a store is especially important for variety-seeking consumers. Variety seeking consumers tend to switch away from the product consumed on the last occasion. Variety-seeking literature demonstrated that consumers adopt this behavior when purchasing food or choosing among hedonic products such as restaurants and music. See Kahn (1995) for a review. Intrapersonal factors (e.g., satiation and the need for stimulation), external factors (e.g., price change, new product introduction), and uncertainty about future preferences promote variety-seeking behavior. On a final note, variety can even negatively affect consumers experience: confusion or complexity due to higher variety may cause dissatisfaction of consumers and decrease sales (Huffman and Kahn 1998).

3. DEMAND MODELS

This section provides a review of demand models as background for assortment planning models. We first present the empirical evidence for consumer driven substitution which is a fundamental assumption in many assortment planning models. The Multinomial Logit model is a discrete consumer choice model, which assumes that consumers are rational utility maximizers and derive customer choice behavior from first principles. Exogenous demand models directly specify the demand for each product and what an individual does when the product he or she demands is not available. The locational choice model is also a utility-based model. Before proceeding, we will define the notation for assortment planning in a single subcategory at a single store. This notation is common throughout this chapter and additional time or store subscripts are introduced when necessary.

N	The set of products in a subcategory, $N = \{1, 2, .., n\}$,
S	The subset of products carried by the retailer, $S \subset N$,
r_j	Selling price of product j,
c_j	Purchasing cost of product j,
λ	Mean number of customers visiting the store per period.

3.1 Consumer driven substitution

We define two types of substitution with a supply side view of the causes of substitution: *Stockout-based* substitution is the switch to an available variant by a consumer when her favorite product is carried in the store, but is stocked-out at the time of her shopping. *Assortment-based* substitution is the switch to an available variant by a consumer when her favorite product is not carried in the store.

The substitution possibilities in retailing can be classified into three groups. (*i*) Consumer shops a store repeatedly for a daily consumable, and one day she finds it stocked out so she buys another. This is an example of stockout-based substitution. (*ii*) Consumer has a favorite product based on ads or her past purchases at other stores, but the particular store she visited on a given day may not carry that product. This is an example of assortment-based substitution. (*iii*) Consumer chooses her favorite from what she sees on the shelf and buys it if it is better than her no purchase option. In this case, there may be other products she may have preferred, but she didn't see them either because the retailer didn't carry them or they are stocked out. This could be an example for either substitution type depending on whether the first choice product is temporarily stocked out or not carried at that store. First two cases fit repeat purchases like food and the third fits one time purchases like apparel.

Let's focus on the options of a consumer who can not find her favorite product in a store, because it is either temporarily stocked out or not carried at all. She can (*i*) buy one of the available items from that category (substitute), (*ii*) decide to come back later for that product (delay), (*iii*) decide to shop at another store (lost customer). If the consumer chooses to substitute, the sale is lost from the perspective of the first favorite product. Table 6-1 summarizes the findings of empirical studies on the consumer response to stockouts. The most recent one, Gruen et al. (2002) examine consumer response to stockouts across eight categories at retailers worldwide and report that 45% of customers substitute, i.e., buy one of the available items from that category, 15% delay purchase, 31% switch to another store, and 9% never buy that item.

The above mentioned papers study the consumer response to stockouts, i.e. stockout based substitution, although none of them explicitly excludes

Table 6-1. Consumer Response to Stockouts in Six Studies of Substitute-Delay-Leave Behavior

	Substitute	Delay	Leave
Progressive Grocer (1968a and 1968b)	48%	24%	28%
Walter and Grabner (1975)	83%	3%	14%
Schary and Christopher (1979)	22%	30%	48%
Emmelhainz et al. (1991)	36%	25%	39%
Zinn and Liu (2001)	62%	15%	23%
Gruen et al. (2002)	45%	15%	40%

assortment-based substitution. Campo et al. (2004) investigate the consumer response to out-of-stocks (OOS) as opposed to permanent assortment reductions (PAR). They report that although the retailer losses in case of a PAR may be larger than those in case of an OOS, there are also significant similarities in consumer reactions in the two cases and OOS reactions for an item can be indicative of PAR responses for that item.

3.2 Multinomial logit

The Multinomial Logit (MNL) model is a utility-based model that is commonly used in economics and marketing literatures. We create product 0 to represent the no-purchase option, i.e., a customer that chooses 0 does not purchase any products. Each customer visiting the store associates a utility U_j with each option $j \in S \cup \{0\}$. The utility is decomposed into two parts, the deterministic component of the utility u_j and a random component ε_j.

$$U_j = u_j + \varepsilon_j.$$

The random component is modeled as a Gumbel random variable. Also known as Double Exponential or Extreme value Type-I, it is characterized by the distribution

$$Pr\{X \leq \varepsilon\} = \exp(-\exp -(\varepsilon/\mu + \gamma)),$$

where γ is Euler's constant (0.57722). Its mean is zero, and variance is $\mu^2\pi^2/6$. A higher μ implies a higher degree of heterogeneity among the customers. The realizations of ε_j are independent across consumers. Therefore, while each consumer has the same expected utility for each product, realized utility may be different. This can be due to the heterogeneity of preferences across customers or unobservable factors in the utility of the product to the individual.

An individual chooses the product with the highest utility among the set of available choices. Hence, the probability that an individual chooses product j from $S \cup \{0\}$ is

$$p_j(S) = \Pr\left\{ U_j = \max_{k \in S \cup \{0\}} (U_k) \right\}.$$

The Gumbel distribution is closed under maximization. Using this property, we can show that the probability that a customer chooses product j from $S \cup \{0\}$ is

$$p_j(S) = \frac{e^{u_j/\mu}}{\sum\limits_{k \in S \cup \{0\}} e^{u_k/\mu}}. \tag{1}$$

See Anderson et al. (1992) for a proof. This closed form expression makes the MNL model an ideal candidate to model consumer choice in analytical studies. See Ben-Akiva and Lerman (1985) for applications to the travel industry, Anderson et al. (1992) for MNL based models of product differentiation, Basuroy and Nguyen (1998) for equilibrium analysis of market share games and industry structure. Moreover, starting with Guadagni and Little (1983), marketing researchers found that MNL model is very useful in estimating demand for a group of products. We will briefly discuss the parameter estimation of MNL model in Section 5.1. For more details on the MNL model and its relation to other choice models, see Anderson et al. (1992) or Mahajan and van Ryzin (1999).

The major criticism of the MNL model stems from its Independence of Irrelevant Alternatives (IIA) property. This property holds if the ratio of choice probabilities of two alternatives is independent of the other alternatives in the choice process. Formally, this property is

for all $R \subset N, T \subset N, R \subset T,$ *for all* $j \in R, k \in R,$

$$\frac{p_j(R)}{p_k(R)} = \frac{p_j(T)}{p_k(T)}.$$

IIA property would not hold in cases where there are subgroups of products in the choice set such that the products within the subgroup are more similar with each other than across subgroups. Consider an assortment with two products from different brands. If brand loyalty is high, adding a new product from the first brand can cannibalize the sales of its sister product more than the rival product. IIA does not capture this important aspect of consumer choice. Another example that illustrates this property is the "blue bus/red bus paradox": Consider an individual going to work and has the same probability of using his or her car or of taking the bus: $\Pr\{car\} = \Pr\{bus\} = 1/2$. Suppose now that there are two buses available that are identical except for their color, red or blue. Assume that the individual is indifferent about the color of the bus he or she takes. The choice set is $\{car, redbus, bluebus\}$. One would intuitively expect that $\Pr\{car\} = 1/2$ and $\Pr\{red\ bus\} = \Pr\{blue\ bus\} = 1/4$. However, the MNL model implies that $\Pr\{car\} = \Pr\{red\ bus\} = \Pr\{blue\ bus\} = 1/3$.

The Nested Logit Model introduced by Ben-Akiva and Lerman (1985) is one way to deal with the IIA property. A two-stage nested process is used for modeling choice, e.g., first brand choice then SKU choice. The choice set N is partitioned into subsets N_l ,$l = 1, .., m$ such that $\cup_{l=1}^{m} N_l = N$ and $N_l \cap N_k = \emptyset$ for any l and k. The individual chooses with a certain probability one of the subsets, from which he or she chooses a variant from that subset. The utility from the choice within subset N_l is also Gumbel distributed with mean $\mu \ln \sum_{j \in N_l} e^{u_j/\mu}$ and scale parameter μ. As a result, the choice process between the subsets follows the MNL model as well and the probability that a consumer chooses variant j in subset N_l is

$$P_j(N) = P_{N_l}(N) * P_j(N_l).$$

Chapter 2 in Anderson et al. describes the Nested Logit in great detail. In the Nested Logit Model, the IIA property no longer holds when two alternatives are not in the same subgroup. However, the use of the Nested Logit requires the knowledge of key attributes and their hierarchy for consumers and makes estimation problems more difficult. Nested Logit model is used in modeling the competition between two-multiproduct firms in several studies (Anderson et al. 1992, Cachon et al. 2006).

Another related shortcoming of the MNL model is related to substitution between different products. The MNL model in its simplest form is unable to capture an important characteristic of the substitution behavior. The utility of the no-purchase option with respect to the utility of the products in S determines the rate of substitution. Consider the following example, where $S = \{1, 2\}$, $\mu = 1$, and $u_0 = u_1 = u_2$. The share of each option is determined by the implication of MNL that the probability of choosing option i is $\exp(u_i)/(\exp(u_0) + \exp(u_1) + \exp(u_2)) = 1/3$ for $i = 0, 1, 2$. Hence, two thirds of the customers are willing to make a purchase from the category. If the second product is unavailable, the probability of her choosing the first product is $\exp(u_1)/(\exp(u_0) + \exp(u_1)) = 1/2$. That is, half of the consumers whose favorite is stocked out will switch to the other product as a substitute and the other half will prefer no-purchase alternative to the other product. In this example, the penetration to the category (purchase incidence) is 2/3 and the average substitution rate is 1/2. These two quantities are linked via u_i's. We can control the substitution rate by varying u_0, but that also determines the initial penetration rate to the category. Hence, it is not possible with this model to have two categories with the same penetration rate but different substitution rates, which we have found severely limits the applicability of this model.

3.3 Exogenous demand model

Exogenous demand models directly specify the demand for each product and what an individual does when the product he or she demands is not available.

There is no underlying consumer behavior such as a utility model that generates the demand levels or that explains why consumers behave as described in the model. As mentioned before, this is the most commonly used demand model in the literature on inventory management for substitutable products. The following assumptions fully characterize the choice behavior of customers.

(A1) Every customer chooses her favorite variant from the set N. The probability that a customer chooses product j is denoted by p_j. $\sum_{j \in N \cup 0} p_j = 1$.

(A2) If the favorite product is not available for any reason, with probability δ she chooses a second favorite and with probability $1 - \delta$ she elects not to purchase. The probability of substituting product j for k is α_{kj}.

When the substitute item is unavailable, consumers repeat the same procedure: decide whether or not to purchase and choose a substitute. The lost sales probability $(1 - \delta)$ and the substitution probabilities could remain the same for each repeated attempt or specified differently for each round.

As a result of (A1) average demand rate for product j is $d_j = \lambda p_j$, and total demand to the category is $\sum_{j \in N} d_j = \lambda(1 - p_0)$.

α_{kj} is specified by a substitution probability matrix that can take different forms to represent different probabilistic mechanisms. Consider the following examples for a four-product category.

Random substitution matrix

$$\begin{bmatrix} 0 & \frac{\delta}{n-1} & \frac{\delta}{n-1} & \frac{\delta}{n-1} \\ \frac{\delta}{n-1} & 0 & \frac{\delta}{n-1} & \frac{\delta}{n-1} \\ \frac{\delta}{n-1} & \frac{\delta}{n-1} & 0 & \frac{\delta}{n-1} \\ \frac{\delta}{n-1} & \frac{\delta}{n-1} & \frac{\delta}{n-1} & 0 \end{bmatrix}$$

Adjacent substitution matrix

$$\begin{bmatrix} 0 & \delta & 0 & 0 \\ \delta/2 & 0 & \delta/2 & 0 \\ 0 & \delta/2 & 0 & \delta/2 \\ 0 & 0 & \delta & 0 \end{bmatrix}$$

Within subgroups substitution matrix

$$\begin{bmatrix} 0 & \delta & 0 & 0 \\ \delta & 0 & 0 & 0 \\ 0 & 0 & 0 & \delta \\ 0 & 0 & \delta & 0 \end{bmatrix}$$

Proportional substitution matrix

$$\begin{bmatrix} 0 & \delta d_2/(\lambda - d_1) & \delta d_3/(\lambda - d_1) & \delta d_4/(\lambda - d_1) \\ \delta d_1/(\lambda - d_2) & 0 & \delta d_3/(\lambda - d_2) & \delta d_4/(\lambda - d_2) \\ \delta d_1/(\lambda - d_3) & \delta d_2/(\lambda - d_3) & 0 & \delta d_4/(\lambda - d_3) \\ \delta d_1/(\lambda - d_4) & \delta d_2/(\lambda - d_4) & \delta d_3/(\lambda - d_4) & 0 \end{bmatrix}$$

The single parameter δ enables us to differentiate between product categories with low and high substitution rates. The adjacent substitution matrix assumes that products are ordered along an attribute space and allows for substitution between neighboring products only. For example, if a customer can't find 1% milk in stock, she may be willing to accept either 2% or skim, but not whole milk. Subgroups substitution matrix allows for substitution within the subgroups only. For example, in the coffee category, consumers may treat decaffeinated coffee and regular coffee as subgroups and not substitute between subgroups.

In the proportional substitution model, the general expression for α_{kj} is

$$\alpha_{kj} = \delta \frac{d_j}{\sum_{l \in N \setminus \{k\}} d_l}. \tag{2}$$

The proportional substitution matrix has properties that are consistent with what would happen in a utility-based framework such as the MNL model. $\alpha_{kj} > \alpha_{kl}$ if $d_j > d_l$. Suppose that a store doesn't carry the whole assortment, i.e., $N \setminus S \neq \emptyset$. Since only one round of substitution is allowed, the realized substitution rate from variant k to other products is $\sum_{j \in S} \alpha_{kj} = \delta \sum_{j \in S} d_j \Big/ \sum_{l \in N \setminus \{k\}} d_l$, which is increasing in the set S. This means that a consumer who can not find her favorite variant in the store is more likely to buy a substitute, as the set of potential substitutes grows.

We next state an assumption commonly made in assortment planning models for tractability.

(A3) No more attempts to substitute occur. Either the substitute product is available and the sale is made, or the sale is lost.

Limiting the number of substitution attempts (A3) is not too restrictive. Smith and Agrawal (2000) show that number of attempts allowed has a smaller effect as more items are stocked, because the probability of finding a satisfactory item by the second try quickly approaches one. Kök (2003) presents an example where effective demands under a three-attempts substitution model with rate δ =0.5 can be approximated almost perfectly with a single-attempt-substitution model with rate δ =0.58.

The exogenous demand model has more degrees of freedom than the MNL model. Since the options in the choice set are assumed to be homogenous, MNL model is unable to capture the types of adjacent substitution, one-product substitution, or within subgroup substitution. In the MNL model the substitution rates depend on the relative utility of the options in $N \cup \{0\}$. This is both an advantage and a disadvantage for the MNL model. The advantage is that it allows one to easily incorporate marketing variables such as prices and promotions into the choice model. The disadvantage is that it cannot differentiate between the initial choice and substitution behavior. Unlike the MNL model, the exogenous demand model can differentiate between categories that have same initial demand for the category but different substitution rates through the choice of p_0 and δ. Therefore, the MNL model cannot treat assortment-based and stockout-based substitutions differently. In contrast, it is certainly possible to use a different δ or different substitution probability matrices for assortment-based and stockout-based substitutions in the exogenous demand model.

3.4 Locational choice model

Also known as the address or the characteristics approach, the locational choice model was originally developed by Hotelling (1929) to study the pricing and location decisions of competing firms. Extending Hotelling's work, Lancaster (1966, 1975) proposed a locational model of consumer choice behavior. In this model, products are viewed as a bundle of their characteristics (attributes) and each product can be represented as a vector in the characteristics space, whose components indicate how much of each characteristic is embodied in that product. For example, defining characteristics of a car include its engine size, gas consumption, and reliability. Each individual is characterized by an ideal point in the characteristics space, which corresponds to his or her most preferred combination of characteristics.

Suppose that there are m characteristics of a product. Let z_j denote the location of variant j in R^m. Consider a consumer whose ideal product is defined by $y \in R^m$. The utility of variant j to the consumer is

$$U_j = k - r_j - g(y, z_j),$$

where k is a positive constant, r_j is the price, and $g : R^m \to R$ is a distance function, representing the disutility associated with the distance from the consumer's ideal point, e.g., Euclidean distance or the rectilinear distance. The consumer chooses the variant that gives him or her the maximum utility. For an extensive discussion of the address approach and its relation to stochastic utility models such as the MNL model, the reader is referred to Chapter 4 in Anderson et al. (1992).

There is one major difference between the locational choice model and the MNL model. In the MNL model, substitution can happen between any two products. In the locational choice model however, IIA property does not hold and substitution between products is localized to products with specifications that are close to each other in the characteristics space. Hence, the firm can control the rate of substitution between products by selecting their locations to be far apart or close to each other.

4. ASSORTMENT SELECTION AND INVENTORY PLANNING

The majority of the papers focus on assortment decisions at a single store. Most papers take a static view of the assortment planning problem, that is the assortment decisions are made once and inventory costs are computed either from a single period model or the steady-state average of a multi-period model. In Sections 4.1 through 4.3, we review four such papers categorized according to the demand model that they are based on. The papers based on the choice models are more stylized but are able to obtain structural properties of the optimal solution. The papers based on the exogenous demand model are more flexible and have more applicability because they allow for more realistic details in modeling, such as nonidentical prices and case packs. In Section 4.4, we review assortment planning papers with supply chain considerations. Section 4.5 discusses a dynamic assortment planning model in which the retailer has a chance to update its assortment throughout the season as it updates its demand estimates every period for products in the assortment. A recent development in the assortment planning literature is the consideration of multiple categories, where consumers are basket shoppers and the assortment decisions across categories are interdependent. In Section 4.6, we discuss two such papers. The first presents an optimization method and the second discusses the long-run impact of variety by considering store choice decisions of consumers.

4.1 Assortment planning with multinomial logit: The van Ryzin and Mahajan model

van Ryzin and Mahajan (1999) formulate the assortment planning problem by using a MNL model of consumer choice. Assume $r_j = r$ and $c_j = c$ for all j. Products are indexed in descending order of their popularity, i.e., such that $u_1 \geq u_2 \geq .. \geq u_n$. *Define* $v_j = e^{u_j/\mu}$. By the MNL share formula, the probability that a customer demands product j is

$$p_j(S) = \frac{v_j}{\sum\limits_{k \in S \cup \{0\}} v_j}. \tag{3}$$

We assume consumers make their product choice (if any) when they observe the assortment, and they do not look for a substitute if the product of their choice is stocked out. Hence, $p_j(S)$ is independent of the inventory status of the products in S. Note that the demand increase in product j due to the decision $S \subseteq N$ is

$$p_j(S) - p_j(N).$$

This demand increase is due to what is termed assortment-based substitution and is comprised of demand from consumer who would have preferred a product in $N - S$ but had to substitute to product j. van Ryzin and Mahajan (1999) also calls this static substitution.

In contrast, in dynamic substitution, consumers observe the inventory levels of all products at the time of their arrival and make their product choice among the products that are available. Hence, dynamic substitution includes both assortment- and stockout-based substitution.

The expected profit of a variant $j \in S$ is

$$\pi_j(S) = (r - c)\lambda p_j(S) - C(\lambda p_j(S)),$$

where $C(\cdot)$ is the operational costs. The cost function is assumed to be concave and increasing to reflect the economies of scale in inventory models such as the EOQ or the newsvendor models.

The objective is to maximize the total category profits by solving

$$\max_{S \subset N} \sum_{j \in S} \pi_j(S).$$

The optimal assortment finds a balance between including a new product and increasing the total demand to the category and cannibalizing the demand of other products' sales and increasing their average cost.

Consider the net profit impact of adding a variant j to assortment S. Define $S_j = S \cup \{j\}$.

$$h(v_j) = \pi_j(S_j) - \left(\sum_{k \in S} \pi_k(S) - \sum_{k \in S} \pi_k(S_j) \right)$$

If the profit of product j is more than the sum of the profit losses of the products in S, then adding j improves profits.

Theorem 1 *The function $h(v_j)$ is quasi-convex in v_j in the interval* $[0, \infty)$.

Since a quasi-convex function achieves its maximum at the end points of the interval, the profit is maximized either by not adding a product to the assortment or by adding the product with the highest v (i.e., the most popular product). This observation leads to the following result that characterizes the structure of the optimal assortment. Define the popular assortment set:

$$P = \{\{\}, \{1\}, \{1, 2\}, .., \{1, 2, .., n\}\}.$$

Theorem 2 *The optimal assortment is always in the popular assortment set.*

This result is intuitive and powerful: it reduces the number of assortments to be considered from 2^n to n. Since only assortment-based substitution is considered, the demand for each product, the optimal inventory level and the resulting profit can be computed for each of the n assortments in the popular assortment set. The above theorems as stated are from Cachon et al. (2005). van Ryzin and Mahajan (1999) originally proved this result for a cost function from the newsvendor model. Specifically, they use the expected costs of a newsvendor model assuming that D is distributed according to a Normal distribution with mean λ and standard deviation σ. The optimal stocking level of product j is the newsvendor stocking quantity:

$$x_j = \lambda p_j(S) + z\sigma\big(\lambda p_j(S)\big)^{\beta},$$

where $z = \Phi^{-1}(1 - c/r)$ and $\beta \in [0, 1)$ controls the coefficient of variation of the demand to product j as a function of its mean. The resulting cost function is

$$C(\lambda p_j(S)) = r\sigma \frac{e^{-z^2}}{\sqrt{2\pi}} \big(\lambda p_j(S)\big)^{\beta}.$$

The authors show that a deeper assortment is more profitable with a sufficiently high price, and a sufficiently high no-purchase preference. In order to compare different merchandising categories, the authors define the fashion of a category using majorization arguments. In a more fashionable category, the utility across products are more balanced, therefore in expectation the market shares of all products are evenly distributed. The paper shows that everything else being equal, the profit of a more fashionable category is lower due to the fragmentation of demand.

This model captures the main trade-off between variety and the increased average inventory costs. The analysis leads to the elegant results that establish the structural properties of the optimal assortment. However, not all assortment planning problems fit the assumption of homogenous group of products with identical prices and costs. The style/color/size combination of shirts in a clothing retailer may be a good example. Even then, the substitutions would occur across styles/colors but not sizes. The assumption that there is a single opportunity to make assortment and inventory decisions can be defended in products with short life cycles, where the season is too short to make changes in the assortment and bring the new products to market before the season is over. Clearly, the main result (Theorem 2) does not hold when products have non-identical price, cost parameters, or different operational characteristics such as demand variance, case pack, and minimum order quantity.

4.1.1 Extensions

Mahajan and van Ryzin (2001a) study the same problem under dynamic substitution. That is, the retailer faces the problem of finding the optimal product selection and stocking levels where customers dynamically substitute among products when inventory is depleted. Consider a customer with the following realization of the utilities: $u_6 > u_4 > u_3 > u_5 > u_0 > u_1 > u_2$. Suppose that the store carries assortment $S = \{1, 2, 3, 4\}$. In the static substitution model, this consumer would choose product 4, buy it if it is available and leave the store if it is not. In the dynamic substitution model, products 4, 3, and 5 are all acceptable to the customer, in that order of preference. Depending on the inventory levels of those products, she will buy the one that is available in the store at the time she visited the store, and won't buy anything only if none of those three products is available. Using a sample path analysis, the authors show that the problem is not even quasi-concave. By comparing the results of a stochastic gradient algorithm with two newsvendor heuristics, they conclude that the retailer should stock more of the more popular variants and less of the less popular variants than a traditional newsvendor analysis suggests. Also, the numerical results support the theoretical insight (Theorem 2) obtained under static substitution. Maddah and Bish (2004) extend the van Ryzin Mahajan model by considering the pricing decisions as well.

Cachon et al. (2005) study the van Ryzin and Mahajan (1999) model in the presence of consumer search, motivated by the following observation: Even when a consumer finds an acceptable product at the retail store, the consumer still faces an uncertainty about the products outside the store's assortment. Therefore, she may be willing to go to another store and explore other alternatives with the hope of finding a better product. In the independent search model, consumers expect each retailer's assortment to be unique, and hence utility of search is independent of the assortment. Examples for this setting include jewelry stores and antique dealers. In the overlapping assortment search model, products across retailers overlap, hence the value of search decreases with the assortment size at the retailer. For example, all retailers choose their digital camera assortments from the product lines of a few manufacturers. In contrast to the no-search model, in the presence of consumer search it may be optimal to include an unprofitable product in the assortment. Therefore, failing to incorporate consumer search in assortment planning results in narrower assortments and lower profits.

Miller et al. (2006) consider the retailer's assortment selection problem with heterogeneous customers and test the impact of different consumer choice models on the optimal assortment. . They develop a sequential choice model in which customers first form Consideration Sets and then make product choices based on the MNL model.

4.2 Assortment planning under exogenous demand models

In this subsection, we review two closely related assortment planning models that consider both assortment-based and stockout-based substitution. Smith and Agrawal (2000) focus on constructing lower and upper bounds to the problem in order to formulate a mathematical program. Kök and Fisher (2007) formulate the problem in the context of an application at a supermarket chain and proposes a heuristic solution to a similar mathematical program. They also provide structural results on the assortments that generate new insights and guidelines for practitioners and researchers.

4.2.1 Smith and Agrawal model

Smith and Agrawal (2000) (hereafter SA) study the assortment planning problem with the exogenous demand model. SA models the arrival process of customers carefully and updates the inventory levels after each customer arrival. Given assortment S, SA sets the stocking level of each product to achieve exogenously determined service levels f_j. Let $g_j(S,m)$ denote the probability that m^{th} customer chooses product j and $A_k(S,m)$ a binary variable indicating the availability of product k when the mth customer arrived. Both clearly depend on the choice of previous customers and the number of substitution attempts made by the customer. For one substitution-attempt-only model,

$$g_j(S,m) = d_j + \sum_{k \notin S} d_k \alpha_{kj} + \sum_{k \in S \setminus \{j\}} d_k \alpha_{kj}(1 - A_k(S,m))$$

The first term is the original demand for product j, the second term is the demand from assortment substitution and the third from stockout substitution. Since exactly determining $g_j(S,m)$ is complex, SA develops lower and upper bounds. The lower bound is achieved by considering only assortment-based substitution and the upper bound by assuming that products achieve f_j in-stock probability even for the first customer, hence overestimating stockout substitution. Specifically,

$$\begin{aligned} &h_j(S) \le g_j(S,m) \le H_j(S) \text{ for all } m, \text{ where} \\ &h_j(S) = d_j + \sum_{k \notin S} d_k \alpha_{kj}, \\ &H_j(S) = d_j + \sum_{k \notin S} d_k \alpha_{kj} + \sum_{k \in S \setminus \{j\}} d_k \alpha_{kj} f_k. \end{aligned} \tag{4}$$

SA shows that these bounds are tight and uses the lower bound $h_j(S)$ to approximate the demand rate. That is, effective demand for product j given assortment S follows a distribution with mean $h_j(S)$. SA provides similar bounds to the demand rate under the repeated-attempts substitution model. Agrawal and Smith (1996) found that Negative Binomial distribution (NBD) fits retail sales data very well. SA shows that when the total number of customers that visit a store is distributed with NBD, the demand for each product would also follow NBD.

The optimization problem is to maximize the total category profits:

$$\max_{S \subset N} \ Z = \sum_{j \in S} \pi_j(S)$$

where the profit function for each product j is the newsvendor profit minus the fixed cost of stocking an item V_j.

$$\pi_j(S) = (r_j - c_j)h_j(S) - c_j E[x_j - D_j | h_j(S)]^+ - (r_j - c_j)E[D_j - x_j | h_j(S)]^+ - V_j,$$

where D_j is the random variable representing the demand for product j, x_j is the optimal newsvendor stocking quantity to achieve the target stocking level $f_j = 1 - c_j/r_j$, e.g., $\Pr\{D_j \geq x_j | h_j(S)\} = f_j$ for a continuous demand distribution. Incorporating salvage value, or holding costs to the newsvendor profit function above is trivial.

This optimization problem is a nonlinear integer programing problem. SA proposes solving the problem via enumeration for small n and a linearization approximation for large n. A single constraint such as a shelf space or a budget constraint can be incorporated into the optimization model. SA proposes a Lagrangian Relaxation approach followed by a one-dimensional search on the dual variable for the resulting mathematical program.

Several insights are obtained from illustrative examples. Substitution effects reduce the optimal assortment size when fixed costs are present. However, even when there are no fixed costs present, substitution effects can reduce the optimal assortment size, because products have different margins. Contrary to the main result of van Ryzin and Mahajan (1999), it may not be optimal to stock the most popular item - a result of the adjacent substitution matrix or the one-item substitution matrix.

4.2.2 Kök and Fisher model

The methodology described in Kök and Fisher (2007) is applied at Albert Heijn, BV, a leading supermarket chain in the Netherlands with 1187 stores and about $10 billion in sales. The replenishment system at Albert Heijn is typical in the grocery industry. All the products in a category are subject to the same delivery schedule and fixed leadtime. There is no backroom, therefore orders are directly delivered to the shelves. Shelves are divided into *facings*. SKUs in a category share the same shelf area but not the same facing, i.e., only one kind of SKU can be put in a facing. Capacity of a facing depends on the depth of the shelf and the physical size of a unit of the SKU. The inventory model is a periodic review model with stochastic demand, lost sales and positive constant delivery lead-time. The number of facings allocated to product j, f_j, determines its maximum level of inventory, $k_j f_j$, where k_j is the capacity of a facing. At the beginning of each period, an integral number of case packs (batches) of size b_j is ordered to take the inventory position as close as possible to the maximum inventory level without exceeding it. Case sizes vary significantly across products and significantly affect returns from inventory. The performance measure is gross profit, which is per-unit margin times sales minus selling price times disposed inventory.

We focus on a single subcategory of products initially for expositional simplicity and then explain how to incorporate the interactions between multiple subcategories. The decision process involves allocating a discrete number of facings to each product in order to maximize total expected gross profits subject to a shelf space constraint:

$$\max_{f_j, j \in N} Z(\mathbf{f}) = \sum_j G_j(f_j, D_j(\mathbf{f}, \mathbf{d}))$$

$$s.t. \quad \sum_j f_j w_j \leq ShelfSpace \tag{AP}$$

$$f_j \in \{0, 1, 2, ..\}, \text{ for all } j$$

where f_j is the number of facings allocated to product j, and w_j is the width of a facing of product j. G_j is the (long run) average gross profit from product j given f_j and demand rate D_j. Due to substitution, effective demand for a product includes the original demand for the product and substitution demand from other products. Hence, $D_j(\mathbf{f}, \mathbf{d})$, the effective demand rate of product j, depends on the facing allocation and the demand rates of all products in the subcategory, i.e., $\mathbf{f} = (f_1, f_2, .., f_n)$ and $\mathbf{d} = (d_1, d_2, .., d_n)$, where d_j is the original demand rate of product j (i.e., number of customers who would select j as their first choice if presented with all products in N). The store's assortment is denoted S and is determined by the facing allocation, i.e., $S = \{j \in N : f_j > 0\}$.

Similar to SA, the effective demand rate function under this substitution model is

$$D_j(\mathbf{f}, \mathbf{d}) = d_j + \left(\sum_{k:f_k=0} \alpha_{kj} d_k + \sum_{k:f_k>0} \alpha_{kj} L_k(f_k, d_k) \right) \tag{5}$$

where the L_k function is the lost sales (average unmet demand) of product k. In our application we estimate $L_k(f_k, d_k)$ via simulation. In (5), $\sum_{k:f_k=0} \alpha_{kj} d_k$ is the demand for j due to assortment-based substitution and $\sum_{k:f_k>0} \alpha_{kj} L_k(d_k, f_k)$ is the demand for j due to stockout-based substitution.

In a stochastic inventory model as described above, G_j is a nonlinear function of the allocated facings to product j. It is a function of the facings of product j (f_j), and the facings of all other SKUs in a subcategory through the D_j function. Hence, (AP) is a knapsack problem with a nonlinear and nonseparable objective function, whose coefficients need to be calculated for every combination of the decision variables. Even if we rule out stockout-based substitution, we need to consider 'in' and 'out' of the assortment values for all products leading to 2^n combinations.

We propose the following iterative heuristic that solves a series of separable problems. The details of the algorithm can be found in Kök and Fisher (2007). We set $D_j(\mathbf{f},\mathbf{d}) = d_j$ for all j and solve (AP) with the original demand rates resulting in a particular facings allocation $\mathbf{f}^0$. At iteration t, we recompute $D_j(\mathbf{f}^{t-1},\mathbf{d})$ given δ for all j according to Equation (5). Note that $\sum_j G_j\left(f_j^t, D_j(\mathbf{f}^{t-1},\mathbf{d})\right)$ is separable now, because $D_j(\mathbf{f}^{t-1},\mathbf{d})$ are computed a priori. We then solve (AP) with $Z(\mathbf{f}^t) = \sum_j G_j\left(f_j^t, D_j(\mathbf{f}^{t-1},\mathbf{d})\right)$ via a Greedy Heuristic. We keep iterating until f_j^t converges for all j. In a computational study, the Iterative Heuristic performs very well with an average optimality gap of 0.5%.

(AP) can be generalized to multiple subcategories of products that share the same shelf space by including several subcategories in the summations in the objective function and the shelf space constraint. Let subscript $i = 1,..,I$ be the subcategory index. The objective function in the multiple subcategory case would be $Z(\mathbf{f}) = \sum_i \sum_j G_{ij}(f_{ij}, D_{ij}(\mathbf{f}_i,\mathbf{d}_i))$, the shelf space constraint can be modified similarly.

Structural Properties of the Iterative Heuristic: The Iterative Heuristic is based on a Greedy Heuristic. Therefore we can find properties of the resulting solution by exploiting the way the Greedy Heuristic works. First we note that the gross profit function for a product depends on demand, margin and operational constraints. Demand level and per-unit margin affect the maximum gross profit a product can generate if sufficient inventory is held. Operational constraints, such as case-pack sizes and delivery leadtime affect the curvature of the gross profit function. For example, a product with a smaller case-pack (batch size) has a higher slope of the gross profit curve for low inventory levels, and therefore can achieve the maximum gross profit with less inventory. These observations lead to the following theorems taken from Kök and Fisher (2007).

Products A and B belong to a subcategory with substitution rate $\delta \geq 0$. They are nonperishable. They are subject to the replenishment system described at the beginning of this subsection. The leadtime is zero. Demand for both products follow the same family of probability distributions. Effective demand for product A (B) has a mean$D_A(D_B)$ and coefficient of variation $\rho_A(\rho_B)$. Unless otherwise stated, $d_A = d_B, \rho_A = \rho_B, r_A = r_B, c_A = c_B$, and $b_A = b_B = 1$.

Theorem 3 *Consider products A and B. Let $\tilde{\mathbf{f}}$ denote the vector of facing allocations for all products in the subcategory other than A and B. If exactly one of the following conditions is met,*

(i) All else is equal and $d_A > d_B$. The demand distribution is one of Poisson, Exponential or Normal distribution.

(ii) *All else is equal and* $r_A - c_A \geq r_B - c_B$.
(iii) $w_A \leq w_B$,

then $f_A \geq f_B$ *in the final solution of the Iterative Heuristic.*

The implications of the first part of this theorem is clear: an allocation algorithm based on demand rates should work fairly well when products are differentiated by demand rates only. This is similar to the property of optimal assortments in the unconstrained problem in van Ryzin and Mahajan (1999). However, the above theorem proves additional results that the product with higher margin, or lower space requirement should be given priority in the assortment.

Theorem 4 *Consider products A and B. Let* $\tilde{\mathbf{f}}$ *denote the vector of facing allocations for all products in the subcategory other than A and B. If exactly one of the following conditions is met,*

(i) *All else is equal and* $\rho_A < \rho_B$,
(ii) *All else is equal,* $b_A \geq 1$, and b_B *is an integer multiple of* b_A,

then the following holds. In the final solution of the Iterative Heuristic, if product B is included in the assortment then so is A (*i.e.,* $f_B > 0 \Longrightarrow f_A > 0$).

Theorem 4 characterizes the impact of the operational characteristics of a product on the assortment choice. When one of the conditions of the Theorem 4 holds, i.e., when *B* has either a larger batch size or higher demand variability, due to limited shelf space, if *A* is not included in the assortment, neither is *B*. Since the maximum value of G_A is higher and the slope is higher for low inventory levels, the profit impact of first facing is higher for *A*, resulting in a higher rank in the ordered input list to the Greedy Heuristic. However, if both products are in the assortment, it is possible to have $f_B > f_A$ in the solution. The reason for this is that G_A reaches its maximum level quickly with the early facing allocations, whereas it takes more facings for *B* to reach its maximum. In such cases, allocation heuristics based on demand rates perform poorly. A reasonable rule of thumb based on these observations would be the following. First high demand rate products shall be included in the assortment, then more facings shall be allocated to the products that have more restrictive operational constraints.

We applied our estimation methodology (to be described in Section 5.2.2) and optimization methodology to the data from 37 stores and two categories. The categories include 34 subcategories or 234 SKUs. (AP) is solved for each category for a given category shelf space. The facing allocations for SKUs also determine the space allocation between subcategories. We compare the category gross profit of the recommended assortments with that of the current assortments at Albert Heijn. The gross profits of the recommended system is 13.8% higher than that of the current assortment. The financial impact of our methodology is a 52% increase in pretax profits of Albert Heijn.

Other work on assortment planning with exogenous demand include Rajaram (2001). He develops a heuristic based on Lagrangian relaxation for the single period assortment planning problem in fashion retailing without consideration of substitution between products.

4.3 Assortment planning under locational choice

Gaur and Honhon (2006) study the assortment planning model under the locational choice demand model. The products in the category differ by a single characteristic that does not affect quality or price such as yogurt with different amounts of fat-content. The assortment carried by the retailer is represented by a vector of product specifications $(b_1, .., b_s)$ where s is the assortment size and $b_j \in [0, 1]$ denotes the location of product j. Each consumer is characterized by an ideal point in [0,1] and chooses the product that is closest to him or her. The coverage interval of product j is defined as the subinterval that contains the most preferred good of all consumers for whom the product yields a nonnegative utility. The first choice interval of product j is defined as the subinterval that contains the most preferred goods of all consumers who choose j as a first choice. To extend Lancaster's model to stochastic demand, the authors assume that customers arrive to the store according to a Poisson process and that the ideal points of consumers are independent and identically distributed with a continuous probability distribution on finite support [0,1]. Only unimodal distributions are considered, implying that there exists a unique most popular product, and that the density of consumers decreases as we move away from the most popular product.

The operational aspects of the problem are similar to the van Ryzin and Mahajan model reviewed in Section 4.1: all products are assumed to have identical costs and selling prices, there is a single selling period, inventory costs are derived from a newsvendor model: excess demand at the end of the period is lost and excess inventory is salvaged. The only difference is that there is a fixed cost associated with including a product in the assortment. This model is closely related to the marketing product line design models in the marketing literature and operations-marketing papers such as de Groote (1994).

Under static substitution (assortment-based substitution), a consumer chooses a first choice product given the assortment but without observing inventory levels and does not make a second choice if the first choice is not available. Under dynamic substitution, the consumer chooses a product (if any) among the available products. This is equivalent to choosing a first choice product from the assortment and then looking for the next best alternative (if any) if the first choice is not available. This is equivalent to stock-out based substitution with repeated attempts.

The paper characterizes the properties of the optimal solution under static substitution and develops approximations under dynamic substitution. We skip the details of the analysis and briefly discuss the results from this paper. The authors show that, under static substitution, the distance between products in the optimal assortment are large enough so that there is no substitution between them. The most popular product, the one that would be located at the mode of the distribution is not included in the assortment when the economies of scale enjoyed by the most popular product is overcome by the diseconomies of scale it created for the other products. This property contrasts with the property of the optimal assortments under the MNL model (Theorem 2). We believe that the difference is not because of the different choice model, but because the problem considered here is a product line design problem at its heart. The authors find that the retailer may choose not to cover the entire market due to fixed costs. An analogous result is obtained under the MNL model as well, but that is purely due to economies of scale created for more popular products by not including some products in the assortment. Whereas in this model, it is optimal to cover the entire market when fixed costs are not present.

The problem is more complex under the dynamic substitution problem, as it is under other demand models. The profits computed under the static substitution assumption provides a lower bound to the dynamic problem, since it does not capture the profits from repeated attempts of the stock-out based substitution. An upper bound is obtained by solving a relaxation of the problem. Namely, the retailer gets to observe the ideal points of all arriving customers before allocating inventory to customers to maximize the profits. This is similar to Bassok, Anupindi and Akella (1999) where consumers do not directly choose a product, but they are assigned a product (if any) either according to an exogenous rule or the retailer's decisions. Clearly, the retailer can generate more profits by doing the allocation itself rather than following the choices of the customers arriving in a random process. The solutions to these bounds are also proposed as heuristic approaches. In a numerical study, the authors make the following observations. Both heuristics generate solutions that are 1.5% within the optimal solution on average. This suggests that the static substitution solution, which is easier to obtain, would serve as a good approximation in most cases. Dynamic substitution has the greatest impact when demand is low, customer distribution in the attribute space is heterogenous, and consumers are willing to substitute more. The retailer provides higher variety under dynamic substitution than under static substitution and locates products closer to each other so that a consumer can derive positive utility from more than one product. The firm offers more acceptable alternatives to the customers whose ideal product is located in areas where consumer density is high.

There are other papers that formulate mathematical models for selecting optimal assortments when customer heterogeneity is represented by locational choice. McBride and Zufryden (1988) deal with manufacturer's product line selection which require specification of product attributes and Kohli and Sukumar (1990) deal with the retailer's problem of choosing an assortment from a set of products.

4.4 Assortment planning in decentralized supply chains

The assortment planning papers reviewed until this section are single location models. There has been some recent work exploring assortment planning issues in two-tier supply chains. Aydin and Hausman (2003) consider the assortment planning problem with MNL (i.e. the van Ryzin and Mahajan model) in a decentralized supply chain with one supplier and one retailer. They find that the retailer chooses a narrower assortment than the supply chain optimal assortment since her profit margins are lower than that of the centralized (vertically integrated) supply chain. The manufacturer can induce coordination by paying the retailer a per-product fee, resembling the slotting fees in the grocery industry, while making both parties more profitable.

Singh et al. (2005) study the effect of product variety on supply chain structures, building on the van Ryzin and Mahajan model. In the traditional channel, the retailers stock and own the inventory, whereas in the drop-shipping channel, the wholesaler stocks and owns the inventory and ships the products directly to customers after the customers place an order at a retailer. Drop-shipping is a common practice in internet retailing: it offers the benefits of risk pooling when there are multiple retailers, but retailers have to pay a per unit fee for drop-shipping. As a result, product variety in the drop-shipping channel is higher than the traditional channel when drop-shipping fees are low and number of retailers is large. The authors derive conditions on the parameters under which the retailers or the wholesaler or both prefer the drop-shipping channel. They also study a vertically integrated firm with multiple retailers and find that a hybrid supply chain structure may be optimal for some parameter combinations: the popular products are stocked at the retailer while the less popular products are stocked at the warehouse and drop-shipped to the customers. The assortment size at the retailer gets smaller as the number of retailers increase or the drop-shipping costs decrease.

Kurtulus and Toktay (2005) compare the traditional category management and category captainship in a setting with two products and deterministic demand under a shelf space constraint. In category captainship, one of the vendors is assigned as the category captain and the pricing and assortment decisions are delegated to her. The argument for category captainship is that the leading manufacturer in a category may have more experience with the category and resources than the retailer. They find that the assortment may be narrower under category captainship, because the noncaptain brand may be priced out of the assortment. Kurtulus (2005) considers the impact of category captainship under three types of contracts in a setting similar to the van Ryzin and Mahajan model. While the resulting assortment is still in the popular assortment set under the target profit and target sales contracts, it is in the least popular assortment set under the target variety contract.

4.5 Dynamic assortment planning

All of the assortment planning papers reviewed in the previous sections consider static assortment planning problems and do not consider revising or changing assortment selection as time elapses. This makes sense for fashion and apparel retailers, because long development, procurement and production lead times constrain retailers to make assortment decisions in advance of the selling season. With limited ability to revise product assortments, academics and industry practitioners focused on optimizing the production quantities in order to delay the production of those products that have high demand uncertainty (e.g., Fisher and Raman 1996). However, innovative firms such as Zara (Spain), Mango (Spain), and World Co. (Japan) created highly responsive and flexible supply chains and cut the design-to-shelf lead time down to 2–5 weeks, as opposed to 6–9 months for a traditional retailer, which enabled them to make design and assortment selection decisions during the selling season. Raman et al. (2001) describes how such short response times are achieved at World Co. through process and organizational changes in the supply chain. Learning the fashion trends and responding with an updated product selection is most critical for these high fashion companies.

Allowing changes in the assortment during a single selling season introduces several new issues. The products put in the store this week can't be removed next week and hence condition the decisions this week; there may be costs associated with adding new products or dropping products from the assortment; it may be optimal to put products in the stores to learn about the demand, even if it isn't optimal to do so given the current knowledge.

Caro and Gallien (2005) formulate the dynamic assortment problem faced by these retailers: At the beginning of each period, the retailer decides which assortment should be offered and gathers demand data for the products carried in the assortment in each period. There is a budget constraint that limits the number of products offered in each period to K. Due to design-to-shelf lead time, an assortment decision can be implemented only after l periods. This problem relates to the classical exploration versus exploitation trade-off. The firm must decide whether to optimize revenues based on the current information (exploitation), or try to learn more about the demand of products not in the assortment with the hope of identifying popular products (exploration).

The authors make several assumptions for tractability. The demand for a product is independent of the demand or the availability of the other products (i.e., there is no substitution between products or correlation in demand). The demand rate for each product is constant throughout the season. There is a perfect inventory replenishment process, therefore there are no lost sales or economies of scale in the operating costs. More importantly, no products carry over from period to period, therefore it is feasible to change the assortment independent of the previous assortment. There are no switching costs. Some of these assumptions are relaxed later.

The demand for product $j \in N$ is from a stationary Poisson process throughout the season. The rate of arrival λ_j is unknown and actual demand is observed only when the product is included in the assortment. The retailer uses a Bayesian learning mechanism: he starts each period with a prior belief that λ_j is distributed according to a Gamma distribution with shape parameter m_j and scale parameter α_j. Suppose that product j is included in the assortment and observed demand is d_j. The prior distribution of λ_j is updated as $Gamma(m_j + d_j, \alpha_j + 1)$. The mean of this distribution is the average sales of product j throughout the periods it is carried. Let $\mathbf{f} = (f_1, .., f_n)$ be a vector of binary variables indicating whether the product is in the assortment and F the set of feasible assortments, $F = \left\{\mathbf{f} : \sum_{j \in N} f_j \leq K\right\}$. Similarly, let $\mathbf{m}, \alpha$, and $\mathbf{d}$ denote the vectors of m_j, α_j, d_j, respectively. Assume that assortment implementation leadtime l is zero.

The dynamic programming formulation is

$$J_t^*(\mathbf{m}, \alpha) = \max_{\mathbf{f} \in F} \sum_{j \in N} f_j r_j E[\lambda_j] + EJ_{t+1}^*(\mathbf{m} + \mathbf{d} \cdot \mathbf{f}, \alpha + \mathbf{f}).$$

Since solution of this dynamic program can be computationally overwhelming, the authors propose a Lagrangian relaxation (of the constraint on the number of products in the assortment) and the decomposition of weakly coupled dynamic programs to develop an upper bound. Performance of two heuristics are compared. The index policy balances exploration by including high expected profit products and exploitation by including products with high demand variance in a single-period look ahead policy. The greedy heuristic selects in each period the K products with the highest expected profits. The index policy is near optimal when there is some prior data on demand available and outperforms the greedy heuristic especially with little prior information about demand or the leadtime. The paper then demonstrates that the heuristics perform well when there are assortment switching costs, demand substitution, and a positive implementation lag.

Another learning method that Zara and other high-fashion companies employ is learning the attributes of the high selling products. That is, if a certain color is hot this season, and products with a special fabric are selling relatively well, the prior distribution of the demand for a product with that fabric-color combination can be updated, even if the product were never included in the assortment before. The attribute-based estimation method by Fader and Hardie (1996) mentioned in Section 5.1 can be instrumental in estimating the demand for new products in this setting.

4.6 Assortment planning models with multiple categories

Although research has primarily focused on single category choice decisions, there is recent research that examines multiple category purchases in a single shopping occasion by modeling the dependency across multi-category items explicitly (see Russell et al. 1997 for a review). Manchanda et al. (1999) find that two categories may co-occur in a consumer basket either due to their complementary nature (e.g., cake mix and frosting) or due to coincidence (e.g., similar purchase cycles or other unobserved factors). Bell and Lattin (1998) show that consumers make their store choice based on the total basket utility. Fixed costs for each store visit (e.g., search and travel costs) provide an intuitive explanation for why consumers basket shop. Bell, Ho and Tang (1998) use market basket data to analyze consumer store choices and explicitly consider the roles of fixed and variable costs of shopping.

Baumol and Ide (1956) study the notion of right level of variety in a very stylized model. The retailer chooses N, the number of different product categories to offer. Consumer utility is increasing in variety, but decreasing in in-store search costs (which increases with N). Therefore for each consumer there is a range of N that makes the store attractive for shopping. The operating cost is the sum of inventory costs per category from an EOQ model and handling costs that is concave increasing in N. The resulting retailer profit function is not well-behaved, therefore profit maximizing level of variety is difficult to characterize and the insights from this model are fairly limited.

There are two papers that consider assortment planning with multiple categories in more detail. Agrawal and Smith (2003) extend the Smith and Agrawal (2000) model and the analysis described in Section 4.2.1 to the case where customers demand sets of products. Cachon and Kök (2007) compare the prices and variety levels in multiple categories under category management to the optimal variety levels in the presence of basket shopping consumers.

The modeling and solution approach in Agrawal and Smith (2003) is very similar to their earlier work. Each arriving customer demands a purchase set. If the initially preferred purchase set is not available, the customer may do one of the following: (*i*) substitute a smaller set that does not contain the missing item, (*ii*) substitute a completely different purchase set, (*iii*) not purchase anything. This behavior is governed by substitution probability matrices. The demand for each set considering the substitution demand from other sets is characterized as in Equation (4). The profit maximization problem is formulated as a mathematical program. For a customer to purchase any set, all the items in the set have to be available. Therefore, the expected profit is much more sensitive to percentage of customers who purchase in sets, the average size of a purchase set, and the substitution structure and parameters. The following observations from numerical examples are quite interesting.

Profits under adjacent substitution structure is much higher than that under random substitution, because under adjacent substitution stocking every other set in the list would result in lower lost sales than that under random substitution. As the percentage of customers who purchases in sets increases (while keeping the total demand constant), the optimal assortment size increases (decreases) if the fixed cost of including a product is low (high). Profits increase with substitution rate δ. Finally, optimizing the category by disregarding the substitution and the purchase sets can result in considerably lower profits than optimal.

Cachon and Kök (2007) work with a stylized model to develop managerial insights regarding the assortment planning process in an environment with multiple categories. Consider two retailers X and Y that carry two categories of goods. Retailer r offers n_{rj} products and sets its margin p_{rj} in category j. The consumer choice model is based on a nested Multinomial Logit (MNL) framework. A consumer's utility from purchasing product i in category j at retailer r is $u_{rji} = v_{rji} - p_{rj} + \varepsilon$ where v_{rji} is the expected utility from the product less the unit cost of the product and ε is i.i.d with Gumbel distribution with zero mean. There are three types of consumers in the market that are characterized by the contents of their shopping baskets: type 1 consumers would like to buy a product in category 1 only, type 2 consumers would like to buy a product in category 2 only, type b consumers are basket shoppers and would like to buy a product from both categories. Consumers buy exactly one unit of one product in every category included in their basket.

The authors show that the choice probability of a non-basket shopper between retailers X, Y and a no-purchase alternative can be written using the nested MNL model as follows:

$$s_{rj} = \frac{A_{rj}}{A_{xj} + A_{yj} + Z_j} \quad \text{for } r = x, y, \text{ and } j = 1, 2,$$

where A_{rj} is the attractiveness function for each alternative (an aggregate function of price and variety level). Using the nested MNL results of Ben Akiva and Lerman (1985), as described in Section 3.2, it can be expressed as

$$A_{rj} = e^{-p_{rj}} \sum_{i=1}^{n_{rj}} e^{v_{rji}}, \text{for } r = x, y.$$

Now, consider a basket-shopping consumer. A basket-shopping consumer chooses retailer r only if she prefers the assortment at r for both categories. As a result, the probability that a basket shopper chooses retailer r is

$$s_{rb} = s_{r1} s_{r2} \text{ for } r = x, y. \tag{6}$$

This is a multiplicative basket-shopping model, as a retailer's share of basket shoppers is multiplicative in its share in each category. An additive model for this problem has been discussed in Kök (2003).

The common practice of category management (CM) is an example of a decentralized regime for controlling assortment because each category manager is charged with maximizing profit for his or her assigned category. Since basket shoppers' store choice decision depends on the prices and variety levels of other categories, one category's optimal decisions depends on the decisions of the other categories. Hence, a game theoretic situation arises. CM can be interpreted as an explicit non-cooperative game between the category managers, since each category manager is responsible exclusively for the profits of her own category. Alternatively, it can be interpreted as an iterative application of single category planning where each category's variety level is optimized assuming all other assortment decisions for the retailer are fixed. Decentralized regimes such as CM are analytically manageable but they ignore (in their pure form) the impact of cross-category interactions. Centralized regimes account for these effects but it is extremely difficult, in practice, to design a model to account for all cross-category effects, to estimate its parameters with available data and solve it.

The authors show that if there are any basket shoppers, CM provides less variety and higher prices than centralized store management. CM can lead to poor decisions because the category manager does not sufficiently account for how his or her decisions influences total store traffic. These results hold both for a single retailer and in duopoly competition. Numerical examples demonstrate that the profit loss due to CM can be significant. The dominant strategy for each retailer is to switch to centralized management.

To address the potential problem with a decentralized approach to assortment planning, we propose a simple heuristic that retains decentralized decision making (category managers optimize their own categories' profit) but adjusts how profits are measured. To be specific, instead of using an accounting measure of a category's profit, the authors define a new measure called *basket profits*. Basket profits can be estimated using point-of-sale data. It enables CM to approximately measure the true marginal benefits of merchandising decisions and lead to near-optimal profits. This analytical approach is an attractive alternative relative to ad-hoc coordination across category managers.

5. DEMAND ESTIMATION

In this section, we briefly discuss the estimation of the demand models specified in Section 3. The estimation method depends on the type of data that is available.

5.1 Estimation of the MNL

5.1.1 With panel data

Starting with the seminal work of Guadagni and Little (1983), an enormous number of marketing papers estimated the parameters of the MNL model to understand the impact of marketing mix variables on demand. These papers use panel data in which the purchasing behavior of households over time are tracked by the use of store loyalty cards. Consider the purchase decision of the household that visited the store in time t. The systematic component of the utility u_{jt} is specified as a linear function of m independent variables including product specific intercepts, price, an attribute of product j, loyalty of the household to the brand of product j (measured as exponentially weighted average of binary variables indicating whether or not the household purchased this brand). Let $x_{jt} = (x_{jt1}, x_{jt2}, .., x_{jtm})$ denote the vector of these attributes for the household's shopping trip at time t, S_t denote the assortment at time t including the no-purchase option, and $\beta = (\beta_1, .., \beta_m)$ denote the vector of common coefficients.

$$u_{jt} = \beta^T x_{jt}, \qquad j = 0, 1, .., n.$$

The outcome of the choice experiment by a household in time t is

$$y_{jt} = \begin{cases} 1, \text{if product } j \text{ is chosen in time } \quad t \\ \qquad hskip - t?0, \text{otherwise} \end{cases}$$

Given u_{jt} it is possible to compute the choice probabilities according to MNL formula (1) . To obtain the maximum likelihood estimates (MLE) for the coefficients, we can write log of the likelihood function by multiplying the probability of observing the choice outcome across all t:

$$\check{L}(\beta) = \sum_t \sum_j y_{jt} \left(\beta^T x_{jt} - \ln \sum_{k \in S_t} e^{\beta^T x_{kt}} \right).$$

McFadden (1974) shows that the log-likelihood function is concave, therefore any nonlinear optimization technique can be used to find the MLE estimate of β. Fader and Hardie (1996) suggest the use of more of the product's attributes and dropping product-specific dummy variables in x_j in the estimation. They argue that this results in a more parsimonious estimation method as the number of coefficients to be estimated would not grow with number of products but with number of significant characteristics. Moreover, this approach enables estimation of the demand for new products.

Extensions of this model such as Chiang (1991), Bucklin and Gupta (1992), and Chintagunta (1993) also investigate whether to buy, and how much to buy decisions of households. In these papers, the whether-to-buy decision is modeled as a binary choice between the no-purchase alternative and the resulting utility from the product choice and quantity decisions in a nested way. Chong et al. (2001) extend the classical Guadagni and Little (1983) model using a nested MNL model, including three new brand-width measures that capture the similarities and the differences among products within and across brands.

Multiplicative Competitive Interactions (MCI) model offers a viable alternative to MNL. Although less popular than MNL, it is used in the marketing area to study market share games (e.g. Gruca and Sudharshan 1991) and it has empirical support. See Cooper and Nakanishi (1988) for a detailed discussion and estimation methods.

5.1.2 With sales transaction data

Consider the demand process in the van Ryzin and Mahajan model, where consumer arrivals follow a Poisson process with rate λ and consumers select an alternative based on the MNL model. Our goal is to estimate λ and β from sales data. Sales transactions are the records of the purchasing time and the product choice for each customer who made a purchase. This is an incomplete data set in the sense that only the arrivals of customers who made a purchase are recorded. Define a period as a very small time interval such that the probability of having more than one customer arrival in a period is zero. Let t denote the index of periods. There is a sales record for a period only if a purchase is made in that period. It is impossible to distinguish a period without an arrival, from a period in which there was an arrival but the customer did not purchase anything. Therefore, the approach described above cannot be used.

The *Expectation-Maximization*(EM) algorithm is the most widely used method to correct for missing data. Proposed by Dempster et al. (1977), the EM method uses the complete-likelihood function in an iterative algorithm. Talluri and van Ryzin (2004) describe an estimation approach based on this method in the context of airline revenue management, but the algorithm is applicable to the retail setting described in Section 4.1. Let P denote the set of periods that there has not been a purchase made and $a_t = 1$ if there has been a customer arrival in period t. The unknown data is $(a_t)_{t \in P}$. We start with arbitrary (λ, β). The E-step replaces the incomplete data with their estimates. That is, we find the expectation of a_t for all $t \in P$ given the current estimates (λ, β). The M-step maximizes the complete-data likelihood function to obtain new estimates. The likelihood function is similar to that in the previous subsection, but includes the arrival probabilities λ. The procedure is repeated until the parameter estimates converge. Greene (1997) shows that the procedure converges under fairly weak conditions. If the expected log-likelihood function is continuous in the parameters, Wu (1983) shows that the limiting value of the procedure would be a stationary point of the incomplete-data log-likelihood function. The advantage of the procedure is that maximizing the complete-data likelihood function is much easier than maximizing an incomplete-data likelihood function.

5.1.3 With sales summary data

The information available in sales data is different from the panel data in several ways, hence requires a different approach. One possibility is the approach in Kök and Fisher (2007), which will be described here. The data typically available for estimating the parameters of a demand model includes the number of customers visiting each store on a given day, sales for each product-store-day, as well as the values of variables that influence demand such as weather, holidays, and marketing variables like price and promotion. At Albert Heijn, the data set included SKU-day-store level sales data through a period of 20 weeks for seven merchandise categories from 37 Albert Heijn stores. For each store-day, the number of customers visiting the store is recorded. For each SKU-day-store, sales data comprised of the number of units sold, the number of customers that bought that product, selling price, and whether the product is on promotion or not. In addition, we have daily weather data and a calendar of holidays (e.g., Christmas week, Easter, etc.). The categories are cereals, bread spreads, butter & margarine, canned fruits, canned vegetables, cookies, and banquet sweets. There were 114 subcategories in these seven categories. The size of subcategories varies from 1 to 29 SKUs, with an average of 7.7 and a standard deviation of 5.7.

The model of consumer purchase behavior is based on three decisions: (1) whether or not to buy from a subcategory (*purchase-incidence*), (2) which variant to buy (*choice*) given purchase incidence, and (3) how many units to buy (*quantity*).[†] This hierarchical model is quite standard in the marketing literature and commonly used with panel data.

The demand for product j is

$$D_j = K(PQ)_j = K\pi p_j q_j \tag{7}$$

where K is the number of customers that visit the store at a given day, $(PQ)_j$ is the average demand for product j per customer, π is the probability of purchase incidence (i.e., the probability that a customer visiting the store buys anything from the subcategory of interest), p_j is the choice probability (i.e., the probability that variant j is chosen by a customer given purchase incidence), and q_j is the average quantity of units that a customer buys given purchase incidence and choice of product j.

The purchase incidence is modeled as a binary choice:

$$\pi = \frac{e^v}{1 + e^v} \tag{8}$$

where v is the expected utility from the subcategory that depends on the demand drivers in the subcategory.

The product choice is modeled with the Multinomial Logit framework, where p_j are given by (1). The average utility of product j to a customer, u_j, is assumed to be a function of product characteristics, marketing and environmental variables.

Let subscript h denote store index, and t denote time index (i.e., day of the observation).

[†] This hierarchical model of choice is similar to Bucklin and Gupta (1992) that models the first two decisions with an additional focus on the segmentation of customers and Chintagunta (1993) that models all three decisions. Both papers work with household panel data, whereas we work with daily sales data.

We compute p_{jht} from the sales data as the ratio of number of customers that bought product j to number of the customers that bought any product in the subcategory at store h on day t. At Albert Heijn, price and promotion are the variables influencing u_j. We fit an ordinary linear regression to the log-centered transformation of (1) (see Cooper and Nakanishi 1988 for details) to estimate $\delta_j^C, \alpha_1^C, \alpha_2^C$, and $\theta_k^C, k = 1, .., n$.

$$\ln\left(\frac{p_{jht}}{\bar{p}_{ht}}\right) = u_j = \delta_j^C + \sum_{k \in N} \theta_k^C I_{jk} + \alpha_1^C \left(R_{jht} - \bar{R}_{ht}\right) + \alpha_2^C \left(A_{jht} - \bar{A}_{ht}\right), \text{for all } j \in S \tag{9}$$

where $\bar{p}_{ht} = \left(\prod_{j \in S} p_{jht}\right)^{1/|S|}$, $I_{jk} = \{1, \text{ if } j = k;\ 0 \text{ otherwise}\}$, R is price, $\bar{R}$ is average price in the subcategory, $A_{jht} = \{1$, if product j is on promotion on day t at store h; 0, otherwise, and $\bar{A}$ is average promotion level in the subcategory. It is straightforward to incorporate variables other than price and promotion into this approach.

We compute π_{ht}, the probability of purchase-incidence for the subcategory, from sales data as the ratio of number of customers who bought any product in S to the number of customers visited the store h on day t. We use the following logistic regression equation to estimate $\alpha_0^\pi, \alpha_1^\pi, \alpha_2^\pi, \alpha_{4t}^\pi, \gamma_k^\pi$, $k = 1, ..6$, and $\beta_l^\pi, l = 1, .., 14$ in (10).

$$\ln\left(\frac{\pi_{ht}}{1 - \pi_{ht}}\right) = v = \alpha_0^\pi + \alpha_1^\pi T_t + \alpha_2^\pi HDI_t + \sum_{k=1}^{6} \gamma_k^\pi D_t^k + \alpha_{4t}^\pi \bar{A}_{ht} + \sum_{l=1}^{14} \beta_l^\pi E_t^l \tag{10}$$

where T is the weather temperature, HDI (Human Discomfort Index) is a combination of hours of sunshine and humidity, D^k are day of the week 0-1 dummies and E^l are holiday 0-1 dummies for Christmas, Easter, etc. Other variables could be used appropriately in a different context.

We compute q_{jht} from sales data as the number of units of product j sold divided by the number of customers who bought product j at store h on day t and and use linear regression to estimate $\alpha_{0j}^Q, \alpha_{1j}^Q, \alpha_{2j}^Q$, and $\beta_{jl}^Q, l = 1, .., 14$ in (11).

$$q_{jht} = \alpha_{0j}^Q + \alpha_{1j}^Q A_{jht} + \alpha_{2j}^Q HDI_t + \sum_{l=1}^{14} \beta_{jl}^Q E_t^l, \text{for all } j \in S \tag{11}$$

In the grocery industry, K_{ht}, the daily number of customers who made transactions in store h on day t is a good proxy for the number of customers who visited the store. We use log-linear regression to estimate $\alpha_{0h}^K, \alpha_{1h}^K, \alpha_{2h}^K, \gamma_k^K, k = 1, .., 6$, and $\beta_{1l}^K, l = 1, .., 14$ in (12).

$$\ln(K_{ht}) = \alpha_{0h}^K + \alpha_{1h}^K T_t + \alpha_{2h}^K HDI_t + \sum_{k=1}^{6} \gamma_k^K D_t^k + \sum_{l=1}^{14} \beta_{1l}^K E_t^l \tag{12}$$

This four stage model of demand estimation has been tested for quality of fit and prediction for multiple stores and subcategories. The average of mean absolute deviation (MAD) across all products, subcategories and stores is 67% in the fit sample and 74% in the test sample. Average bias of our approach is 0% and -9% in fit and test samples, respectively. The current method used at Albert Heijn is estimating $(PQ)_j$ for each SKU directly via logistic regression with similar explanatory variables. The MAD of this method is 72% and 94% and average bias is -43% and -30% in the fit and test samples, respectively.

5.2 Estimation of substitution rates in exogenous demand models

5.2.1 Estimation of stockout-based substitution

Anupindi et al. (1998) estimate the demand for two products and the substitution rates between them using data from vending machines. They assume that consumers arrive according to a Poisson process with rate λ and choose product A (B) as their first choice product with probability p_A (p_B) and substitute according to an asymmetric substitution matrix $\begin{bmatrix} 0 & \alpha_{AB} \\ \alpha_{BA} & 0 \end{bmatrix}$. The demand for product A when B is not available is Poisson with rate $\lambda(p_A + p_B\alpha_{BA})$.

They consider two information scenarios. In the first one, so-called perpetual inventory data, each sales transaction and the exact time that each product runs out of stock (if they do) is observed. In this case, it is not difficult to write down the log-likelihood function and maximize it to obtain the MLE estimates. They show that the timing of the stockouts and the sales volume before and after those times are sufficient statistics. Therefore, it is not necessary to trace each sales transaction. This result of course would not hold if the arrival process were a nonstationary process.

In the second information scenario, so-called periodic review data, the stockout times of the products are not observed, but whether or not they are in-stock at the time of replenishment is known. We encounter an incomplete data problem, and again we can use the EM algorithm briefly discussed in Section 5.1.2 to correct for the missing data (i.e., the stockout times). To be able to generalize the methodology to more than two products, it is necessary to make further assumptions.The authors restrict the substitution behavior to a single-attempt model, i.e., no repeated attempts are allowed and they estimate the parameters for a problem with six products. Their results show that naive demand estimation based on sales data is biased, even for items that rarely stockout. They also find significant differences in the substitution rates of the six brands.

Anupindi et al. (1998) estimate stationary demand rates (i.e., do not consider a choice process) and a substitution matrix. Talluri and van Ryzin (2004) estimate demand rate and the parameters of the MNL choice model (λ, β) but do not consider a substitution matrix. Kök and Fisher (2007) generalize these two approaches and propose a procedure that simultaneously estimates the parameters of the MNL model, on which the consumer's original choice is based, and a general substitution probability matrix.

5.2.2 Estimation of assortment-based substitution

Some retailers do not track inventory data. Some others do, but there is empirical evidence that the inventory data may not be accurate (e.g. DeHoratius and Raman 2004). Hence, sales data may be the only source of information in some cases. Here we review the methodology proposed by K ök and Fisher (2007) to estimate substitution rates using sales data. We assume that substitution structure (i.e., the type of the matrix) is known, and we only need to estimate the substitution rate δ. We demonstrate the method for the proportional substitution matrix, that is assume α_{kj} is given by (2).

The methodology can be explained briefly as follows. Suppose that a store carries assortment $S \subset N$ with 100% service rate (i.e., no stockout-based substitution takes place). We observe D_j for products $j \in S$ from sales data. Notice that at a store that has full assortment (i.e., $S = N$), no substitution takes place, hence $D_j = d_j$ for all j. We can therefore estimate d_j for $j \in N$ from sales data of a similar store that carries a full assortment. We can conclude that the substitution rate is positive for this subcategory if $\Sigma_{j \in S} D_j > \Sigma_{j \in S} d_j$. Let $y(S) = \Sigma_{j \in S} D_j$. Given $\mathbf{d}$, substitution rate δ, and assortment S, we compute what each product in S would have sold at this store using Equation (5), and the total subcategory sales denoted $\hat{y}(S, \delta)$. The error associated with a given δ is the difference between the observed and theoretical subcategory sales at a store (i.e., $y(S) - \hat{y}(S, \delta)$). We find the substitution rate δ that minimizes the total error across all available data from multiple stores and different time periods. The details of the procedure can be found in the paper.

As Campo et al. (2004) point out, there are significant similarities in consumer reactions to a permanent assortment reduction and to stockouts. Therefore, the substitution rate estimated for assortment based substitution can be also used for stockout-based substitution if that cannot be estimated. Another advantage of this methodology is that it enables us to estimate the demand rates of products in a store including those that have never been carried in that particular store.

The next step after the estimation of the substitution rate is the computation of the true demand rates. This involves two tasks. (*i*) deflating the demand rate of the variants already in the assortment S_h, and (*ii*) estimating a positive demand rate for the variants that are not in S_h. Clearly, if $S_h = N$, no computation is necessary. Figure 6-1 presents an example of observed demand rates and the computed true demand rates for a subcategory with ten products.

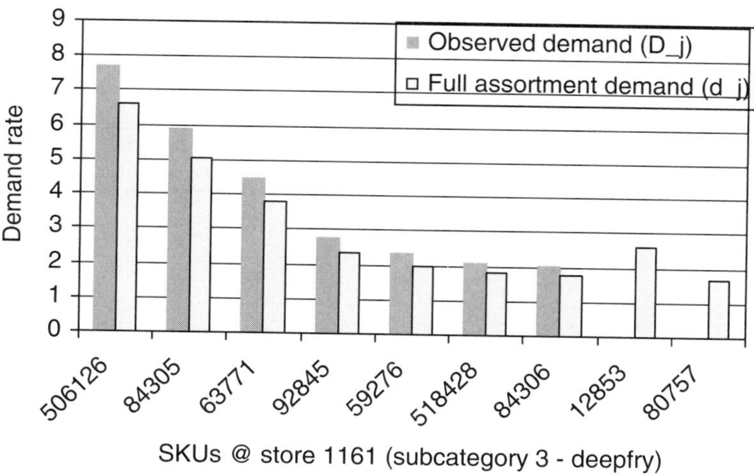

Figure 6-1. Estimates of Observed and Original Demand Rates for a Subcategory

6. ASSORTMENT PLANNING IN PRACTICE

The goal of this section is to describe assortment planning practice as illustrated by the processes used by a few retailers with whom we have interacted: Best Buy, Borders Books, Tanishq and Albert Heijn. Levy and Weitz (2004), Chapter 12, also provides a description of retail assortment planning.

6.1 Best Buy

Most retailers divide their products into various segments, usually called categories and sub categories. The assortment planning process begins by forecasting the sales of each segment for a future planning period ranging from a several month season to a fiscal year. Then scarce store shelf space and inventory purchase dollars are allocated to each segment based in part on the sales projections. Finally, given these resource allocations, the number of SKUs to be carried in each segment is chosen. As such, assortment planning in practice is essentially a strategic planning and capital budgeting process.

Best Buy offers a good example of this process. The following description of the planning process is based on Freeland (2004). In their planning process, conventional still cameras and digital still cameras are two of the product segments. The starting point for a forecast of next year's sales is last year's sales adjusted for trend. Figure 6-2 shows sales of digital and traditional cameras through 2002. A logical forecast for 2003 would be less than 2002 sales for traditional cameras and more than 2002 sales for digital cameras.

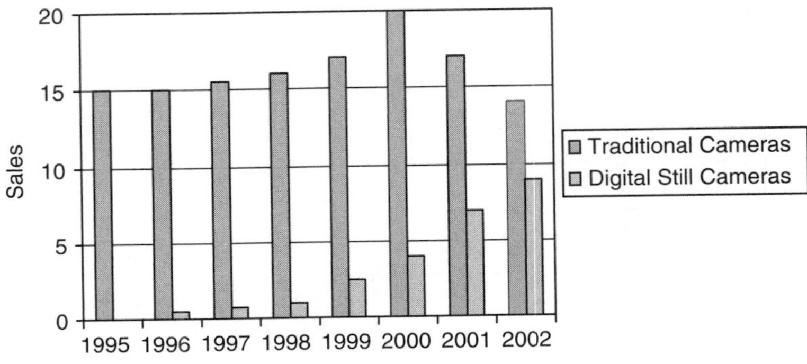

Figure 6-2. Historical Sales of Traditional and Digital Cameras

The forecasts based on sales history are then adjusted based on information from trade shows, vendors, observations of competitor moves and reviews of new technology. The goal of assimilating these inputs is to identify changes in sales for a product category that might not be apparent from a straight forward extrapolation of sales history.

The next step is to set goals for each segment for sales, margin and market share based on the sales forecast, to allocate shelf space and inventory purchase dollars and then to determine how many SKUs to carry in each product segment. A critical input in deciding how many SKUs to carry is the importance to the customer of a broad selection in a particular category. Figure 6-3 was created by Best Buy to show the factors that influence sales and the importance of these factors for different types of products. For example, an accessory item such as a surge protector is often an impulse buy whose sales would be significantly increased by placing it on display near the check out register or in some other high traffic area. However, the customer is not particularly sensitive to price and doesn't require a broad selection. By contrast, placing a refrigerator next to the cash register to drive sales would be silly, because this isn't an impulse purchase for customers. However, they do value a broad selection and low prices. Another way of interpreting the data in this table is that Best Buy believes customers shopping for accessories are very willing to substitute if they don't find exactly what they are looking for, but refrigerator and movie customers are relatively unwilling to substitute.

This matrix is used to guide the number of SKUs to be carried in each product category. Other things being equal, a greater number of SKUs would be carried for those products where selection has a high impact on sales.

Once the number of SKUs to be carried in a product segment has been determined, it is left to the buyer for that segment to determine exactly which SKUs to carry. As an example, in flat panel TV's, Best Buy might carry 82 different SKUs. By contrast, the number of potential SKUs is much larger, comprising of 8 diagonal widths (e.g. 19" , 25" , 32" , 35" . 40" , etc.), 5 screen

Category	Promo	Labor	Impulse	Price	Selection
Computer	High	High	Low	High	Medium
Refrigerator	Medium	High	Low	Medium	High
Accessories	Low	Low	High	Low	Low
Movies	High	Med	High	High	High

Figure 6-3. The Impact of Sales Drivers for Various Types of Products

types (plasma, LCD, projection, etc.), 7 resolutions (analog, 480i, 720p, 1080i, etc.) and 9 major vendors (Sony, Panasonic, Pioneer, etc.) for a total of $8 \times 5 \times 7 \times 9 = 2520$ potential SKUs. It is left to the buyer through a largely manual process to determine which 82 out of these 2520 SKUs will be carried by Best Buy. The buyer incorporates a number of factors into the choice of SKUs. For example, it is highly desirable to carry products from several vendors so that Best Buy can benefit from competition when negotiating with vendors on price.

The Best Buy example suggests that practice and academic research are complementary, in that practice ends with delegating to the buyer the decision of which products to carry from the universe, and this is precisely the problem that has been emphasized in the academic literature.

6.2 Borders

Two interrelated issues in assortment planning are the division of decision rights between corporate and stores and the degree to which the assortment varies by store. Figure 6-4 below depicts alternatives of these two factors.

By far the most common approach is for corporate headquarters to decide on a single common assortment that is carried by all stores of the chain, except that in smaller stores, the breadth of the assortment may be reduced by removing some of the least important SKUs. A relatively small number of retailers (Bed Bath & Beyond would be an example) allow their store managers considerable authority in deciding which SKUs to carry in their stores. Usually, a portion of the assortment is dictated by corporate, and the remainder is chosen by store management from a corporate approved list of options. Obviously a

Figure 6-4. Approaches to Assortment Planning

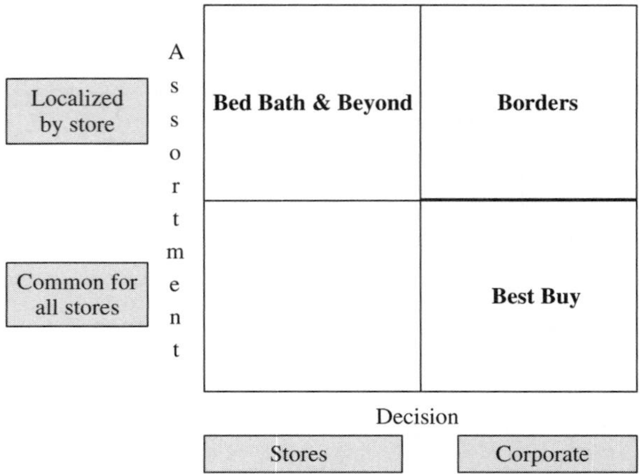

result of this approach is that the assortment is different in all stores, and is hopefully tuned to the tastes of that store's customers.

Borders Books is one of the few retailers that have developed a central approach to creating a unique assortment for each store. They segment their products into about 1000 book categories and define the assortment at a store by the number of titles carried in each category. To choose these parameters they rely on a measure called Relative Sales per Title (RST) that equals the sales in a category over some history period divided by the number of titles carried in the category over the same period. If RST is high for a store-category in a recent period, then they increase the number of titles in that category, and conversely, reduce the titles in low RST categories. For example, a rule could be to divide the 1000 categories in a store into the upper, middle and lower third of RST values and then increase number of titles carried in upper 3rd by Δ and reduce lower 3rd by Δ, where Δ and the frequency of adjusting the assortment are parameters of the process that determine how quickly and aggressively the assortment is adjusted based on history. Their overall process also takes seasonality into account, but that is outside the scope of this survey article.

6.3 **Tanishq**

Tanishq, a division of Titan Industries Ltd. (India's largest watch maker) is India's leading branded jewelry manufacturer and retailer in the country's \$10 billion jewelry market. Tanishq jewelry is sold exclusively through a company controlled retail chain with over 60 boutiques spread over 39 cities. This network of boutiques is supplied and supported by a strong distribution network.

Assortment planning is a key activity at Tanishq involving significant challenges. First, jewelry is a complex product category with a very broad offering to choose from (more than 30,000 active SKUs) making assortment selection non-trivial. Second, given the small to medium size of most of the retail outlets, there were inventory limitations; as a consequence, getting the assortment decision right was critical. Significant differences in customer profile across its 60 boutiques and the frequent introduction of new products added further layers of complexity to the assortment planning process.

Traditionally, each store placed its own order, subject to guidelines on total inventory drawn up by the supply chain team at the corporate headquarters. This was done since the store associates were the ones closest to the customers and hence believed to have the best understanding of their preferences. This was true to a large extent, as the jewelry buying process in the Indian market was highly interactive, with store associates playing a significant role in guiding the customer through the product offerings based on their preferences (e.g. price range, design). Consequently, the store associates had a fairly accurate knowledge of customer choices, their willingness to substitute across product attributes, and reasons that led them to reject certain product variants.

However, there were issues with this model. First, store associates were already burdened with monthly sales targets and hence had little time to do full justice to the ordering process. Second, their knowledge was limited only to product variants that the store had stocked in the past. Hence, they were missing out on potential product opportunities. This necessitated the need to modify the existing assortment planning process and address those shortcomings.

Tanishq accomplished this by moving from a store-centric model to a hybrid model involving both the store associates and a central supply chain team. The supply chain team at the corporate headquarters had the best access to sales and inventory data from all stores. They had detailed information about market trends and were in the best position to analyze historical data to detect selling patterns, and best selling variants at the state, regional, and national levels. This, combined with the local, store specific knowledge of the store associates, resulted in a more refined process for Tanishq.

The first step was the identification of product attributes relevant to the customers' choice process. This was done by the central supply chain team, based on inputs from the store associates. For example, the product category of rings was defined by the following attributes: theme, collection, design, gem type and size.

The next step was the determination of an appropriate assortment strategy for each product category. Again, this was carried out by the central supply chain team. They analyzed historical sales and inventory data in order to understand differences in sales mix across stores by attribute, to identify best sellers, and to develop an understanding of basic selling patterns.

The assortment strategy for each product category was developed based on a simple 2×2 matrix of percent contribution to sales vs. sales velocity (see Figure 6-5). For example, in the case of a product category like Daily Neckwear, which has high percent sales contribution as well as high velocity, the high volume SKUs were put on replenishment, with inventory levels decided based on simple

EOQ models. For the rest of the category, norms were drawn for overall inventory level and product attribute mix at each store (e.g. at Store A, overall inventory of Neckwear should be $2 million and the mix should be: Themes – 50% traditional, 30% contemporary, and 20% fashion; Gem – 40% large, 30% medium, and 30% small).

Based on the assortment strategy, the supply chain team developed a preliminary assortment plan for each store, with suggested products and inventory levels. With a bulk of the products put on SKU level replenishment, the work of store associates has been considerably reduced.

For the products not on SKU-level replenishment, the store associates were at liberty to modify the products selected and order quantities based on their knowledge of localized customer preferences. This was subject to the overarching inventory and product attribute mix guidelines drawn by the central team. This is done through a visual interface that provides the store associates a dynamic picture of how the modified order is stacking up against corporate guidelines.

Through the adoption of a hybrid model, Tanishq was thus able to customize its product offering to suit each store's clientele, while at the same time automating a bulk of the assortment planning process.

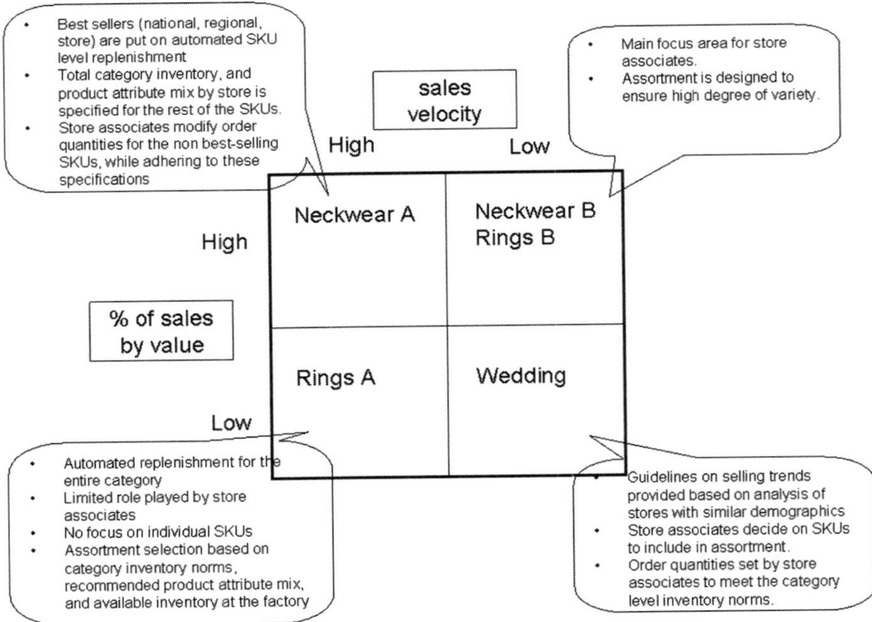

Figure 6-5. Assortment Strategy Based on Percent Sales vs. Sales Velocity Matrix

6.4 Albert Heijn

Albert Heijn, BV is a leading supermarket chain in the Netherlands with 1187 stores and about \$10 billion in sales.[‡] In the grocery indùstry, supermarkets often carry more than 30,000 stock keeping units (SKUs). At the top level of the hierarchy, SKUs are divided into three groups: chilled products, dry goods, and groceries. Each group then is divided into merchandising categories, such as wines, bread spreads, butter & margarine. A subcategory is defined as a group of variants such that the difference between products within a subcategory is minimal, but the difference between subcategories is significant. For example, the subcategories in the butter & margarine category include deep-fry fat, regular butter, healthy butter, and margarines. We assume that substitution takes place within a subcategory but not across subcategories. The assortment planning models reviewed in this chapter focused on the selection and inventory/space allocation within a subcategory given a fixed shelf space and other constraints. Albert Heijn follows a hierarchical approach to assortment planning. First, store space is allocated to categories. Then product selection and facing allocation to products are carried out, subject to the shelf space constraint. In this subsection, we describe the details of this hierarchical approach.

Albert Heijn solves the following optimization problem to allocate shelf space between categories for each store.

$$\max \quad \left\{ \sum_i P_i(x_i) : \sum_i x_i \leq \textit{Store Shelf Space}; x_i \geq 0, \forall i. \right\}$$

$P_i(x_i)$ is the category gross profit when x_i meters of shelf space is allocated to category i. The function P_i is assumed to have a logarithmic form whose parameters are estimated using data from multiple stores $(x_i, P_i(x_i))$. The optimization is done by a Greedy Heuristic – allocating one meter of shelf space at each step to the category with the highest incremental gross profit. Note that this shelf space allocation approach is similar to Corstjens and Doyle (1981), except that cross-space elasticities are not included in the formulation (i.e., category gross profit depends only on the category shelf space).

(Contrast this with the shelf space allocation approach at Borders Bookstores. Borders grouped 300,000 titles into 300 categories and allocated shelf space to categories on the premise that, "Except for best sellers, a customer is interested not in title but category". Category popularity is assessed by computing RST (Relative sales per title = Category sales/Number of titles). Shelf space is periodically reassigned from low RST to high RST. Following the principle of " Survival of the Fittest" , categories "fight" for shelf space. Store managers are allowed to pick titles to be stocked within each category, thereby decentralizing a part of the decision process. Assuming that the number of titles is a proxy for

---

[‡] Albert Heijn, BV is a subsidiary of Ahold Corporation, which owns many supermarket chains around the world with about 8,500 stores and \$50 billion in sales.

category shelf space, RST is equivalent to $P_i(x_i)/x_i$. The Borders approach is similar to that of Albert Heijn except that rather than allocating the last meter of shelf space based on the marginal return, Borders allocates space based on average return from a category.

At Albert Heijn, it is the category manager's responsibility to choose the number of products and their shelf space allocation in each category, given a fixed shelf space. Category managers use several heuristics and their expertise about the category in order to make these decisions. Firstly, Albert Heijn wants to be known as the high variety, high quality supermarket in the Netherlands. One of the guidelines to achieve this strategical mandate is to carry 10% more variety than the nearest competitor. The minimum number of SKUs in a subcategory, the minimum number of facings in a subcategory, the minimum and maximum number of facings for particular SKUs are also specified by category managers. If there is a need to reduce variety in a subcategory, the likely candidate is the subcategory with the highest substitution rate. To introduce new products periodically, m worse products are discarded and m new products are included in the assortment. Given the product selection, facings are allocated to products proportional to their demand rates.

Inventory management operates within the given facing allocations for a selection of products. For non-perishable items, the assigned facings are filled as much as possible at all times, even in the non-peak-load periods. That is achieved by ordering an integral number of case packs such that the inventory position is as close as possible to and less than the maximum inventory level that would fit in the allocated facings. For perishable items that have a shelf life of a few days or less (e.g., produce), the inventory control is done in a more dynamic way. Albert Heijn uses a real-time system that estimates the demand for each product in the assortment based on the sales in the last few hours, and places an order to maximize each product's expected revenues minus cost of disposed inventory.

6.5 Comparison of academic and industry approaches to assortment planning

This section compares and contrasts the approaches taken by academia and industry to assortment planning. Industry has taken a more strategic and holistic approach, while academics use a more operational and detail oriented approach. In some respects these approaches are nicely complementary in that the aspects of assortment planning that have received least attention in practice have received the most attention in academia, and academic research has the potential to fill a void in retail practice.

For most retailers, the process of assortment planning starts at the strategic level. The breadth of product categories carried and the depth of products offered in each of them is a function of the retailer's position in the competitive landscape. For example, a retailer like Best Buy would carry a rarely demanded product such as a 10 mega-pixel camera, just to maintain consumer perception of Best Buy as offering the latest technologies. In other words, the assortment would carry products which are otherwise unprofitable, but are a strategic necessity. While academic research does acknowledge such phenomenon (Cachon et al. 2005), there is little research that focuses on incorporating these strategic considerations while optimizing the assortment.

The other strategic aspect that retailers are concerned with is the role of a product category in their mix. Going back to the Best Buy example, it might be the case that Best Buy offers a very extensive assortment of HDTV's, more than what might be the optimal number when looked upon in isolation, for they are the main traffic drivers for the store. In other words, customers prefer to shop at Best Buy as they see extensive variety on offer in key categories, and as a result end up buying at Best Buy. There is little academic research (except Cachon and Kök 2007) that models this aspect of an assortment. On the other hand, the pricing version of this phenomenon (loss leaders and advertising features to drive traffic into the store and the razor-blade model) is extensively studied in the marketing literature.

One common theme across all the industry examples is that retailers recognize the fact that not all categories should be treated the same. The major drivers of sales in each category are different. While product variety may be the most important factor in a consumers store choice and purchasing decisions for one category, promotions, in-store service experience, and impulse buying (aisle displays) may be more critical for another category. For example, Dhar et al. (2001) find that increasing the breadth and depth of the assortment does not have a positive effect on the performance of high penetration, high frequency categories like coffee and cereals.

Most retailers consider product selection as one among several levers (like promotions, pricing, etc.) that influence sales. Hence, they find it critical to integrate assortment planning decisions with the other influencing parameters. For example, if an apparel retailer is advertising a certain line of clothing heavily, then the variety that needs to be offered is higher than what might have been required without the attention due to advertising. Hence, retailers make assortment decisions in conjunction with other key factors that influence sales.

Retailers are well aware of the dynamic nature of the problem. At many retailers, the initial assortment developed by the buyers is tested across a sample of stores to get an early read, prior to the actual selling season. The test results are used to understand trends on winners and losers and gaps in the portfolio so as to redesign the assortment. As there are several other factors such as promotions, pricing, display, etc. which affect sales on an ongoing basis, the assortment is reviewed from time to time and appropriate changes are made. Academic papers, with the exception of Caro and Gallien (2005), consider static assortments. Even in mature categories, the frequent introduction of new products make it a necessity to revise the assortments. In practice, categories in different stages of their life cycles or categories with seasonal products require different assortment planning approaches. Growth potential is another strategic consideration that influences a retailer's assortment. For example, a dying product category like VCRs might not have the variety that a growing category like DVDs would.

The Tanishq example illustrates how assortment planning and replenishment can be attribute-focused rather than product-focused. For non-best sellers, Tanishq chooses a certain theme and gem size distribution as the defining properties of the target assortment. This approach is sensible, especially for categories in which attributes of the products are critical in driving traffic and influencing consumers' choice behavior. The attribute-focused approach is common in apparel retailing as well. Levy and Weitz (2004) describe the assortment plan for a jeans category where the size distribution, colors and styles are the main attributes that define the assortment. The total inventory budget is then allocated to products given the required distribution of the assortment over these attributes. Academic assortment planning models are mostly product-focused.

Customization of the assortment at the store level has gotten scant attention from retailers and no attention from academics. The Tanishq example illustrates a hybrid approach, where either the assortment or the guidelines for the assortment of the categories are planned at the corporate level, and for some categories store associates tinker with the assortment given the guidelines. Albert Heijn also follows the hybrid approach in that the store assortments are chosen from a chain-wide assortment. Borders Books is the best example we know of a retailer that aggressively customizes assortments at the store level.

Retailers take supply chain considerations into account in assortment planning. For example, Best Buy considers vendor relations, vendor performance and the number of products in other categories from a vendor while developing the assortment plans. However, there is very limited discussion of assortment planning from a supply chain view in the academic literature.

We performed a search on Google for "retail assortment planning" and found more than 700 references. Most of these references are to the product description of software providers and consulting firms, indicating a strong industry interest in the topic. Some academic papers come up in the search as well. One interesting observation that complements the discussion above is that there is a huge disconnect between the two groups: the language or the terminology of each group is substantially different and neither group acknowledges the existence of the other.

7. DIRECTIONS FOR FUTURE RESEARCH

Four research avenues emerge as important future research directions based on our discussion in this chapter. First, more empirical work is needed in understanding the impact of assortment variables on consumers' store choice and purchasing behavior. Second, most of the existing theoretical models have not been implemented as part of industry applications (or their theoretical predictions have not been empirically tested). The field would benefit from such applications and empirical tests, as a validation of the assumptions in the increasingly complicated assortment planning models being formulated in the academic literature.. Third, it seems that there are significant opportunities in generalizing the existing theoretical work to handle more complex problems faced by the retailers. One example would be to allow customization of the assortment by store. Fourth, incorporating the empirical findings on consumer behavior and perception of variety in assortment optimization models seems a worthy area of research. Below we describe some possible research topics from these four avenues in no particular order.

Demand arrival is assumed to be exogenous in most academic models. Understanding the drivers of store traffic through market share or store choice models, and incorporating those in assortment planning is a possible research direction. Lower prices, for example, would increase store traffic, but on the other hand, lower margins would lead to narrower assortments. Retailers recognize these interactions but make these decisions sequentially and in rudimentary ways. The joint pricing and assortment planning problem has not been studied in depth. Aydin and Ryan (2000) study optimal pricing under MNL model but do not consider operational costs. Cachon et al. (2006) are interested in the impact of competitive intensity on the variety level and prices.

Academic models take a static view of the assortment planning problem, whereas in practice, assortment decisions in a category can be made several times throughout the season. The problems that industry faces include not only multi-period problems, but also managing the assortment for multiple generations of products, as in the digital versus traditional camera example. The dynamic assortment problem provides a rich set of research questions. Assortment planning with demand learning through tests in sample stores is another topic worthy of investigation.

Assortment planning models assume that there is a well defined set of candidate products, for which the consumer choice behavior is known perfectly. It may be interesting to take an attribute view of this problem, where consumers are interested in particular attributes rather than products. Mostly, a category is assumed to be composed of homogenous products that are potential substitutes from a consumer's perspective. Assortment planning for vertically differentiated products (i.e., varying quality) or subgroups of products that are more likely than others to be substitutes can be studied to generalize the existing results on properties of optimal assortments. There is a significant body of literature in marketing on consumers' perception of variety as mentioned in Section 2.4. Incorporating some of those concepts in assortment planning may increase the applicability of the theoretical models.

Consumers are usually assumed to be a homogenous group. However, marketing literature places particular emphasis on understanding consumer segments. Estimation papers attempt to identifythe latent consumer segments, and products are carefully positioned to achieve price discrimination between consumer segments in the product line design literature. Similarly in retail assortment planning, the consideration of multiple consumer segments may lead to optimal assortments that are composed of clusters of products that target these different segments.

Consumer purchase decisions across product categories may not always be independent. For example, a consumer's decision to buy a red colored sheet might depend on his being able to find a matching pillow. Explicitly incorporating this basket effect of consumer behavior while optimizing the assortment is an interesting research avenue. Agrawal and Smith (2003) and Cachon and Kök (2007) are first examples of this.

Estimating model parameters such as substitution probabilities, is another area that needs further research. There is an extensive body of literature in marketing (conjoint analysis) and econometrics that deal with parameter estimation for a wide variety of consumer choice models. However, there is little application of these in the assortment planning literature. For academic research to impact the industry, it is critical to invest research time in this area and to come up with innovative techniques to estimate the parameters which form the backbone of the several optimization models.

It is usually assumed that each individual buys a single unit of a single product in a category. This may not be true, even among substitutable products. For example, one shopper may buy multiple units of multiple flavors of yogurt in the same purchase occasion. This behavior violates the assumptions of standard choice models like the MNL, and it might be interesting to develop alternate models and study the properties of the resulting assortment. It would also be worthwhile to study the structure of the optimal assortment for product categories in situations when consumers are variety-seeking, causing the inventory-variety trade-off to take a different form.

Clearly, it is necessary to develop methods to understand the role of categories and to measure the intangible factors (such as the strategic importance of a category, the impact of assortment breadth or inventory levels on attractiveness of a store). The relation of assortment and inventory decisions with other levers such as pricing, promotions, and advertising has not been studied empirically. Joint optimization of some of these variables may lead to interesting results. It may be possible to draw from the literature on economics of product differentiation and the marketing/operations literature on product line design, both of which have extensively studied these variables and their impact on industry structures or product variety.

Assortment planning in multi-store, multi-tier supply chains is a completely open research area. Singh et al. (2005) and Aydin and Hausman (2003) are the only cases in the literature that incorporate supply chain considerations into assortment planning. The pros and cons of the hierarchical approach, the benefits of localization, and the execution problems associated with them have not been studied empirically or analytically. Balancing the benefits of customizing assortments by store with the increased cost of complexity is increasingly seen by retailers as a significant source of competitive advantage.

An extremely interesting research question here is how to strike the balance, find the sweet spot between a "one size fits all" and "each store is its own" philosophies.

Incentive conflicts between the levels of the hierarchy may be a hurdle in deployment of the corporate assortment plans to the store level. Corporate level plans that are built based on strategic considerations may be imperfectly executed because the store managers' incentives are based on more short term objectives. The conflict of incentives between store managers, buyers, and vendors in a decentralized supply chain is yet another potential research area. For example, it is not clear how a category level assortment plan and the vendor-managed inventory agreements should be reconciled.

In conclusion, it seems to us that academics could make a tremendous contribution to retailing in the area of assortment planning. Retailers have developed practices that enable them to incorporate the complexities of the world in which they live, but they realize their approaches are too much based on art and judgment and that they could benefit from more rigorous use of the huge quantities of data available to them. If academics would be willing to work with individual retailers to understand their true complexity, they could make an enormous contribution in adding rigor and science to the retailer's planning process, much as academics have done in other areas like finance, marketing and strategy.

REFERENCES

ACNielsen. 1998. *Eighth Annual Survey of Trade Promotion Practices*. Chicago, IL. ACNielsen.

Agrawal N., S.A. Smith. 1996. Estimating negative binomial demand for retail inventory management with unobservable lost sales. *Naval Research Logistics*. **43** 839–861.

Agrawal N., S.A. Smith. 2003. Optimal retail assortments for substitutable items purchased in sets. *Naval Research Logistics*. **50**(7) 793–822.

Alptekinoglu, A. 2004. Mass customization vs. mass production: variety and price competition. *Manufacturing & Service Operations Mgmt*. **6**(1) 98–103.

Anderson, S.P., A. de Palma., J.F. Thisse. 1992. *Discrete Choice Theory of Product Differentiation*. The MIT Press, Cambridge, MA.

Anupindi R., M. Dada, S. Gupta. 1998. Estimation of consumer demand with stockout based substitution: An application to vending machine products. *Marketing Science*. **17** 406–423.

Avsar, Z.M., M. Baykal-Gursoy. 2002. Inventory control under substitutable demand: A stochastic game application. *Naval Research Logistics*. **49** 359–375.

Aydin, G., W.H. Hausman. 2003. The role of slotting fees in the coordination and assortment decisions. Working paper. University of Michigan.

Aydin, G., J. K. Ryan. 2000. Product line selection and pricing under the multinomial logit choice model. Working paper. Stanford University.

Bassok, Y., R. Anupindi, R. Akella. 1999. Single-period multiproduct inventory models with substitution. *Operations Research*. **47** 632–642.

Basuroy, S., D. Nguyen. 1998. Multinomial logit market share models: Equilibrium characteristics and strategic implications. *Management Science*. **44**(10) 1396–1408.

Baumol, W.J., E.A. Ide. 1956. Variety in retailing. *Management Science*. **3** 93–101.

Bell, D.R., T-H Ho, C.S. Tang 1998. Determining where to shop: Fixed and variable costs of shopping. *Journal of Marketing Research*. **35** 352–369.

Bell, D.R., J.M. Lattin. 1998. Shopping behavior and consumer preference for store price format: Why large basket shoppers prefer EDLP. *Marketing Science*. **17** 66–88.

Ben-Akiva, M., S.R. Lerman. 1985. *Discrete choice analysis: Theory and application to travel demand*. Cambridge, MA. The MIT Press.

Boatwright, P., J.C. Nunes. 2001. Reducing assortment: An attribute-based approach. *Journal of Marketing*. **65**(3) 50–63.

Borin, N.,P. Farris. 1995. A sensitivity analysis of retailer shelf management models. *J Retailing*. **71** 153–171.

Broniarczyk, S.M., W.D. Hoyer, L. McAlister. 1998. Consumers' perception of the assortment offered in a grocery category: The impact of item reduction. *J. of Mrkt Res*. **35** 166–76.

Bucklin, R.E., S. Gupta. 1992. Brand choice, purchase incidence, and segmentation: An integrated modeling approach. *Journal of Marketing Research*. **29** 201–215.

Bultez, A., P. Naert. 1988. SHARP: Shelf allocation for retailers profit. *Marketing Science*. **7** 211–231.

Cachon, G.P., A.G. Kök. 2007. Category management and coordination of categories in retail assortment planning in the presence of basket shoppers. *Management Science*. **53** (6) 934–951.

Cachon, G.P., C. Terwiesch, Y. Xu. 2005. Retail assortment planning in the presence of consumer search. *Manufacturing & Service Operations Mgmt*. **7** (4) 330–346.

Cachon, G.P., C. Terwiesch, Y. Xu. 2006. On the Effects of Consumer Search and Firm Entry in a Multiproduct Competitive Market. *Marketing Science*. Forthcoming.

Campo, K., E. Gijsbrechts, P. Nisol. 2004. Dynamics in consumer response to product unavailability: do stock-out reactions signal response to permanent assortment reductions? *Journal of Business Research*. **57** 834–843.

Caro, F., J. Gallien. 2005. Dynamic assortment with demand learning for seasonal consumer goods. Working paper. Sloan School of Management.

Chen, F., J. Eliashberg, P. Zipkin. 1998. Customer preferences, supply-chain costs, and product line design. In *Product Variety Management: Research Advances*, T.-H.Ho and C.S.Tang (Eds.), Kluwer Academic Publishers.

Chen, Y., J.D. Hess, R.T. Wilcox, Z.J. Zhang. 1999. Accounting profits versus marketing profits: A relevant metric for category management. *Marketing Science*. **18**(3) 208–229.

Chiang, J. 1991. A simultaneous approach to the whether, what and how much to buy questions. *Marketing Science*. **10** 297–315.

Chintagunta, P.K. 1993. Investigating purchase incidence, brand choice and purchase quantity decisions of households. *Marketing Science*. **12** 184–208.

Chong, J-K., T-H. Ho, C.S. Tang. 2001. A modeling framework for category assortment planning. *Manufacturing & Service Operations Mgmt*. **3**(3) 191–210.

Cooper, L.G., M. Nakanishi. 1988. *Market-Share Analysis: Evaluating Competitive Marketing Effectiveness*. Kluwer Academic Publishers.

Corstjens, M., P. Doyle. 1981. A model for optimizing retail space allocations. *Management Science*. **27** 822–833.

de Groote, X. 1994. Flexibility and Marketing/Manufacturing Coordination. *International Journal of Production Economics*. **36** 153–167.

Dempster, A.P., N.M. Laird, D.B. Rubin. 1977. Maximum likelihood from incomplete data via the EM algorithm. *J. Roy. Statist. Soc. B*. **39** 1–38.

DeHoratius, N., A. Raman. 2004. Inventory record inaccuracy: an empirical analysis. Working paper. Graduate School of Business. University of Chicago.

Desai, P., S. Radhakrishnan., K. Srinivasan. 2001. Product differentiation and commonality in design: balancing revenue and cost drivers. *Management Science*. **47** 37–51.

Dhar, S.K., S.J. Hoch, N. Kumar. 2001. Effective category management depends on the role of the category. *Journal of Retailing*. **77**(2) 165–184.

Dobson, G., S. Kalish. 1993. Heuristics for pricing and positioning a product line. *Management Science*. **39** 160–175.

Downs, B., R. Metters, J. Semple. 2002. Managing inventory with multiple products, lags in delivery, resource constraints, and lost sales: A mathematical programming approach. *Management Science*. **47** 464–479.

Dreze, X., S.J. Hoch, M.E. Purk. 1994. Shelf management and space elasticity. *Journal of Retailing*. **70** 301–26.

Eliashberg, J., R. Steinberg. 1993. Marketing-production joint decision-making. In J. Eliashberg and G.L. Lilien, Eds. *Handbooks in OR & MS*. **5**.

Emmelhainz, L., M. Emmelhainz, and J. Stock. 1991. Logistics Implications of Retail Stock-outs. *Journal of Business Logistics*. **12**(2) 129–141.

Fader, P.S., B.G.S. Hardie. 1996. Modeling consumer choice among SKUs. *Journal of Marketing Research*. **33** 442–452.

Fisher, M. L., A. Raman. 1996. Reducing the cost of demand uncertainty through accurate response to early sales. *Operations Research*. **44** 87–99.

Freeland, K., 2004. Assortment Planning. Presentation in COER Conference. Consortium for Operational Excellence in Retailing, The Wharton School, University of Pennsylvania.

Gaur, V., D. Honhon. 2006. Assortment planning and inventory decisions under a locational choice model. *Management Science*. **52**(10) 1528–1543.

Greene, W.H. 1997. *Econometric Analysis*. Prentice Hall, Englewood Cliffs, NJ.

Gruca, T.S., D. Sudharshan. 1991. Equilibrium characteristics of multinomial logit market share models. *Journal of Marketing Research*. **28** (11) 480–482.

Gruen, T.W., Corsten, D.S., Bharadwaj, S. 2002. *Retail out-of-stocks: A worldwide examination of extent, causes and consumer responses*. Grocery Manufacturers of America.

Guadagni, P.M., J.D.C. Little. 1983. A logit model of brand choice calibrated on scanner data. *Marketing Science*. **2** 203–238.

Hadley, G; T.M. Whitin. 1963. *Analysis of Inventory Systems*. Prentice Hall.

Hoch, S.J., E.T. Bradlow, B. Wansink. 1999. The variety of an assortment. *Mrkt Sci*. **25** 342–55.

Hotelling, H. 1929. Stability in competition. *Economic Journal*. **39** 41–57

Huffman, C., B.E. Kahn. 1998. Variety for sale: Mass customization or mass confusion? *Journal of Retailing* **74** 491–513.

Irion, J., F. Al-Khayyal. J. Lu. 2004. A Piecewise Linearization Framework for Retail Shelf Space Management Models. Working paper. Georgia Institute of Technology.

Kahn, B.E. 1995. Consumer variety-seeking in goods and services: an integrative review. *Journal of Retailing and Consumer Services*. **2** 139–48.

Kohli, R., R. Sukumar. 1990. Heuristics for product line design. *Management Science*. **36**(3) 1464 –1478.

McFadden, D., 1974. Conditional logit analysis of qualitative choice behavior. in P. Zarembka (ed.) *Frontiers in Econometrics*. Academic Press, NY.

Kök, A.G. 2003. Management of product variety in retail operations. Ph.D. Dissertation. The Wharton School, University of Pennsylvania.

Kök, A.G., M.L. Fisher. 2007. Demand estimation and assortment optimization under substitution: methodology and application. *Operations Research*. **55** (6) 1001–1021.

Kurt Salmon Associates. 1993. Efficient consumer response: Enhancing consumer value in the grocery industry. *Food Marketing Institute Report # 9–526*. Food Marketing Institute.

Kurtulus, M. 2005. Supply chain collaboration practices in consumer goods industry. Ph.D. Dissertation. INSEAD.

Kurtulus, M., B.Toktay. 2005. Category captainship: Outsourcing retail category management. Working paper. INSEAD.

Lancaster, K. 1966. A new approach to consumer theory. *Journal of Political Economy*. **74** 132–57.

Lancaster, K. 1975. Socially optimal product differentiation. *American Economic Review*. **65** 567–585.

Lancaster, K. 1990. The economics of product variety: A survey. *Marketing Science*. **9** 189–210.

Levy, M., B.A. Weitz. 2004. *Retailing Management*. McGraw-Hill/Irwin, NY. 398–400.

Lippman S.A., K.F. McCardle. 1997. The competitive newsboy. *Operations Research*. **45** 54–65.

Maddah, B., E. K. Bish. 2004. Joint pricing, assortment, and inventory decisions for a retailer's product line. Working Paper, Virginia Polytechnic Institute and State University.

Mahajan, S. and van Ryzin, G.J. 1999. Retail Inventories and Consumer Choice. Chapter 17 in Tayur, S. et al. (eds.). *Quantitative Methods in Supply Chain Management*. Kluwer, Amsterdam.

Mahajan, S., G. van Ryzin. 2001a. Stocking retail assortments under dynamic consumer substitution. *Operations Research*. (**49**)3 334–351.

Mahajan, S., G. van Ryzin. 2001b. Inventory competition under dynamic consumer choice. *Operations Research*. (**49**)5 646–657.

Manchanda, P., A. Ansari, S. Gupta. 1999. The "shopping basket": A model for multi-category purchase incidence decisions. *Marketing Science*. **18**(2) 95–114.

McBride, R.D. and F.S. Zufryden. 1988. An integer programming approach to the optimal product line selection problem. *Marketing Science*. **7**(2) 126–140.

McGillivray, A.R., E.A. Silver. 1978. Some concepts for inventory control under substitutable demand. *INFOR*. **16** 47–63.

Miller, C.M., S.A. Smith, S.H. McIntyre, D.D. Achabal. 2006. Optimizing retail assortments for infrequently purchased products," working paper, Santa Clara University Retail Workbench.

Moorthy, S. 1984. Market segmentation, self-selection, and product line design. *Marketing Science*. **3** 288–307.

Mussa, M., S. Rosen. 1978. Monopoly and product quality. *Journal of Economic Theory*. **18** 301–317.

Netessine, S., N. Rudi. 2003. Centralized and competitive inventory models with demand substitution. *Operations Research*. **51** 329–335.

Netessine, S., T.A. Taylor. 2005. Product line design and production technology. Working paper, Wharton School.

Noonan, P.S. 1995. When consumers choose: a multi-product, multi-location newsboy model with substitution. Working paper. Emory University.

Parlar, M. 1985. Optimal ordering policies for a perishable and substitutable product: A Markov decision model. *Infor*. **23** 182–195.

Parlar, M., S.K. Goyal. 1984. Optimal ordering policies for two substitutable products with stochastic demand. *Opsearch*. **21** (1) 1–15.

Progressive Grocer. 1968a. The Out of Stock Study: Part I. October. S1–S16.

Progressive Grocer. 1968b. The Out of Stock Study: Part II. November. S17–S32.

Quelch, J.A., D. Kenny. 1994. Extend profits, not product lines. *Harvard Bus Rev*. **72** 153–160.

Raman, A., A. McClelland, M.L. Fisher. 2001. Supply Chain Management at World Co. Ltd. Harvard Business School Case # 601072.

Rajaram, K. 2001. Assortment planning in fashion retailing: methodology, application and analysis. *European Journal of Operational Research* **129** 186–208.

Rajaram, K., C.S. Tang. 2001. The impact of product substitution on retail merchandising. *European Journal of Operational Research* **135** 582–601.

Russell, G.J., D.R. Bell, et al. 1997. Perspectives on multiple category choice. *Marketing Letters*. **8**(3) 297–305.

Schary, P., M. Christopher. 1979. The Anatomy of a Stockout. *Journal of Retailing*. **55**(2) 59–70.

Simonson, I. 1999. The effect of product assortment on buyer preferences. *J of Ret*. **75** 347–70.

Singh, P., H. Groenevelt, N. Rudi. 2005. Product variety and supply chain structures. Working paper. University of Rochester.

Smith, S.A., N. Agrawal. 2000. Management of multi-item retail inventory systems with demand substitution. *Operations Research*. **48** 50–64.

Song, J.-S. 1998. On the order fill rate in multi-item, base-stock systems. *Operations Research*. **46**. 831–845.

Song, J.-S., P. Zipkin. 2003. Supply chain operations: assemble-to-order systems. *Handbooks in Operations Research and Management Science*, Vol. XXX: *Supply Chain Management*, S. Graves and T. De Kok, eds, North-Holland.

Talluri, K., G. van Ryzin. 2004. Revenue management under a general discrete choice model of consumer behavior. *Management Science*. **50** 15–33.

Urban, T.L. 1998. An inventory-theoretic approach to product assortment and shelf space allocation. *Journal of Retailing*. **74** 15–35.

van Herpen, E., R. Pieters. 2002. The variety of an assortment: An extension to the attribute-based approach. *Marketing Science*. **21**(3) 331–341.

van Ryzin, G., S. Mahajan. 1999. On the relationship between inventory costs and variety benefits in retail assortments. *Management Science*. **45** 1496–1509.

Walter, C., J. Grabner. 1975. Stockout models: Empirical tests in a retail situation. *Journal of Marketing*. **39**(July) 56–68.

Wu, C.F.J. 1983. On the convergence properties of the EM algorithm. *Ann. Statist*. **11** 95–103.

Zinn, W., P. Liu. 2001. Consumer response to retail stockouts. *Journal of Business Logistics*. **22**(1) 49–71.

Chapter 7
MANAGING VARIETY ON THE RETAIL SHELF: USING HOUSEHOLD SCANNER PANEL DATA TO RATIONALIZE ASSORTMENTS

Ravi Anupindi[1], Sachin Gupta[2], and M.A. Venkataramanan[3]
[1]*Associate Professor of Operations Management, Department of Operations and Management Sciences, Stephen M. Ross School of Business, University of Michigan, Ann Arbor, MI, USA*
[2]*Henrietta Johnson Louis Professor of Management and Professor of Marketing, Johnson Graduate School of Management, Cornell University, Ithaca, NY, USA*
[3]*Lawrence D. Glaubinger Professor of Business, Kelley School of Business, Indiana University, Bloomington, IN, USA*

Abstract We propose a model for the rationalization of retail assortment and stocking decisions for retail category management. We assume that consumers are heterogeneous in their intrinsic preferences for items and are willing to substitute less preferred items to a limited extent if their preferred items are not available. We propose that the appropriate objective function for a far-sighted retailer includes not only short-term profits but also a penalty for disutility incurred by consumers who do not find their preferred items in the available assortment. The retailer problem is formulated as a constrained integer programming problem. We demonstrate an empirical application of our proposed model using household scanner panel data for eight items in the canned tuna category. Our results indicate that the inclusion of the penalty for disutility in the retailer's objective function is informative in terms of choosing an assortment to carry. We find that customer disutility can be significantly reduced at the cost of a small reduction in short term profits. We also find that the optimal assortment behaves non-monotonically as the weight on customer disutility in the retailer's objective function is increased.

Key Words Retail Assortment · Substitution · Heterogeneity · Scanner Data · Integer Programming · Optimization

N. Agrawal, S.A. Smith (eds.), *Retail Supply Chain Management*,
DOI: 10.1007/978-0-387-78902-6_7,

1. INTRODUCTION

Two fundamental retailer decisions are which items to stock in a category (the assortment decision) and how much to stock of each item (the inventory decision). While these decisions have always been key to retailer profitability, they have received renewed attention because of industry initiatives labeled Efficient Consumer Response (ECR). Category Management, a component of ECR, emphasizes the need to recognize the inter-relatedness (e.g., substitutability) of items within a category when making decisions. Thus, categories need to be managed as strategic business units, with an emphasis on total category performance. Point-of-sale information can potentially play a critical role in providing insights into consumer behavior to help develop sound category strategies.

Retailers recognize that wider assortments help their business by catering to the needs of multiple consumer segments (Coughlan, Anderson, Stern & El-Ansary, 2006), as well as by offering variety to variety-seeking consumers. However, there are limits to the value of variety. Adding items with small differences offers little in the way of "real" variety to the consumer (Boatwright & Nunes, 2001), yet adds to costs of operations such as administrative costs and cost of warehouse space. The sharp growth of warehouse clubs and deep discount drug stores in recent years is attributed, in part, to their cost advantages arising from their limited variety offering. The resultant loss of market share has re-focussed attention of supermarkets on the need to manage variety. It is believed that there is substantial potential for lowering supermarket operating costs without hurting business by making store assortments more efficient; see, for example, a report by the Food Marketing Institute (1993).

Managing retail space entails solving two types of problems. The first is allocating space to categories, called the *inter-category* space allocation problem. The second is allocating space to items within a category or the *intra-category* space allocation problem. This second problem is often referred to as the assortment problem. Ideally, assortment decisions need to incorporate a variety of factors. On the demand side, one needs to consider the (heterogenous) customer purchase behavior including substitution patterns when their preferred items are not available (either temporarily due to stock-out or permanently due to limited assortment), the stochastic nature of demand arising due to the uncertainty inherent in consumer choice, the effect of product display on sales, etc. On the supply side, retailers face a finite shelf-space constraint for a category and incur fixed costs to include items in the assortment. Further, since limited assortments may have longer term consequences on profitability, a retailer needs to balance current profits with implications of the assortment on future profits. Finally, such a model for decision making should be driven by actual data and the solution strategy should be scalable to address the large problem sizes that any realistic assortment decision would entail.

In this chapter, we outline a modeling framework that incorporates some of the above features to assist the retailer in determining the optimal subset of items to carry in a category, from the set currently carried, and the quantity to stock of each item. We propose the use of household purchase data collected via scanners to estimate intrinsic preferences of consumers and to infer their substitution patterns. Such information is key to ensuring that the assortment carried caters to heterogeneous consumers' tastes, while avoiding unnecessary and expensive duplication.

Previous research on the retailer's assortment problem has typically not modeled consumer substitution behavior explicitly. Empirical evidence from several studies suggests that in packaged goods markets, consumers are often willing to substitute a less preferred item for their (non-available) preferred item. A Food Marketing Institute survey reports that only 12–18% of shoppers said they would not buy an item on a shopping trip if their favorite brand-size was not available; the rest indicated they would be willing to buy another size of the same brand, or switch brands. A number of other studies (Emmelhainz, Stock & Emmelhainz, 1991; Carpenter & Lehmann, 1985; Urban, Johnson & Hauser, 1984; Gruen, Cortsen & Bharadwaj, 2002) support a similar conclusion. A 1993 study by Willard Bishop Consulting Ltd. and Information Resources, Inc. found that when duplicative items were removed, 80% of consumers saw no difference (Business Week (1996)). Other evidence suggests that consumers make about two-thirds of their purchase decisions about grocery and health-and-beauty products while they are in the store (Nielsen Marketing Research (1992)). Thus, it is important to take account of substitution behavior of consumers when rationalizing assortments.

It is likely that consumers who do not find their preferred item in the store assortment are not fully satisfied, whether or not they buy another item in the category. The decision to rationalize assortments needs to take account of the potential adverse impact on customer retention. Traditional formulations of the assortment problem typically assume that the retailer is a myopic profit maximizer. Such formulations disregard the longer-term adverse impact on profits of not satisfying consumers' demand for their preferred items. In our proposed formulation, the retailer's objective function is a weighted sum of profits and a penalty for disutility caused to consumers who do not find their preferred items in the assortment. The rationale for including a penalty is that dissatisfied customers may take their future business elsewhere, thereby hurting longer term profits, even if they purchase less preferred items in the current period. Our proposed model can be used by a retailer to balance short term profits and customer disutility when choosing assortments.

Another contrast of our proposed approach with previous research lies in our accommodation of differences in item preferences between consumers. Most previous work assumes an aggregate demand model. Aggregate demand specifications do not allow us to distinguish between the extent of disutility or dissatisfaction caused by not stocking a particular item to, for example, more versus less loyal groups of consumers. Clearly this distinction is relevant for a retailer who cares about retaining customers in the longer run. The existence of consumer heterogeneity has been established by a number of previous empirical studies. Our proposed model allows for completely idiosyncratic patterns of substitution, as well as disutility due to non-stocking, between consumers.

To demonstrate an empirical application of the proposed model, we estimate consumer preferences for eight items in the canned tuna category using household scanner panel data, a commonly available source of market research information. A hierarchical Bayesian approach is used to estimate an interval scaled measure of each household's utility for the eight items, and the household's price and promotion sensitivity. The retailer's decision problem is then solved as an integer programming problem. Although the problem is large in terms of the number of decision variables and constraints, we show that it can be solved efficiently. Our solution reveals that a significant reduction in customer disutility can be accomplished at the cost of a small reduction in the current period profits.

Our model should be considered as an illustrative first step. While we have captured the richness of customer heterogeneity, substitution behavior, and the current vs. future profit tradeoff, we also have made simplifying assumptions on other aspects of this complex problem. In section 6 we outline several ways to enhance our proposed model to incorporate these remaining aspects, which we hope will inform further research in this important field.

The rest of the chapter is organized as follows. In Section 2 we review related research. We discuss the consumer model in Section 3. In Section 4 we develop an optimization framework for the assortment decision, discuss special cases and some properties of the model. In Section 5, we demonstrate an empirical application of our proposed model using household panel data. We conclude in Section 6 with a brief summary and a discussion of extensions and further research.

2. LITERATURE REVIEW

Two broad streams of literature are relevant to this study – one in marketing, the other in operations management. Early research in marketing deals with issues of retail shelf space allocation and is empirical in nature. Corstjens and Doyle (1981) proposed a model to optimize space allocation across categories, given an overall store space constraint. Direct and cross space elasticities were measured via a multiplicative sales response model using cross-sectional data. Their model does not explicitly include the assortment decision, although allocation of zero space to an item may be interpreted as exclusion of the item. However, as pointed out by Borin et al. (1994), the multiplicative sales response model predicts zero sales for a given category if the space of any of the store's other categories is set to zero. Bultez and Naert (1988) and Bultez et al. (1989) model the intra-category space allocation problem. Space elasticities are measured experimentally with item sales as the criterion variable. However, the assortment decision is not explicitly modeled. Borin et al. (1994) incorporate both the space allocation and assortment decisions in a retailer model. However, this study does not empirically estimate the demand model. Instead, parameter values are assumed. More recently, Van Dijk et al. (2004) use observed variation in shelf-space allocation across stores to infer shelf-space elasticities.

The focus of these studies is on allocation of a scarce resource – space – given that different items show varying responsiveness to space. Thus, emphasis is placed on methods and data for measurement of space elasticities (own and cross) and on algorithms to solve the retailer profit maximization problem efficiently. By contrast, our focus is on estimating consumers' brand preferences to infer their willingness to substitute, thereby determining the optimal assortment of items to stock. In the present study we do not tackle issues of responsiveness of demand to space allocations, but leave that for future research. The primary emphasis in our work is motivated by the empirical observation that in most consumer packaged goods categories, consumers can often be (imperfectly) satisfied by one of several items. This characteristic of consumer behavior is used in determining optimal assortments.

Recent empirical findings in the marketing literature provide strong support for the idea that assortment reductions may be profitable for retailers. Broniarczyk et al. (1998) conduct controlled lab experiments as well as field experiments in which assortments were reduced in five categories in convenience stores. They measure consumer perceptions of variety, which are shown to mediate store choice. A key finding is that elimination of low-selling items had little or no impact on shoppers' perceptions of variety, as long as favorite items were available and category shelf space was held constant.

Boatwright and Nunes (2001) analyze data from a natural experiment conducted by an online grocer, in which 94% of the categories experienced dramatic reductions in the number of SKUs offered. Sales increased an average of 11% across the 42 categories examined.[1] An important finding especially relevant to our work is that customers who lose their favorite item when the assortment is reduced are significantly less likely to purchase in the category on a future purchase occasion.

Borle et al. (2005) use household panel data of the same online grocer that Boatwright and Nunes study to analyze the effects of assortment reductions in several categories on overall store sales. They find that although the effect is positive in several categories, overall store sales are reduced due to decreases in the number of store visits and the size of the shopping basket. To our knowledge, this is the first study that demonstrates that customer retention,

---

[1] Part of the increase is attributed to enhanced utility due to reduced clutter in the category. Our model does not allow for such an effect.

i.e., customers' repeat store visit behavior, is adversely affected by reductions in category assortments.

Sloot et al. (2006) distinguish between short and long term sales effects of a 25% item reduction in the assortment in one category. They find that while short-term category sales suffer a sharp reduction, long-term category sales display only a weak negative effect.

The findings of both Broniarczyk et al. (1998) and Boatwright and Nunes (2001) highlight that the impact of assortment reductions is heterogeneous across consumers, depending on the extent of loyalty exhibited towards the lost item. Borle et al. (2005) show conclusively that assortment reductions may reduce a shopper's probability of returning to this store on the next shopping visit. Although our data do not permit us to directly model the effects of assortment availability on consumers' store choice decisions, in our assortment optimization model we formalize the idea by including in the retailer's objective function the disutility incurred by consumers as a result of not finding their preferred items in the available assortment. This disutility is idiosyncratic to each consumer, and serves as a proxy for the reduced profits resulting from the lower probability of consumers choosing this retailer in future.

In the operations literature, work on assortment problems was motivated by the textile industry where decisions regarding which sizes (e.g., in-seam lengths for slacks) to carry had to be made. Pentico (1974) considers the single dimension assortment problem with probabilistic demands, with assumptions about substitution behavior of consumers. Pentico (1988) extends the earlier work to two-dimensional assortment problems with deterministic demands. Other related work deals with determining optimal stock levels for multiple items given stochastic demands and a pattern of substitution based on non-availability; see, for example, Bassok, Anupindi and Akella (1997) and the references therein. In this work, however, substitution is determined by the supplier firm and not by the buyer or consumer.

van Ryzin and Mahajan (1999) study a stochastic single period assortment planning problem under a Multinomial Logit (MNL) Choice model. A consumer's choice depends on the variants that the store carries and they assume that consumers do not substitute in the event of a stock-out. Using a newsvendor framework with identical exogenous retail prices across all variants, they show that the optimal assortment always consists of a certain number of the most "popular" products. They also illustrate that retail prices and profits increase when consumer preferences are more "fashion" oriented. In a follow-up paper, Mahajan and van Ryzin (2001) incorporate both assortment-based as well as stock-out based substitution behavior and present a stochastic sample path optimization method to solve for the optimal assortment. In contrast to these papers that assume a MNL model of choice, Gaur and Honhon (2006) use a locational choice model to study the assortment problem.

Smith and Agrawal (2000) study the assortment planning problem using a general probabilistic model of demand allowing for substitution behavior.

Using a substitution matrix, they estimate the derived demand for a given assortment. They then present a methodology to determine the assortment and stocking levels jointly when retailers incur a fixed cost for carrying an item in stock as well as the classical inventory and shortage costs for excess inventory and shortage at the end of the period.

Some recent papers have focused on jointly addressing demand estimation as well as assortment planning. Chong et al. (2001) present a category assortment planning problem. Consumer choice is represented as a combination of a category-purchase-incidence model and a brand-share model. While the former predicts the probability of an individual consumer's purchase in a category on a given shopping trip, the latter predicts which brand will be purchased. The optimization problem then determines the optimal number of facings for the various products to maximize profits, subject to a shelf space constraint. The authors illustrate their methodology using data from five stores in eight food categories.

Kok and Fisher (2004) present a demand estimation as well as an assortment optimization model. Using cross-sectional data across stores that carry different assortments, they estimate the substitution behavior of a homogeneous set of customers. Using a probabilistic model of choice, they posit an assortment optimization model and develop heuristics to determine the number of facings of a particular product that a retailer should carry. They apply their method to a supermarket chain in the Netherlands and demonstrate that their methodology for assortment planning potentially leads to a 50% increase in profits.

Miller et al. (2006) propose an approach to optimize retail assortments with demand specified as a multinomial logit model. Consumers' utilities for products are estimated via a conjoint approach wherein consumer heterogeneity is allowed. In an empirical application they find that there is a significant negative impact on profits when heterogeneous consumers are assumed to be homogeneous.

Like the papers just discussed, our chapter focuses on a joint demand estimation and assortment planning problem. Demand is modeled at the household level using a discrete choice framework, specifically a probit model. Households are modeled as heterogeneous in unobserved utility function parameters, and the heterogeneity distribution is estimated using household scanner panel data. Thereby, posterior estimates of households' preferences are derived.

The formulation of our optimization model is similar to the one studied by Dobson and Kalish (1988; 1993) in the context of positioning and pricing a product line. They present welfare and profit maximization formulations for positioning and pricing respectively. Our formulation is also similar to McBride and Zufryden (1988) who apply integer programming techniques to the optimal product line selection problem. Their model formulation recognizes heterogeneity in consumer preferences. Our approach of incorporating consumer disutility into the retailer's objective function is, however, more general than that of Dobson and Kalish (1993) or McBride and Zufryden (1988). The idea of penalizing the objective function for lost goodwill due to non-availability of stock is not new. In stochastic inventory theory (Arrow, Karlin & Scarf, 1958; Lee & Nahmias, 1994) a penalty cost for shortages is routinely included in the objective function. However, to our knowledge, this chapter is the first to operationalize the penalty based on disutilities estimated from market-place data.

A key point of distinction between our paper and most of the literature discussed previously is with respect to the model of consumer heterogeneity. The classical multinomial logit (MNL) model as used in van Ryzin and Mahajan (1999) and Mahajan and van Ryzin (2001) allows for heterogeneity between consumers only via the stochastic term in the random utility. However, these differences between consumers are unobservable to the firm a priori, since the *expected* utility of a product is identical across consumers. This is why the model is sometimes referred to as the "homogeneous" MNL model. By contrast, we explicitly incorporate differences between consumers in the expected utility via a distributional assumption on the utility function parameters. The distribution of these parameters is then empirically estimated and can be used when determining the optimal assortment. Our approach is similar in theory to conjoint models (e.g., Miller et al. (2006)) in which idiosyncratic utility functions are estimated.

3. CONSUMER MODEL

Our model of the retailer's decision problem of which items to carry and how much to carry, discussed at length in the next section, assumes that each consumer chooses that item from the available assortment which maximizes the consumer's utility. Solving this problem requires empirical estimates of consumers' preferences. We discuss in this section our approach to estimate consumer preferences.

Traditionally, data on consumer preferences have been collected via surveys as stated preferences (ordinal- or interval-scaled), or trade-offs that individuals would be willing to make on particular attributes (e.g., conjoint studies). An alternative approach is revealed preference data as obtained from reported or observed brand choices of consumers in actual purchase situations. For most product categories in the grocery industry these data are readily available from syndicated sources (e.g., household panels of Nielsen and Information Resources Inc.). The primary advantage of stated preference data is the ability to measure preferences for items currently not stocked (in particular, for new products). The major disadvantages of stated preference data relative to brand choice data are potentially lower validity of the data, and often substantially higher cost of data gathering.

Since the focus of our empirical work is on assortment decisions for supermarket product categories, we consider a model to estimate preferences that can be applied to observed brand choices of consumers – a multinomial probit model of brand choice. The probit model can be derived by assuming that the utility a consumer obtains from purchasing an item in the category is composed of a deterministic component and a stochastic component. The stochastic component represents unobserved (to the researcher) components of utility. In the typical formulation of the brand choice model, the utility of item $j, j = 1, 2, ..., J$ to consumer i on occasion t is given by U_{ijt}, thus: $U_{ijt} = \tilde{V}_{ijt} + \varepsilon_{ijt}, \varepsilon_{ijt} \sim N(0, \Sigma)$ where

$$\tilde{V}_{ijt} = \tilde{\alpha}_{ij} - \beta_i p_{ijt} + \tilde{\gamma}_i X_{ijt} \tag{1}$$

where for consumer i and item j, $\tilde{\alpha}_{ij}$ is the *intrinsic* utility or valuation, p_{ijt} is the price of the item on occasion t, X_{ijt} represents other attributes of the item (such as in–store promotions) on that occasion, and β_i and $\tilde{\gamma}_i$ are parameters. The assumption that the stochastic term has a multivariate normal distribution leads to the multinomial probit model of brand choice. We use a diagonal covariance structure $\varepsilon_{ijt} \sim N(0, \Sigma)$ where Σ is a $J \times J$ diagonal matrix, coupled with the identifying restriction that the first diagonal element is one. The choice of diagonal covariance structure simplifies the calculation of choice probabilities, while obviating the restrictive IIA property associated with a scalar covariance, as well as with a multinomial logit model.

Note that the parameters of the utility function are individual specific, thus allowing for heterogeneity in both the intrinsic preferences and the effects of price and other attributes. As we demonstrate subsequently, this characteristic of the model has important implications for the optimal assortment decision of the retailer. The objective of model estimation is to recover the unknown parameters of the deterministic component of the utility function. Data required to estimate the model are observations of consumer choices as well as prices and in-store promotional conditions on each purchase occasion. Such information is typically available in household scanner panel data.

We model heterogeneity by specifying a series of conditional distributions in a Hierarchical Bayesian fashion. The reader is referred to Imai and van Dyck (2005) and McCullogh and Rossi (1994) for details of the estimation approach. A key benefit of using this approach is that it yields posterior estimates of utility function parameters at the individual level. These estimated utility functions are inputs into the retailer's optimization problem.

To obtain item-specific intrinsic utilities, we assume that prices are determined exogenously.[2] Furthermore, for simplification they are assumed to remain constant at their observed mean level p_j. We also assume the in-store promotion variables are fixed at their average levels X_j, again for simplification. Since utility is linear in prices, we divide utilities by the estimated price coefficient β_i (Kalish & Nelson, 1991) to obtain a \$-metric utility, thus:

[2] In section 6 of the chapter, as future research, we discuss the possibility of extending the model to determine optimal prices as well.

$$V_{ij} = \alpha_{ij} - p_j + \gamma_i X_j \tag{2}$$

where $\alpha_{ij} = \tilde{\alpha}_{ij}/\beta_i$,$\gamma_i = \tilde{\gamma}_i/\beta_i$ and p_j is the (constant) price of item j.[3]

The difference in \$-utility between two items may be considered the cost of substituting one item for the other for the consumer; see Krishna (1992) and Bawa and Shoemaker (1987) for a similar notion of substitution costs. An alternative interpretation of this difference is the reduction in price of the less preferred item necessary to make the consumer indifferent between the two items.

We assume that a consumer is willing to substitute lower utility items when higher utility items are not carried in the retail assortment. This assumption is strongly supported by empirical studies (Urban, Johnson & Hauser, 1984; Emmelhainz, Stock & Emmelhainz, 1991). The order of substitution is described by the rank-ordering of estimated preferences for items. When such substitution occurs, however, the consumer is assumed to incur a disutility equal to the difference in \$-metric of intrinsic utility between the most preferred item in the category and the item bought (i.e., the substitute item).

Empirical evidence also suggests that consumers may be willing to incur disutility due to downward substitution only upto a point. Below this point they may be unwilling to substitute and may choose to either postpone purchasing in the category or purchase at a different store (Borle, Boatwright, Kadane, Nunes & Shmueli, 2005). In an ideal setting, one would estimate the utility of a no-purchase decision and expect that consumers will be willing to substitute items as long as the utility of these items is above the utility for no-purchase. However, in the form they are currently available, household scanner panel data do not allow empirical estimation of the no-purchase threshold of households. Thus, in the subsequent empirical illustration we posit alternate mechanisms for operationalizing the no-purchase decision; we outline some options in Section 4.2.

Since the vector of intrinsic brand utilities is unique to each consumer, our consumer model allows completely idiosyncratic patterns of substitution. Not only is the highest preference brand allowed to be different across consumers, consumers who have a given brand as the most preferred may substitute a different brand in the event the most preferred item is not carried in the assortment. Such heterogeneity in substitution behavior between consumers has been documented in empirical studies (Emmelhainz, Stock & Emmelhainz, 1991). Furthermore, since we obtain an interval-scaled measure of preference, consumers who have exactly the same rank-ordering of brand preferences may incur differing amounts of disutilities due to non-availability of the most preferred item. This allows us to capture differences in intensities of brand

---

[3] The transformation of utilities by dividing by the price coefficient also serves to remove the influence of the unidentified scale factor that confounds the vector of parameter estimates (Swait & Louviere, 1993).

preferences between consumers (e.g., loyals vs. switchers) that are relevant for the assortment and inventory decision.

To summarize, our model of the process consumers follow to choose an item to purchase in a category after entering the store is as follows. Consumers have preferences for various items in a category; these preferences vary from consumer to consumer. A consumer observes the available assortments (and the prices of items) and picks the highest utility item from those available or choses not to purchase. The exact operationalization of the no-purchase decision is discussed in the next section.

To use the consumer demand model in the retailer optimization problem, we revert to the utility measures V_{ij} in (2) (at constant prices) and use the estimated utilities $\hat{V}_{ij}$. Disutilities form an important component of the retailer's objective function in our model, as detailed in the subsequent section. Ideally, we should use the random utility function U_{ijt} shown earlier. However, since U_{ijt} contains both a deterministic and a stochastic component, its use will lead to a potentially complex stochastic programming formulation. While accurate, this formulation does confound the impact of heterogeneity and probabilistic choice on the assortment decision. Instead, to focus exclusively on the heterogenous model of consumer behavior, we use only the deterministic component of the utility given by V_{ijt}. Our modeling choice is not without precedence; see Dobson and Kalish (1988; 1993) and McBride and Zufryden (1988). We comment on alternative approaches that could incorporate stochastic choice in the concluding section.

4. THE RETAILER ASSORTMENT AND STOCKING PROBLEM

In this section, we describe a model to solve the retailer's assortment and stocking problem. We first develop a basic model that incorporates profits and disutility. We then discuss some special cases and properties of the formulation.

4.1 Basic formulation

The retailer's problem can be defined as follows: We are given a set of N items indexed by j. There is a fixed cost of stocking each item. Consumers belong to one of the s index segments,[4] $s \in \{1, \cdots, S\}$. There exists a (monetary)

[4] The consumer model in Section 3 was developed assuming each consumer is a separate segment, i.e., the number of consumers in each segment is one. Other models of brand choice that provide estimates for "segments" of consumers could be employed, such as formulations of Kamakura and Russell (1989) and Chintagunta, Jain and Vilcassim (1991).

utility measurement, V_{sj}, for every segment s for every item j (see Section 3). As noted previously, for solving the retailer's optimization problem we assume that prices and promotional activities are held constant at their average levels. As a consequence, item utilities are time invariant. A consumer (segment) chooses from all available items the one that maximizes its utility.[5] The retailer's problem is to select an assortment and determine the stock for items in the assortment to maximize profits. The profit function can be written as:

$$PR(\mathbf{x}, \mathbf{y}) = \sum_j \left[\sum_s (p_j - c_j) x_{sj} n_s - K_j y_j \right] \tag{3}$$

where p_j is the per unit (regular) price of item j, c_j is the per unit variable cost of stocking item j, x_{sj} is a 0–1 variable which takes on a value of one if segment s customers are assigned to item j and zero otherwise (a decision variable),[6] n_s is the number of consumers in segment s, K_j is the fixed cost of stocking item j, and y_j is a 0-1 decision variable which takes the value one if item j is stocked and zero otherwise. Finally, $\mathbf{x}$ is a $S \times N + 1$ matrix of x_{sj} and $\mathbf{y}$ is an $N + 1$-vector of y_j. We let no-purchase decision be a "product" that is always available, thus expanding the product space to $N + 1$; further, $p_0 = c_0 = K_0 = 0$ and $y_0 = 1$.

Typically a retailer may do assortment planning for its stores twice a year; thus the planning horizon for assortments is about six months. In our formulation, we have not specified any planning horizon explicitly. The data can be scaled to accommodate any planning horizon. We need to, however, consider the fixed costs – which include costs relating to sourcing, supplier selection, negotiations, etc., – appropriate for the planning horizon. Due to fixed costs of carrying an item in the assortment, not all items may be stocked. As a consequence, the following situations are possible:

1. A customer segment buys a less preferred item because its most preferred item is not available.
2. A customer segment does not purchase at all because no satisfactory item is available.

In either case the customer incurs a disutility. We postulate that such disutility adversely affects the customer's likelihood of repurchasing at this store, thereby affecting long–run profits.[7] We propose the following measure of customer disutility:

---

[5] We assume, for simplification, that each consumer buys exactly one unit in each restocking period. This assumption can be relaxed by weighting each consumer by the number of units bought. In general, the number of units bought by a consumer within any stocking period may be uncertain. Incorporating this uncertainty will result in a stochastic programming formulation. We elaborate upon this idea in the discussion of future work in section 6.

[6] In the optimization model, the item "assigned" to a consumer will be the one that maximizes the consumer's utility. Thus, consumers will in effect self-select their best alternative from the available assortment.

[7] Notice that this disutility is due to non-stocking of items and not due to stock-out of an item.

$$DU(\mathbf{x}) = \sum_s n_s \left[\sum_k \{(V_{sj_1} - V_{sk})x_{sk}\} + (V_{sj_1} - V_{sj_0})(1 - \sum_k x_{sk}) \right] \quad (4)$$

where, $V_{sj_1} = \max_j\{V_{sj}\}$, and V_{sj_0} is the no-purchase utility, as discussed later in subsection 4.2.

For those customers who are assigned an item k, the disutility is the difference between the utility of item k and their most preferred item.[8] Similarly, customers who do not purchase are also dissatisfied. The disutility incurred by these customers is the difference in utility between their highest utility and their utility for no-purchase. Clearly, customers who find their most preferred item in the assortment do not incur any disutility.

We propose that the overall objective function for a retailer should be a weighted combination of profits as measured by (3) and disutility as measured by (4). The extent to which a retailer should weight consumer disutility will depend on the product category. Customer dissatisfaction with some categories is likely to have a larger adverse impact on store choice. In the context of pricing, for example, Harris and McPartland (1993) classify categories into "traffic generators" (i.e., affect store choice) and others. We model this by taking a convex combination of the profit and disutility functions. Thus the objective function of the retailer is:

$$\Pi(\mathbf{x}, \mathbf{y}, w_c) = (1 - w_c)PR(\mathbf{x}, \mathbf{y}) - w_c DU(\mathbf{x}) \quad (5)$$

where $0 \leq w_c \leq 1$. w_c may be interpreted as a control or policy parameter whose value is to be subjectively determined by the decision maker.[9]

The optimization problem of the retailer is then written as follows:

$$(P1) \quad \max_{\mathbf{x},\mathbf{y}} \Pi(\mathbf{x}, \mathbf{y}, w_c)$$

such that,

$$\sum_k V_{sk} x_{sk} \geq V_{sj} y_j \qquad \forall s, j \quad (6a)$$

[8] Dissatisfaction measured as sum across segments of the differences in utilities implies that a large number of small disutilities is equivalent to a small number of large disutilities; e.g., two segments with one unit of disutility each is equivalent to one segment (of same size) with two units of disutility. This may not be desirable since larger differences in utilities signify consumers *loyal* to certain brands, and smaller differences in utilities signify *switchers*. A non–linear (say, e.g., exponential) function of difference in utilities will allow us to distinguish between *loyals* and *switchers*.

[9] A similar objective function (weighted combination of profits and consumer utility) was also considered by Little and Shapiro (1980) in the context of pricing nonfeatured products in supermarkets. Similarly, there is extensive literature on bi-criterion optimization problems; see, for example, French and Ruiz-Diaz (1983).

$$\sum_j x_{sj} \leq 1 \qquad \forall s \tag{6b}$$

$$x_{sj} \leq y_j \qquad \forall s, j \tag{6c}$$

$$x_{sj} = 0, 1 \qquad \forall s, j \tag{6d}$$

$$y_j = 0, 1 \qquad \forall j \neq 0 \tag{6e}$$

$$y_0 = 1 \tag{6f}$$

Constraints (6a) ensure that of the items stocked, a customer is assigned his/her most preferred item. Constraints (6b) ensure that segment s is assigned to at most one item; finally, constraints (6c) ensure that only items that are offered are chosen by the customers.

At first glance it may appear that incorporating consumer disutility through $DU(\cdot)$ in the objective function makes constraints (6a) redundant. The constraints are redundant (or trivially satisfied) only when a retailer sets $w_c = 1.0$. Otherwise, in the absence of constraints (6a) it is possible that a retailer may assign a less preferred item (with a higher contribution margin) to a consumer even though a more preferred item (with a lower contribution margin) is stocked, albeit for a different consumer. Such an assignment is problematic from an implementation viewpoint in the context of supermarkets since a consumer walks into a store and necessarily picks his most preferred item if it is available. Constraints (6a) ensure that the retailer incorporates this fact into its decision making.

4.2 Modeling no purchase

As discussed previously, a customer may decide to not purchase in the category if its preferred item is not stocked. Since scanner data do not report non-purchasing on account of unavailability in the assortment, we model this outcome and assume its value.[10] There are at least two ways one could model no purchase in the optimization problem. For a customer segment s, first rank order the utilities V_{sj} in decreasing order to write:

$$V_{sj_1} \geq V_{sj_2} \geq \cdots \geq V_{sj_N}.$$

Then,

[10] Category purchase incidence is frequently modeled using scanner data (e.g. Bucklin, and Gupta (1992)). However, the consumers' decision is considered to be one of choosing to buy one of the items in the assortment at today's prices and promotions, versus postponing the purchase decision to a future occasion when prices may be better, and relying meanwhile on available household inventory for consumption. Thus, the impact of assortment unavailability is not modeled.

1 For all customer segments s, assume that customers do not purchase if their most preferred d (exogenously specified) items are not stocked (see Smith and Agrawal (2000) for a similar operationalization). We call d the *depth of no purchase*. Clearly $d \in [1, N]$. An alternate interpretation of d is that it captures the (store) switching cost of a consumer; a large d implies high switching cost. Intuitively, a large d implies that a customer is willing to substitute less preferred items when more preferred items are not stocked rather than not purchase, regardless of the magnitude of disutility incurred. Under this operationalization, we set the no-purchase utility $V_{sj_0} = V_{sj_{d+1}}$ if $d < N$ and $V_{sj_0} = V_{sj_N} - \epsilon$ (for some $\epsilon > 0$) if $d = N$

2 Alternately, let T be an exogenously specified threshold level of disutility that signifies no purchase. Suppose there exists an item j_{k+1} for segment s, such that $V_{sj_1} - V_{sj_{k+1}} \geq T$. Then we infer that a customer in segment s will not purchase if items j_1 through j_k are not available in the assortment. Under this operationalization, we set the no-purchase utility $V_{sj_0} = V_{sj_{k+1}}$.

While either formulation is easily incorporated in our model, in this chapter we use the former approach to model no-purchase. Later, we will analyze the sensitivity of the assortment solution to the depth of no purchase, d. To incorporate the depth of no purchase into problem (P1), we modify constraint (6a) as follows. For each customer segment, s, define an order set consisting of d elements $N_s^d = \{j_1, j_2, \ldots, j_d, j_0 | V_{sj_1} \geq V_{sj_2} \geq V_{sj_d} \geq V_{sj_0}\}$. We then rewrite (6a) as:

$$\sum_{k=0}^{d} V_{sj_k} x_{sj_k} \geq V_{sj_i} y_{j_i} \qquad \text{for } j_i \in N_s^d \text{ and } \forall s \qquad (6a')$$

Furthermore, to ensure that a customer is assigned a product within their first d choices or no-purchase, we need to modify constraint (6c) to:

$$\sum_{j=0}^{d} x_{sj} \leq 1 \qquad \forall s \qquad (6b_1')$$

$$\sum_{j=d+1}^{N} x_{sj} \leq 0 \qquad \forall s \qquad (6b_2')$$

4.3 **Reformulation**

In this section we reformulate problem (P1), specifically contraint ($6a'$) which facilitates solution of (P1) as a linear program when integrality constraints on x_{sj} are relaxed. We observe that constraint set (6b)–(6e) is of the same form as that for an uncapacitated plant / warehouse location problem (Cornuejols, Fisher & Nemhauser, 1977). We now reformulate constraint set (6a′) that results in a tighter formulation for (P1). Observe that (6a′) ensures that a customer segment is assigned its most preferred product amongst the ones stocked. Thus it merely depends on the rank order of products for any given consumer segment and not on the interval scaled utilities as measured by V_{sj}. We exploit this structure to replace (6a′) with

$$1 - \sum_{k=i+1}^{d} x_{sj_k} \geq y_{j_i} \qquad \text{for } j_i \in N_s^d \text{ and } \forall s. \tag{6a''}$$

We also relax the constraints on x_{sj} in (6d) as follows:

$$x_{sj} \leq 1. \tag{6d'}$$

Proposition 4.1 *Problem (P1) with (6a″) set of constraints is at least as tight a formulation as (P1) with (6a′) set of constraints. Furthermore the relaxation of integrality constraints to (6d′) still guarantees an integer solution for* x_{sj}.

A proof is provided in the appendix.

Thus the new constraint set (6a″) achieves the same results as (6a′), i.e., ensuring that of the items stocked a customer segment is assigned its most preferred item. Furthermore, this reformulation does not increase the number of constraints. Finally, the relaxation guarantees an integer solution. In the sequel we will use (P1) with (6a″) and (6d′).

4.4 Discussion of the optimization model and some special cases

Readers familiar with the literature on plant location will see that problem (P1) has an embedded uncapacitated plant location model (when $w_c = 0$, and constraints (6a) are relaxed). This problem is extensively researched by Cornuejols et al. (1977) and they show that the problem is NP-hard. Hence problem (P1) is also NP-hard. Our computational study shows that similar to the uncapacitated plant location model (Erlenkotter, 1978), the solution to problem (P1) is easily obtained for problem sizes (relatively small) of interest in this study. Large scale models comprising several products in a product line and a larger number of customer segments will call for development of heuristics.

We now consider a few special cases of Problem (P1). First, we consider the situation when a retailer places zero weight on the disutility incurred by the consumers due to his assortment decision; we shall identify a retailer with $w_c = 0.0$ as a *myopic retailer* who maximizes just short-term profits.

To highlight the need to model "no purchase", consider the myopic retailer who solves (P1) with $w_c = 0.0$ and with a depth of no purchase $d < N$. Recall that as d increases, consumers are more willing to substitute to the available items in the assortment and less willing to not purchase. We then observe that in a model without a no-purchase decision, a myopic retailer will stock only one product. Effectively, we solve problem P1 with $w_c = 0.0$ and $d = N$; that is, the retailer does not care about disutilities incurred by the consumers and all consumers purchase some product. This implies that the total demand is unaffected by the choice of items available. Then the retailer carries just one product $j* = argmax_j\{(p_j - c_j)n_s - K_j\}$ which maximizes his profit.

We would like to be able to study the behavior of the assortment decision with respect to parameters like weight on disutility (w_c), depth of no–purchase (d), contribution margins ($p_j - c_j$), etc. In general, (P1) is a complex optimization problem and usually does not permit many comparative statics results. Analytically, we were unable to get any general sensitivity results with respect to p_j, w_c and d. The main difficulty appears to be the very general formulation of the heterogeneity of consumers. Any change in these parameters affects the substitution pattern through change in the interval scaled utilities and hence the demand patterns. The obvious case is when profit margins increase due to decrease in marginal costs. This increases the contribution margin and with fixed p_j, d and w_c, the retailer will find it optimal to increase his assortment sizes, since for d it may help him satisfy more consumers and/or decrease disutility if $w_c > 0$.

5. COMPUTATIONAL STUDY

5.1 Description of household scanner panel data

The data were collected by the AC Nielsen Company and are available for a two year period. A panel of households provided information on their purchasing in several categories. These data were supplemented with data on prices, in-store displays, and feature advertising collected from the supermarkets in the city. We include purchases of the eight largest brand-sizes of canned tuna made by 1097 panelist households in our estimation sample. These eight items account for approximately 90% of category volume. Brand names are disguised to meet confidentiality requirements of the data provider.

In Table 7-1 we provide descriptive statistics of the data. Besides shelf price, we include in–store displays and retailer feature advertising in the choice model.

Table 7-1. Descriptive Statistics of Data

Item	Average Price (cents/oz.)	Display (% occasions)	Feature (% occasions)
1	12.3	3.9	25.9
2	21.8	0	1.7
3	12.0	4.0	29.9
4	11.5	8.7	24.4
5	15.1	0	0
6	24.2	0	0
7	11.3	4.3	24.4
8	9.8	4.2	13.7

Table 7-1 indicates that there is considerable variation in shelf prices and promotional activity between brands, highlighting the need to control for the effects of these variables when measuring intrinsic brand preference or valuation.

Bayesian posterior estimates of the demand model parameters are obtained for each household using the approaches of Imai and van Dyck (2005) and McCullogh and Rossi (1994). Table 7-2 contains the mean value of the estimated posterior estimates. The coefficients of price, display, and feature, have the expected signs.

We use the estimated \$-metric intrinsic preferences for items V_{ij} to infer patterns of primary demand and likely substitution between items. We computed optimal assortments under two separate assumptions about consumers' willingness to substitute. First we assume that consumers are willing to make one substitution. That is, they will not purchase in the category if their first preference and second preference brands are not available (i.e., $d = 2$). Therefore, we focus on the top two brands for each consumer. Note that customers who do not find their most preferred brand but do find their second-most-preferred brand still incur a disutility, which our decision model incorporates. Next, we also solved for the optimal assortment under the assumption that consumers are willing to substitute twice (i.e., $d = 3$). In the subsequent discussion we describe the solution under the $d = 2$ assumption in detail and thereafter briefly talk about the $d = 3$ case.

Table 7-3 shows the cross classification of the first and second preference brands for the sample of 1097 consumers.[11] Row total $N_{i.}$ indicates the number of consumers whose first preference brand is brand i. Similarly, column total $N_{.j}$

Table 7-2. Mean Value of Household Parameter Estimates of Probit Demand Model

Mean Brand Specific Constants	
Item 1	0.815
Item 2	2.350
Item 3	0.494
Item 4	1.030
Item 5	0.267
Item 6	2.885
Item 7	–0.273
Price (\$/oz.)	–26.882
Display	0.597
Feature	0.163

[11] Note that only the rank ordering of preferences is used to construct Table 7-3 to illustrate the nature of substitution between items. The retailer optimization problem uses interval-scaled values of preferences.

Table 7-3. Cross classification of First and Second Preference Brands (Cell entries are in %)

First Preference Product	Second Preference Product 1	2	3	4	5	6	7	8	No. of Consumers
1	0	0	71.7	3.3	1.7	0	10.0	13.3	60
2	0	0	0	0	0	0	0	0	0
3	69.7	1.4	0	1.9	0	0	17.1	9.9	211
4	3.4	0	6.8	0	1.1	0	78.4	10.2	88
5	12.5	6.3	3.1	0	0	0	28.1	50.0	32
6	0	0	0	0	0	0	0	0	0
7	2.3	0	14.7	55.4	2.5	0	0	25.1	354
8	14.5	0	21.3	2.8	13.6	0	47.7	0	352
Number of Consumers	213	5	177	212	59	0	288	143	1097

is the number of consumers whose second preference brand is brand j. Each cell entry in the table denotes the percentage of $N_{i.}$ consumers who have brand j as their second preference brand.

The row totals are indicative of primary demands for items. For example, it is clear that items 3, 7 and 8 are the first-preference products of a large number of consumers, while none of the consumers in our sample prefer items 2 and 6. Similarly, items 1, 4, and 5 have relatively weak primary demand. Column totals indicate whether items are acceptable as substitutes. Item 1, for example, is the brand of second choice for a large number of consumers (213) as compared with its primary demand (60). A similar preference pattern is evident for items 3 and 4. Item 8 has the opposite kind of preference pattern, with large number of consumers (352) preferring it in first place while only 143 prefer it in second place. Large cell entries indicate items that are more substitutable. For example, we see that 71.7% of consumers who have item 1 as their first preference have item 3 as their second preference. Conversely, 69.7% of those who prefer item 3 are willing to accept item 1. There is some evidence of asymmetries in patterns of substitution between brands. For instance, the entry in row 5 and column 8 is 50.0% while that in row 8 and column 5 is only 13.6%. These data further confirm the existence of substantial heterogeneity in patterns of substitution between consumers.

5.2 Solution technique for assortment problem

We used LINDO, a commercial linear programming package, to solve the reformulated optimization model. The problems are generated from the preference, price, and cost data using a program written in C. This program allows the decision maker to vary the weight w_c (weight on consumer welfare and profit objectives) and d (depth of no purchase) to evaluate various solutions.

For our computational study we solved 80 instances of the problem. We varied the weight w_c from 0.01 to 0.99 with $d = 2$ (40 problems) and d = 3 (40 problems) for two different fixed costs. On average the problem took 32 seconds of cpu time, with times ranging from 20 seconds to 48 seconds. Based on our computational times it seems appropriate to solve this problem to obtain the optimal solution using a commercial package. Specialized implementation and heuristics may be necessary for larger problems if the computational times become prohibitive.

5.3 Optimal assortment

To solve the retailer optimization problem (P1), we need estimates of fixed costs (K_j), contribution margins ($p_j - c_j$), and of w_c, the weight placed by the retailer on customer disutility relative to current period profits. We did not have access to real cost and contribution data for the market for which consumer data were available. For the empirical illustration, we assume values of these parameters as follows. Retail contribution margins are assumed to be 30% of the average retail price of the item. Thus, items can be ordered in terms of margin based on the average prices shown in Table 7-1. We examine two different levels of fixed costs in our illustrations: $1 per stocking period and $5 per stocking period. These levels of fixed costs ensure that at least one item is unprofitable to carry based on its primary demand. We explore the impact of varying w_c (over the space 0 to 0.99 in small steps) on the optimal assortment, profits and customer disutility.[12]

Case 1: Fixed Cost is $1 per item per stocking period

In Table 7-4 we show changes in the optimal assortment of items, customer disutility, and optimal profits as the weight on disutility in the objective function (w_c) is increased from 0 to 0.99. Note that items 2 and 6 are never included in the optimal assortment, regardless of the value of w_c, because of the pattern of first and second preferences discussed previously. When $w_c = 0$, the problem reduces to the pure profit maximization problem of a myopic retailer. Thus, the retailer should carry only those products whose contribution margin exceeds the fixed cost. The demand for a product, given an assortment, is the sum of its primary demand, and spillover demand from items not carried. The solution to the pure profit maximization problem is to carry four items (item numbers 1, 3, 5, and 7). Table 7-1 shows that products 1, 3, and 5 are the highest margin products (after products 6 and 2). Although item 4 has higher margin than item 7, item 7 is included in the optimal assortment instead of item 4 because of its large primary demand (354 consumers) relative to item 4 (88 consumers). When

[12] For the illustration here we assume that the total market consists of the 1097 consumers in our sample.

Table 7-4. Optimal Assortment and Resulting Disutility and Profits (Fixed cost = $1)

Weight on Disutility (w_c)	Disutility	Profit	# customers not served	Optimal Assortment
0.00	85.83	34.50	19	1,3,5,7
0.01	85.83	34.50	19	1,3,5,7
0.03	69.62	34.17	0	1,3,4,5,7
0.04	20.38	32.64	13	3,7,8
0.05	4.45	31.93	7	3,4,7,8
0.20	2.27	31.60	0	3,4,5,7,8
0.30-0.99	0.00	30.72	0	1,3,4,5,7,8

a weight of 0.03 is placed on disutility we find that item 4 is also included in the assortment now. As noted previously, item 4 has low primary demand, but is acceptable as a substitute by a large number of customers. 19 customers who were previously not served at all now find an acceptable product to buy. Moreover, with this assortment profits are slightly lower, but disutility is significantly reduced. This suggests that profit as a function of assortment carried is quite flat near the maximum. The introduction of a second criterion (i.e., disutility) into the objective function helps us to select the assortment that delivers close to maximum profits while reducing disutility. If customer disutility influences future store traffic and hence long-run profits, the results presented help the decision maker balance short-run with long-run profits.

As w_c is increased further, we find that the number of items in the optimal assortment decreases and then increases. At $w_c = 0.040$ the optimal assortment shrinks from {1,3,4,5,7} to {3,7,8}. The inclusion of item 8 is probably explained by its large primary demand (352 customers), which implies that when it is omitted from the assortment, large disutility is incurred. Further, half of the customers who prefer item 5 find item 8 acceptable. At $w_c = 0.050$ the optimal assortment expands to include item 4 once again. At $w_c = 0.30$ the optimal assortment expands to include all six products, other than items 2 and 6.

Note that we observe two kinds of non-monotonicities in the optimal behavior with increases in w_c. One, the number of items in the optimal assortment expands and then shrinks. Two, certain items (such as 4 and 1) enter the optimal assortment, then get dropped, and then get re-included. Such non-monotonic behavior of the optimal assortment reinforces the need for a decision support model for retail assortment decisions.

Case 2: Fixed Cost is $5 per item per stocking period

In Table 7-5 we show the optimal assortment and associated profits and disutility. Note that in the pure profit maximization case, 155 customers are not served and disutility incurred is quite high. Placing a weight of 0.03 on disutility expands the optimal assortment to include product 8 in addition to items 3 and 7. As a consequence, profits drop. However, the number of customers served increases significantly and disutility drops sharply.

Table 7-5. Optimal Assortment and Resulting Disutility and Profits (Fixed Cost = \$5)

Weight on Disutility (w_c)	Disutility	Profit	# customers not served	Optimal Assortment
0.00	95.92	22.72	155	3,7
0.01	95.92	22.72	155	3,7
0.03	20.38	20.64	13	3,7,8
0.30	4.45	15.93	7	3,4,7,8
0.70-0.99	0.00	6.72	0	1,3,4,5,7,8

A distinguishing feature of the optimal assortment in Case 2, relative to Case 1, is that with increase in w_c the number of items in the optimal assortment always increases. Furthermore, once an item enters the optimal assortment it stays in the assortment with increases in w_c . We conjecture that the high fixed cost may cause such monotonic behavior of the optimal assortment.

Results in the $d = 3$ case are entirely consistent with the results for the $d = 2$ case with some differences that are intuitive. For reasons of space we do not show detailed results. At each level of w_c, we find that optimal profits are at least as large in the $d = 3$ case since consumers are assumed to be more willing to substitute to less-preferred products. As a result, the spillover demand to any product from items not carried is no lower in this case than in the $d = 2$ case. Further, disutility is at least as large in the $d = 3$ case. When the fixed cost per item is \$1, the optimal assortment changes non-monotonically with increases in w_c. When the fixed cost is \$5, on the other hand, the optimal assortment changes monotonically.

To deduce further inferences, we ran the model for both cases of fixed costs considered previously ($K = 1$ and 5) and equal margins across all products, set equal to average margin of eight products using depths $d = 2$ and $d = 3$. The optimal solutions exhibited monotone changes to the optimal assortment for all w_c values. While this is true for our particular data set, we are able to construct a three-product, three-customer instance to provide a counter-example (see data in Table 7-6) for this monotone behavior.

In this counterexample, we find that when $w_c = 0$, the optimal profits are 2.2, the disutility is 9, and the optimal assortment has only product 2. As w_c grows to 0.1379, the assortment consists of product 1 only, and for higher values of w_c the optimal assortment consists of products 1 and 3.

Table 7-6. A three-product example

Customer	Utility		
	Product 1	Product 2	Product 3
1	5	2	1
2	5	2	1
3	1	2	5
Fixed Cost	2	2	2
Margin	1.4	1.4	1.4

The results show that it is very hard to predict the structure of the optimal assortment, especially when we consider a data-driven problem setting.

6. SUMMARY, EXTENSIONS, AND FUTURE WORK

We propose a model for the optimal assortment and stocking decisions for retail category management. In particular, we address the question of rationalization of the retail assortment, i.e., determining the optimal subset of items to retain from the set of items currently carried. We assume, based on empirical evidence reported in the literature, that consumers are willing to partially substitute less preferred items if their preferred items are not available. We also assume that consumers are heterogeneous in their intrinsic preferences for items and in their price sensitivities, an assumption strongly supported empirically.

We propose that the appropriate objective function for a far-sighted retailer should include not only short-term profits but also a penalty for the disutility incurred by consumers who do not find their preferred items in the available assortment. The rationale for including such a penalty is that dissatisfied consumers are less likely to return to the store in the future. We propose a measure for disutility that recognizes differences between consumers in their intensity of dissatisfaction.

The retailer problem is formulated as an integer programming problem. We show that the problem is large but can be solved efficiently to obtain an optimal solution. We demonstrate an empirical application of our proposed model using household scanner panel data for eight items in the canned tuna category. Our results indicate that the inclusion of the penalty for disutility in the retailer's objective function is informative in terms of choosing an assortment to carry. We find that customer disutility can be significantly reduced at the cost of a small reduction in short term profits.

An immediate extension of the current work is to develop heuristics to solve the optimization problem since problem sizes in categories with a large number of items may be very large and computational times to find optimal solutions might be prohibitive. Furthermore, we realize that there is uncertainty due to errors in the utility function parameter estimates, which our optimization model assumes to be fixed. The problem formulation can be modified to allow for uncertain parameter estimates and use a stochastic programming approach to solve the assortment problem.

The approach described in this chapter is an illustrative first-step that attempts to close some of the modeling gaps in the literature. As outlined in the introduction, the complete assortment planning problem needs to consider several other factors. Next we discuss briefly several directions to extend the proposed model in future research.

1. *Shelf Space Constraints:* Typically, retailers have shelf space constraints which limit the amount of stock that can be carried within a category. These constraints can be incorporated within the context of our problem (P1). A complexity that now arises is the occurrence of stock-outs. Since customers have heterogeneous preferences for items, the dynamics of their arrival process also needs to be accounted for.
2. *Incorporating Demand Uncertainty*: In the current model, we assumed that utilities of each consumer segment are deterministic. In fact, from the retailer's perspective utilities are stochastic. Including stochastic utilities results in a mixed–integer stochastic programming problem.
3. *The Pricing Problem*: The basic formulation outlined in this chapter can be extended to study the joint pricing and assortment decisions. However, maximization over prices makes (P1) a non–linear optimization problem which can be solved using procedures outlined in Adams and Sherali (1990), for example. Alternately, heuristic procedures could be explored.
4. *The Display Effect or the Effect of Facings on Sales*: The literature on shelf space management has been concerned with the relationship between shelf space allocations and sales due to the influence of product display on demand. The number of facings allocated to an item also determines the quantity stocked of this item (usually an integer multiple of the number of facings). Thus, the problem of determining the optimal assortment and inventory is inter–related with the shelf–space allocation problem. Extending the model presented in this chapter to incorporate the display effect presents two challenges: one, the problem of measuring the effect of product display on demand, and two, the optimization problem changes considerably since we will now have to decide on number of facings which will be an integer variable.
5. *Joint Fixed Costs*: Product lines for a retailer typically consist of several SKU's being supplied by the same manufacturer or wholesaler. Therefore, multiple products in a category may require common resources (contact, vendor management, etc.). The Dobson and Kalish (1993) formulation assumes independent fixed costs, and therefore it can overstate the fixed costs associated with incremental introduction of products that share fixed costs with incumbent products. In case of shared fixed costs, a firm can take the savings available into account when introducing products that require common resources. One approach is to define product classes, similar to manufacturing classes used by Morgan, Daniels, and Kouvelis (2001). We hypothesize that inclusion of common fixed costs (relative to the assumption of independent fixed costs) will increase the number of products offered, profits, as well as consumer satisfaction.

Acknowledgment We are grateful to the A.C. Nielsen Company for generously providing the data used in this paper, to Edward Malthouse for his help with setting up the data, to Qiang

Liu for help with data analysis, and to Pradeep Chintagunta, Maqbool Dada and Yehuda Bassok for valuable comments on an earlier version of the paper.

Appendix

Proof of Proposition 4.1

Without loss of generality, we will illustrate this for the general case rather than the special case of fixed depth of search d.

First consider the constraint (6a″). The constraint for $j = 1$ will be

$$1 - (x_{s2} + x_{s3} + \ldots + x_{sK} + x_{s0}) \geq y_1$$

However, from (6c) we know that

$$x_{s1} + x_{s2} + \ldots + x_{jK} + x_{s0} \leq 1$$

Actually, given a "no purchase" option, the above is an equality; i.e.,

$$x_{s1} + x_{s2} + \ldots + x_{jK} + x_{s0} = 1$$

Using this we rewrite $1 - (x_{s2} + x_{s3} + \ldots + x_{sK} + x_{s0}) \geq y_1$ as simply $x_1 \geq y_1$. Similarly, we can write (6a″) for $j = k$ as

$$x_{s1} + x_{s2} + \ldots + x_{sk} \geq y_k.$$

Using this, for any arbitrary customer segment s that prefers K products in the ordinal order (without loss of generality) the constraint sets (6a′) and (6a″) are

(6a′)		(6a″)	
$V_{s1}x_{s1} + V_{s2}x_{s2} + \cdots + V_{sk}x_{sk} \geq V_{s1}y_1$	(1′)	$x_{s1} \geq y_1$	(1″)
$V_{s1}x_{s1} + V_{s2}x_{s2} + \cdots + V_{sk}x_{sk} \geq V_{s2}y_2$	(2′)	$x_{s1} + x_{s2} \geq y_2$	(2″)
$\vdots$		$\vdots$	
$V_{s1}x_{s1} + V_{s2}x_{s2} + \cdots + V_{sk}x_{sk} \geq V_{sk-1}y_{k-1}$	((k-1)′)	$x_{s1} + x_{s2} + \cdots + x_{sk-1} \geq y_{k-1}$	((k-1)″)
$V_{s1}x_{s1} + V_{s2}x_{s2} + \cdots + V_{sk}x_{sk} \geq V_{sk}y_k$	(k′)	$x_{s1} + x_{s2} + \cdots + x_{sk} \geq y_k$	(k″)

Consider normalized constraint (1′) and (1″):

$$x_{s1} + \left(\frac{V_{s2}}{V_{s1}}\right)x_{s2} + \cdots \left(\frac{V_{sk}}{V_{s1}}\right)x_{sk} \qquad \textit{and} \qquad x_{s1} \geq y_1.$$

Since (1″) and (1′) are identical in x_{s1} dimension and (1′) has $k - 1$ extra variables (degrees of freedom), constraint (1″) is tighter than constraint (1′).

Using similar arguments one can show that constraints (2″) to ((k − 1)″) will be tighter than (2′) to ((k − 1)′). Constraint (k″) may be identical to (k′). The argument can be repeated for other segments. Thus problem (P1) with (6a″) is a tighter formulation than (P1) with (6a′).

To see that relaxation of x_{sj} still leads to an integer solution, first consider (1″). If $y_1 = 0$, then $x_{s1} = 0$ using (6c). If $y_1 = 1$, then $x_{s1} = 1$. Now consider (2″). Suppose $y_1 = 0$. If $y_2 = 0$ then $x_{s2} = 0$; otherwise ($y_2 = 1$), $x_{s2} = 1$. However, if $y_1 = 1$, then (6b) ensures that $x_{s2} = 0$. Following this argument, we can show that x_{sj} is integer.

REFERENCES

Adams, P. W. and H. D. Sherali (1990). Linearization Strategies for a Class of Zero–One Mixed Integer Programming Problems. *Operations Research*, **38**, 217–226.

Arrow, K., S. Karlin, and H. Scarf (1958). *Studies in the Mathematical Theory of Inventory and Production*. Stanford University Press.

Bassok, Y., R. Anupindi, and R. Akella (1997). Single Period Multi–product Inventory Models with Substitution. *Operations Research*, **47**, 632–642.

Bawa, K. and R. W. Shoemaker (1987). The Effects of a Direct Mail Coupon on Brand Choice Behavior. *Journal of Marketing Research*, **24**, 370–76.

Boatwright, P. and J. C. Nunes (2001). Reducing Assortment: An Attribute Based Approach. *Journal of Marketing*, **65**, 50–63.

Borin, N., P. Farris, and J. Freeland (1994). A Model for Determining Retail Product Category Assortment and Shelf Space Allocation. *Decision Sciences*, **25**(3), 359–384.

Borle, S., P. Boatwright, J. B. Kadane, J. C. Nunes, and G. Shmueli (2005). Effect of Product Assortment Changes on Customer Retention. *Marketing Science*, **24**(4), 612–22.

Broniarczyk, S. M., W. D. Hoyer, and L. McAlister (1998). Consumers' Perceptions of the Assortment Offered in a Grocery Category: The Impact of Item Reduction. *Journal of Marketing Research*, **35**, 166–176.

Bucklin, R. E. and S. Gupta (1992). Brand Choice, Purchase Incidence, and Segmentation: An Integrated Modeling Approach. *Journal of Marketing Research*, **29**(9), 201–15.

Bultez, A., E. Gijsbrechts, P. Naert, and P. V. Abeele (1989). Asymmetric Cannibalism in Retail Assortments. *Journal of Retailing*, **65**(2), 153–192.

Bultez, A., and P. Naert (1988) S.H.A.R.P.:Shelf Allocation for Retailer's Profit. *Marketing Science*, **7**(3), 211–231.

Business Week (1996). Marketing: Making It Simple. September 9.

Carpenter, G. and D.R. Lehmann (1985). A Model of Marketing Mix, Brand Switching and Competition. *Journal of Marketing Research*, **22**, 318–329.

Chintagunta, P., D.C. Jain, and N.J. Vilcassim (1991). Investigating Heterogeneity in Brand Preferences in Logit Models for Panel Data. *Journal of Marketing Research*, **28**, 417–28.

Chong, J., T-H Ho, and C. Tang (2001). A Modeling Framework for Category Assortment Planning. *Journal of M&SOM*, **3**(3), 191–210.

Cornuejols, G., M. Fisher, and G. Nemhauser (1977). Location of Bank Accounts to Optimize Float : An Analytic Study of Exact and Approximate Algorithms. *Management Science*, **23**(8), 789–810.

Corstjens, M. and P. Doyle (1981). A Model for Optimizing Retail Space Allocations. *Management Science*, **27**(7), 822–833.

Coughlan, A. T., E. Anderson, L. W. Stern, and A. I. El-Ansary (2006). *Marketing Channels*. Prentice Hall, Englewood Cliffs, New Jersey.

Dobson, G. and S. Kalish (1988). Positioning and Pricing a Product Line. *Marketing Science*, **7**(2), 107–125.

Dobson, G. and S. Kalish (1993). Heuristics for Pricing and Positioning a Product–line Using Conjoint and Cost Data. *Management Science*, **39**, 160–175.

Emmelhainz, M. A., J. R. Stock, and L. W. Emmelhainz (1991). Consumer Responses to Retail Stockouts. *Journal of Retailing*, **67**(2), 139–147.

Erlenkotter, D. (1978). A Dual-based procedure for Uncapacitated Facility Location. *Operations Research*, **26**(6), 992–1009.

Food Marketing Institute (1993). Variety or Duplication : A Process to Know Where You Stand. Published by The Research Department, Food Marketing Institute, Washington, D.C.

French, S. and F. Ruiz-Diaz (1983). A Survey of Multi–Objective Combinatorial Scheduling. In S. French et al. (Eds.), *Multi–Objective Decision Making*. Academic Press, New York.

Gaur, V. and D. Honhon (2006). Assortment Planning and Inventory Decisions Under a Locational Choice Model. *Management Science*, **52**(10), 1528–1543.

Gruen, T., D. S. Cortsen, and S. Bharadwaj (2002). *Retail out-of-stocks: A worldwide examination of extent, causes and consumer responses*. Grocery Manufacturers of America.

Harris, B. and M. McPartland (1993). Category Management Defined: What it is and why it works. *Progressive Grocer*, **72**(9), 5–8.

Imai, K. and D. A. van Dyck (2005). A Bayesian Analysis of the Multinomial Probit Model Using the Marginal Data Augmentation. *Journal of Econometrics*, **124**(2), 311–334.

Kalish, S. and P. Nelson (1991). A Comparision of Ranking, Rating, and Reservation Price Measurement in Conjoint Analysis. *Marketing Letters*, **2**(4), 327–335.

Kamakura, W. A. and G. Russell (1989). A Probabilistic Choice Model for Market Segmentation and Elasticity Structure. *Journal of Marketing Research*, **26**, 379–390.

Kok, A. and M.L. Fisher (2007). Demand Estimation and Assortment Optimization under Substitution: Methodology and Application. *Operations Research*, Nov/Dec, **55**(6), 1001–1021.

Krishna, A. (1992). The Normative Impact of Consumer Price Expectations for Multiple Brands on Consumer Purchase Behavior. *Marketing Science*, **11**(3), 266–286.

Lee, H. L. and S. Nahmias (1994). Single Product Single Location Models. In S. Graves, A. R. Kan, and P. Zipkin (Eds.), *Logistics of Production and Inventory*, Handbook in Operations Research and Management Science. North–Holland.

Little, J. D. and J. Shapiro (1980). A Theory for Pricing Nonfeatured Products in Supermarkets. *Journal of Business*, **53**(3), S199–S209.

Mahajan, S. and G. van Ryzin (2001). Stocking Retail Assortments Under Dynamic Consumer Substitution. *Operations Research*, (49), 334–351.

McBride, R. and F. S. Zufryden (1988). An Integer Programming Approach to Optimal Product Line Selection. *Marketing Science*, **7**, 126–140.

McCullogh, R. and P. E. Rossi (1994). An Exact Likelihood Analysis of the Multinomial Probit Model. *Journal of Econometrics*, **64**, 207–240.

Miller, C., S. A. Smith, S. H. McIntyre, and D. D. Achabal (2006). Optimizing Retail Assortments for Infrequently Purchased Products. Working Paper. Retail Management Institute, Santa Clara University.

Morgan, L. O., R. L. Daniels, and P. Kouvelis (2001). Marketing/Manufacturing Trade-Offs in Product Line Management. *IIE Transactions*, **33**, 949–962.

Pentico, D. (1974). The Assortment Problem with Probabilistic Demands. *Management Science*, **21**, 286–290.

Pentico, D. (1988). A Discrete Two-Dimensional Assortment Problem. *Operations Research*, **36**(2), 324–332.

Nielsen Marketing Research (1992). *Category Management: Positioning Your Organization to Win*. NTC Business Books, Lincolnwood, IL.

Sloot, L., D. Fok, and P. C. Verhoef (2006). The Short- and Long-Term Impact of an Assortment Reduction on Category Sales. *Journal of Marketing Research*, **43**, 536–48.

Smith, S. and N. Agrawal (2000). Management of Multi–Item Retail Inventory Systems with Demand Substitution. *Operations Research*, (48), 50–64.

Swait, J. and J. Louviere (1993). The Role of the Scale Factor in the Estimation and Comparision of Multinomial Logit Models. *Journal of Marketing Research*, **30**, 305–314.

Urban, G. L., P.L. Johnson, and J.R. Hauser (1984). Testing Competitive Market Structures. *Marketing Science*, **3**(2). Spring.

van Dijk, A., H. J. van Heerde, P. S. H. Leeflang, and D. R. Wittink (2004). Similarity Based Spatial Methods to Estimate Shelf-Space Elasticities. *Quantitative Marketing and Economics*, **2**, 257–277.

van Ryzin, G. and S. Mahajan (1999). On the relationship between inventory costs and variety benefits in retail assortments. *Management Science*, **45**, 1496–1509.

Chapter 8
OPTIMIZING RETAIL ASSORTMENTS FOR DIVERSE CUSTOMER PREFERENCES

Stephen A. Smith
Department of Operations and MIS, Leavey School of Business, Santa Clara University, Santa Clara, CA 95053, USA

1. INTRODUCTION

Assortment selection is one of the most important and difficult decisions that retailers face. Assortment are typically chosen subjectively, often before any sales have been observed for some candidate products. Compared to price or advertising decisions, assortment decisions are more difficult to adjust later on. For multi-featured items such as consumer electronics and durable goods, the large number of product options, together with limited display space and financial constraints all contribute to the complexity of this decision. Consumer preferences for the various product attributes may also be heterogeneous, which requires assessing tradeoffs between the products that appeal to diverse customer segments. Because of these complexities, intuitively chosen retail assortments seem likely to be suboptimal.

This paper develops an operational methodology for selecting optimal retail assortments based on an underlying multinomial logit (MNL) choice model for each customer's selection of product and retailer. A formulation is developed for optimizing the retailer's expected profit across customers with heterogeneous preferences. The formulation can also include a variety of additional merchandising constraints, such as display space, price point coverage or brand offerings.

Choice models have been successfully applied in consumer package goods to predict customers' response to assortment changes, based on observing repeat purchase behavior. The increased use of the Internet as a shopping guide for more complex, less frequently purchased products provides an opportunity to obtain detailed preference information for broader classes of merchandise. A commercial data base of consumer preferences for attributes and features of DVD players, which was obtained through interactive Internet sessions, is used to illustrate the methodology. Consumer surveys or past buying behavior of individuals might also be used as alternative sources for the preference information needed for this assortment optimization methodology.

N. Agrawal, S.A. Smith (eds.), *Retail Supply Chain Management*,
DOI: 10.1007/978-0-387-78902-6_8,

The methods in this paper provide a basis for several strategic retailer decisions including: (1) determining the optimal set of SKUs to offer and their estimated selling proportions; (2) how the retailer's relative market strength affects the contents of the optimal assortment; (3) how changing the contents of the assortment affects the probability that customers choose a given retailer and (4) how the customers' preference structure affects the optimal assortment and the corresponding expected profits. In analyzing the sample data set, it was found that accounting for preference heterogeneity and customers' use of consideration sets both had significant impacts the retailer's expected profits.

Literature Review

Kok, *et al.* (Assortment planning, Chapter 6) provide a comprehensive survey of recent papers in retail assortment planning, and thus this paper's literature review will focus on a few papers that are particularly relevant for the optimization model developed here. Several recent papers have developed models for assortment optimization based on a newsvendor type model for inventory cost. Van Ryzin and Mahajan (1999), Cachon and Kok (2007) and Cachon, et al. (2005) use a multinomial logit (MNL) model in which customers have homogeneous expected utilities. In Mahajan and van Ryzin (2001), customers are heterogeneous with regard to utility and their paper explicitly models the substituted demand that results from random stockouts of the retailer's inventory, but optimizing the assortment requires solution heuristics that are based on the set of possible inventory trajectories over the season. Guar and Honhon (2006) used a Lancaster type of model of substitution for products distributed along a single attribute dimension, and analyzed the impacts of static and dynamic substitution under this preference structure. Honhon, et al. (2006) consider assortment optimization with dynamic substitution for more general deterministic preference structures. This leads to a dynamic programming formulation, for which they develop solution heuristics. Smith and Agrawal (2000) used a probability of substitution matrix across products to optimize assortments in combination with an approximate newsvendor inventory model. Kok and Fisher (2007) develop a heuristic for optimizing the allocation of shelf facings and inventory levels for a supermarket based on a particular substitution structure that also considers stockouts. Chong, *et al.* (2001) developed a more general hierarchical market model for retail assortment planning for repeat purchase items, but due to the complexity of the resulting objective function, used a local improvement heuristic for optimization.

Only two of the above papers address the issue of retailer choice. Cachon, et al. (2005) investigates how three different consumer models for the value of additional search at alternative retailers can affect the optimal assortment. Cachon and Kok (2007) develop a more general category management model based on the retailer choice probabilities obtained from the nested logit model, but require mean utilities that are homogeneous across customers.

Product line optimization models have used mathematical programming formulations to solve a related problem. In this setting, a manufacturer decides which set of products to produce, where each potential product is viewed as a collection of adjustable product attributes. Chen and Hausman (2000) considered product line selection based on the MNL choice model, with homogenous customer preferences. Green and Kreiger (1985), McBride and Zufryden (1988), Dobson and Kalish (1988, 1993) and Kohli and Sukumar (1990) consider heterogeneous customer utilities, but assume deterministic product choices. Green and Krieger treat discrete price options as product attributes, as in this paper, while Dobson and Kalish treat product prices as separate decision variables. With the exception of Chen and Hausman, these mathematical programming formulations are computationally difficult to solve, in part because they assume strict utility maximization by customers. Some product line selection papers developed solution heuristics (Kohli and Sukumar 1990; Dobson and Kalish 1993) or suggested clustering of customer preferences to reduce the problem size (Green and Kreiger 1985) so that iterative search methods can be applied. These product line optimization methods do not model retailer choice, nor do they include inventory management costs.

Summary of Results

This paper provides an operational assortment optimization model that includes general heterogeneous consumer preferences as well as the customer's choice of retailer within the MNL framework. It is shown that the input parameters required for modeling product choice and retailer choice can be estimated separately, which facilitates their use in an operational model for assortment optimization. Assuming homogeneous mean utilities, Van Ryzin and Mahajan (1999) showed that the optimal assortments form nested sets as the assortment size increases. For heterogeneous mean utilities, this paper shows that this property no longer holds, but that nested optimal assortment sets do occur for two limiting cases: (1) a monopoly retailer and (2) perfect competition among retailers. An optimization formulation is developed, which can include linear retailer constraints on the contents of the assortment, such as brand coverage and display space limitations. Finally, a commercial data base of preferences for DVD players is analyzed to illustrate the sensitivity of the expected profit and optimal assortment to the customer preference structure. The results for this data set illustrate the importance of including preference heterogeneity and customers' use of consideration sets in assortment optimization, as well as the sensitivity of the retailer's profit to assortment size.

2. MODEL DESCRIPTION

This paper focuses on the assortment decision for a particular retailer r, whose objective is to maximize the expected profit over a fixed time period, e.g., the Fall season. It is assumed that other retailers do not react competitively to this decision. The retailer's assortment is defined by a binary vector $y = y_1, y_2, \ldots, y_n$, where $y_j = 1$ if the retailer's assortment includes product j and 0 otherwise. Then let

$$D_j(y) = \text{the random demand for product } j,$$

which depends on y as well as other factors that affect demand. We now develop a choice model that determines the probability distribution for $D_j(y)$.

2.1 Modeling the Consumer's Purchase Decision

First, suppose that customers are classified according to n distinct customer types indexed by $i = 1, \ldots, n$. It is assumed that customers of the same type assign the same expected values to various choice alternatives, but their actual purchase decisions also reflect random variations.

Actual purchases are the result of a sequential process that can be diagramed as follows:

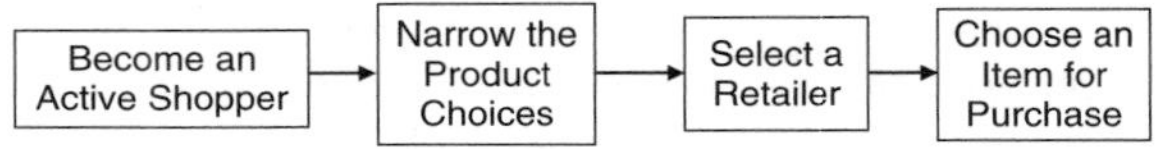

The choice decisions in each of these steps can be described in terms of the iPACE model for retail shopping decisions that has been developed in the marketing literature, where iPACE stands for information, Price, Assortment, Convenience and Entertainment, (see e.g., Hanson and Kalyanam 2006, Chapter 13.) By becoming an active shopper, the customer is sufficiently interested in the product category to gather information. Using a variety of sources, which may include both Internet research and store visits, customers assesses their utilities for the available products and the relative values of purchasing from the alternative retailers. This process allows the customer to narrow the set of choices to a "consideration set" of products. The customer selects a retailer based on the retailer's assortment, as well as the assessed convenience and entertainment values of shopping at that retailer. Finally, the customer makes a product selection from the choice set, which is defined as the intersection of the consideration set and the chosen retailer's assortment. Although this description is sequential, these decisions do not necessarily need to be made in any specific order. For example, the customer might choose the most preferred product first, and then select the retailer from which to purchase. The key assumption is that it is the combined utility of the retailer and the chosen product that jointly determine the decision. This paper assumes that these decisions are made normatively by customers, based on maximizing expected utility.

From the perspective of a particular retailer r, the customer may choose the "no purchase" option for two reasons: (1) no product in the consideration set has positive net value, i.e., the choice set is empty or (2) the combined value of shopping and purchasing from this particular retailer's assortment either does not exceed the product's price, or is less than the combined value obtained from another retailer.

The mathematical models for each of these steps can be summarized as follows. The assortment decision is made for a fixed period of time, e.g., one season, and the time dependent parameters correspond to the length of this season. A random number N_i of customers of type i will become "active shoppers," i.e., they will gather information and make a purchase decision this season for this product category. We assume that N_i is a Poisson random variable with rate parameter λ_i. For the N_i shoppers, define

$$q_{ij}(y) = P\{\text{type } i \text{ chooses product } j \text{ from this retailer} \mid \text{assortment} y\}.$$

This implies that $D_j(y)$, the random demand for product j defined previously, has a Poisson distribution with mean

$$\mu_j(y) = \sum_i \lambda_i q_{ij}(y). \tag{1}$$

The remaining customer decisions, which determine $q_{ij}(y)$, are based on the following utility model. The underlying choice model is a multinomial logit (MNL) in which customer i's combined utility for product j and retailer r is a random variable of the form

$$U_{ij}^r = U_{ij} + V_{ir} + \varepsilon_{ijr}, \tag{2}$$

where ϵ_{ijr} = Gumbel distributed error terms with mean 0 and scale parameter $\xi_{i,}$,

U_{ij} = the expected utility obtained from purchasing product j,
V_{ir} = the additional utility obtained from purchasing from retailer r.

For this paper's analysis, the product price is included in U_{ij} as a fixed attribute, rather than a decision variable. For many retailers, this is justified based on operational practice. At the individual product level, tactical pricing decisions such as temporary markdowns are typically made later by the retailer during the selling season, as part of promotional and advertising activities. Strategic pricing decisions, such as how to price relative to competitors, are typically made less frequently and at a higher level than one product category. For assortment planning purposes, the product price is therefore the estimated average price for the season. While a combined model that could simultaneously optimize product prices and the retail assortment is conceptually superior to separate decision models, it could not feasibly include all the other aspects of customers' purchasing decisions that are analyzed here.

Additive MNL models of the form (2) are frequently used for two dimensional choice decisions. (See, e.g., Ben Akiva and Lerman 1985 for further discussion.) In the context of this application, the error terms ϵ_{ijr} can capture both the customer's imprecise knowledge of his or her own utilities, as well as the retailer's imperfect knowledge of customers' utilities. It is common practice to rescale the utilities for each customer i so that $\xi_i = 1$. This is possible because dividing all utilities with subscript i by the same scalar ξ_i does not change which utility is the maximum. That is, probability statements about the maximum utility for customer i are not affected by this rescaling.

Narrowing the Product Choices

Narrowing the product choices is a "prescreening" step that does not change the fundamental structure of the underlying logit model. When there are many product alternatives to consider, marketing researchers have found that customers typically use some criteria to narrow their choices to a "consideration set" of products, which are then investigated in more detail. (See, e.g., Roberts and Lattin 1991 ; Andrews and Srinivasan 1995, Siddarth, *et al.* 1995). In a normative framework, customer i would form a consideration set by eliminating all products with expected utility less than some threshold u_i, where the threshold is based on his or her cost of considering additional alternatives. Thus we define

u_i = customer i's minimum acceptable expected utility for considering a product,
$X_{ij} = 1$ if $U_{ij} \geq u_i$ and 0 otherwise, for all i, j.

Consideration sets can have a significant impact on the assortment optimization, as the numerical analysis in Section 3 illustrates.

Determining $q_{ij}(y)$

The definition of conditional probability implies that

$$\begin{aligned} q_{ij}(y) &= \mathrm{P}\{\text{ customer } i \text{ purchases product } j \text{ from retailer } r \mid y\} \\ &= P\{\text{customer } i \text{ purchases product } j \mid \text{purchases from retailer } r,\ y\} \\ &\quad * P\{\text{ customer } i \text{ purchases from retailer } r \mid y\} \end{aligned} \tag{3}$$

This equation does not necessarily imply that the customer chooses the retailer first, but this decomposition allows a separable estimation of the required model parameters, as will be discussed later.

Given that customer i selects retailer r's assortment for a purchase, his or her choice set is defined as the intersection of the consideration set and retailer r's assortment, i.e,

$$S_{ri} = \{j \mid y_j X_{ij} = 1\}, \quad \text{for all } i.$$

Given any choice set S_{ri}, the probability of selecting item $j \in S_{ri}$ is the standard MNL probability, which in this case is

$$P\{\text{customer } i \text{ purchases product } j \mid \text{chooses retailer } r,\ y\} = \frac{e^{U_{ij}}}{\sum_{k \in S_{ri}} e^{U_{ik}}}. \tag{4}$$

Ben Akiva and Lerman (1985, p. 282) show that the maximum utility that customer i obtains from the choice set S_{ri} has a Gumbel distribution, with mean

$$V_{ir}^{*} = \ln\left(\sum_{j\in S_{ri}} e^{U_{ij}}\right),$$

and the same scale parameter as the individual utilities. Thus, the total utility of purchasing from retailer r's assortment is Gumbel distributed with mean $v_{ir} = V_{ir} + V_{ir}{}^{*}$. The analogous result holds for all other retailers' assortments, which we index by ρ. Therefore, the maximum utility that customer i could obtain from shopping at other retailers also has a Gumbel distribution with mean

$$v_{io} = \ln\left(\sum_{\rho\neq r} e^{V_{i\rho}+V_{i\rho}^{*}}\right),$$

and the same scale parameter as the individual utilities. This allows the retailer choice probability to be written as a binary logit probability

$$P\{\text{customer } i \text{ selects retailer } r|y\} = \frac{e^{v_{ir}}}{e^{v_{ir}} + e^{v_{io}}} = \frac{\sum\limits_{j\in S_{ri}} e^{U_{ij}}}{e^{a_{ir}} + \sum\limits_{j\in S_{ri}} e^{U_{ij}}} \tag{5}$$

with $a_{ir} = v_{io} - V_{ir}$. ;

The second fraction results if we multiply top and bottom by $\exp\{-V_{ir}\}$. From this point onward, we focus on the particular retailer r and simply write a_i for a_{ir}.

Multiplying the two probabilities in (4) and (5) using the assortment y for retailer r and the X_{ij} for customer i to define the choice set S_{ri}, we obtain the formula

$$q_{ij}(y) = \frac{y_j X_{ij} e^{U_{ij}}}{e^{a_i} + \sum\limits_{k} y_k X_{ik} e^{U_{ik}}}, \tag{6}$$

after cancelling the term $\sum_{j\in S_{ri}} e^{U_{ij}}$. A key result in (6) is that a_i is a constant that is independent of retailer r's assortment decision y.

The size of a_i indicates the relative strength of retailer r's competitors for customer type i. The value of a_i can be obtained in various ways. One possible method is to assume that customer i knows the contents of all the retailers' assortments and chooses the best retailer by maximizing the total utility as described above. Alternatively, the customer might simply decide whether to continue shopping at other retailers based on an estimated value a_i which corresponds to the estimated maximum utility obtained from other retailer's products plus the difference in the value between buying from an alternative retailer versus retailer r. For assortment optimization using (6), retailer r does not need to know which behavioral model applies to customer i, since a_i is simply a parameter to be estimated, as discussed below.

Kahn and Lehmann (1991) and others have suggested adding terms to V_{ir} to capture the additional customer value associated with properties of the assortment that increase its "breadth," such as the total number of products or the number of brands offered. The structure of the optimization model in this paper does not allow these additional variables to be included in the retailer's objective function. But features such as the total number of products or the number of brands in the assortment can be included as constraints for the assortment optimization model, with their corresponding values being added as constant terms to V_{ir}. This allows a sensitivity analysis to be done with respect to these assortment constraints.

Estimation and Empirical Testing

The form of the choice probabilities in (4) and (5) allows the utilities U_{ij} and the parameters a_i to be estimated separately. For frequently purchased items such as consumer package goods, the utilities can be estimated from the observed market shares of products in any given assortment. Ben Akiva and Lerman (1985, p. 188) obtain maximum likelihood estimates for the utilities of the form $U_j = \boldsymbol{\beta}^{\mathrm{T}} \boldsymbol{x}_j$, where $\boldsymbol{x}_j$ is a vector of attribute settings for product j, and $\boldsymbol{\beta}$ is a vector of parameters to be estimated. In the data base used for the calculations in this paper, the utility estimates were obtained through conjoint analysis of customer response data.

An estimate for a_i can be obtained using (5) from the observed fraction f_i of customers of type i who choose retailer r for *any given* assortment. Since f_i, which equals the lefthand side of (5) is known, we can solve for the corresponding a_i as follows

$$a_i = \left(\frac{1}{f_i} - 1\right) \sum_{j \in S_{ri}} e^{U_{ij}}.$$

This formula requires utility estimates for each product, which can be obtained from (4) as discussed previously. Thus, equations (4) and (5) allow the $\{U_{ij}\}$ and $\{a_i\}$ to be estimated separately, possibly from different existing assortments.

Purchasing behavior for consumer package goods based on multi-stage logit models has been studied empirically for a variety of model forms. For example, Cintagunta (1993) provides a summary of articles that include empirical studies of three stages of consumer purchase decision making: (a) whether or not to purchase from this retailer (b) item choice from a retailer and (c) purchase quantity. See also Roberts and Latin (1997) for a literature review. In forecasting demand for consumer package goods, "purchase incidence," which is defined as the probability that the customer makes a shopping trip to a given retailer that results in a purchase from the category, plays a role that is similar to retailer choice in this paper. [See, e.g., Bucklin and Lattin 1991 for a discussion of using the binary logit model for purchase incidence.]

Elasticity Comparisons

Formula (6) shows that adding another product to the assortment *increases* the probability that customer i purchases from this retailer, but *decreases* the probability that each other product in the assortment is selected. The magnitudes of these affects depend on a_i, as shown below.

Let $Q_i(y)$ = P{customer i purchases from this retailer}, where

$$Q_i(y) = \sum_j q_{ij}(y) = \frac{P_i(y)}{e^{a_i} + P_i(y)}, \text{with } P_i(y) = \sum_j y_j X_{ij} e^{U_{ij}}.$$

Interpreting partial derivatives as changes in y_k from 0 to 1, we can define the two elasticities

$$\frac{1}{Q_i(y)} \frac{\partial Q_i(y)}{\partial y_k} = \frac{e^{a_i} X_{ik} e^{U_{ik}}}{P_i(y)[e^{a_i} + P_i(y)]} \quad \text{and} \quad \frac{1}{q_{ij}(y)} \frac{\partial q_{ij}(y)}{\partial y_k} = -\frac{X_{ij} e^{U_{ij}}}{[e^{a_i} + P_i(y)]^2}.$$

The first and second elasticities show, respectively, that:

(1) the percentage increase in total sales to customer i from adding product k is *greater* when a_i is *larger*.
(2) the percentage of cannibalization of product j's sales due to adding product k is *smaller* when as a_i is *larger*.

Both of these results imply that including additional products in the assortment is more advantageous to the retailer when competition is strong. However, if these additional products also have high fixed costs, this may not be the most profitable approach for the retailer.

2.2 Retailer's assortment optimization

The profit function $\Pi_j(D_j(y))$ for each product j is based on a newsvendor type model. The expected profit $\Pi(y)$ for the planning period as a function of y can be written as the sum of the expected profits for the various products

$$\Pi(y) = \sum_j E[\Pi_j(D_j(y))].$$

Even though the random variables $\Pi_j(D_j(y))$ are not independent, their expectations are still additive. A fixed cost F_j of stocking product j can also be included. A more general form results if one assumes that there are nonlinear cost interactions between products.

The expected profit for product j over a fixed time period as a function of the assortment y can be written as

$$E[\Pi_j(D_j(y))] = \max_{S_j}\{m_j\mu_j(y) - c_{uj}E[D_j(y) - s_j]^+ - c_{oj}E[s_j - D_j(y)]^+ - y_jF_j\}, (7)$$

where $E[x]^+$ denotes the expected value of $\max\{0, x\}$,

s_j = the base stock level for product j for the time period
m_j = unit profit margin for product j
$\mu_j(y)$= expected demand during the time period = $E[D_j(y)]$
c_{uj} = "understock" cost per unit
c_{oj} = "overstock" cost per unit
F_j = fixed cost of stocking product j.

In terms of the usual financial input quantities, m_j = selling price – unit cost, c_{uj} = shortage loss – unit cost and c_{oj} = unit cost – salvage value. From (6), we see that $y_j = 0$ implies $D_j(y) = 0$ with probability 1, which implies that the expected profit is 0. That is, no shortage cost c_{uj} results from not including a given item in y, but there is a loss of expected utility for the retailer's assortment, which increases the likelihood that the customer will choose another retailer. This is because when a customer's most preferred item is missing, the customer either substitutes another item from this retailer's assortment or chooses another retailer. The demand that results from substitutions for items not in the retailer's assortment is captured in $\mu_j(y)$, while substitutions from stockouts are ignored, as discussed below. The probability of choosing another retailer is included in the "no purchase" option.

Using the newsvendor critical ratio formula, the optimal base stock level s_j^* satisfies

$$s_j^* = \arg\min_S\left\{s|P\{D_j(y) \leq s\} \geq \alpha_j = \frac{c_{uj}}{c_{uj} + c_{oj}}\right\}.$$

The overstock cost c_{oj} above can have a variety of interpretations. For continuing products that will be offered in subsequent seasons, it is the unit holding cost for the season, while for "seasonal" products, it is the unit cost minus the expected salvage value per unit for any excess inventory at the end of the season.

There are various fixed costs F_j that can be associated with stocking items in a product category. For larger items such as furniture, it is common to display one unit in the store and hold additional inventory elsewhere, for example. In this case, F_j would include the required floor space for display. For smaller items, there may be a shelf facing with one item viewable, and the remaining items stored behind it. In both these cases, F_j would include the fixed cost of the required display space in the store when the item is in the assortment.

Incremental Demand Arising from Substitution

Kok, *et al.* (Assortment planning, Chapter 6) define two kinds of substitution-based demand: (1) assortment based substitution in which a customer switches to another product when a more preferred product is not carried in the assortment and (2) stockout-based substitution in which the customer substitutes another product if a more preferred alternative is in the assortment but out of stock. This paper captures assortment-based substitution through the MNL choice model discussed previously, but ignores stockout-based substitution. Some recent papers have modeled stockout-based substitutions, but this generally leads to very complex optimizations, and thus solution heuristics are required. Mahajan and van Ryzin (2001) and Guar and Honhon (2006) and Honhon, *et al.* (2006) assume that customers maximize utility over the items that are currently available, i.e., they treat the retailer's assortment as dynamic. These approaches are quite general, but require heuristic solutions for most customer preference structures. The other assortment optimization models discussed previously have either not treated this stockout-based substitution or have bounded its effects.

This paper assumes that the customer chooses the retailer based on the complete assortment y, and that product demands which encounter stockouts of products in the retailer's assortment become lost sales. That is, the demand arising from stockout-based substitutions is ignored. Smith and Agrawal (2000) argue using bounds, that the absolute percentage error in expected demand that results from ignoring demand from stockout based substitutions is bounded by $(1 - \alpha)(1 - L)$, where α is the target service level and L is the probability that the customer is unwilling to substitute. The actual percentage error may be much lower than this bound. For example, for the normal distribution with a service level $\alpha = 0.9$, approximately 96% of demand will be served, and with $\alpha = 0.95$ approximately 98% of demand will be served before a stockout occurs. The remaining unserved demand is important only to the extent that customers are willing to switch to another product from the same retailer. Given that there are alternative retailer choices for most items, customers who choose another retailer instead of substituting a different product will be correctly captured by the lost sales assumption. Thus, for retail products that have fairly high service levels such as 0.9 or 0.95, it seems acceptable to assume that this component of the demand results in lost sales.

Two Variants of the Objective Function

Products that may have purchase quantities larger than one can be handled in a variety of ways. One method is to use a compound Poisson distribution for demand, where customers arrive according to a Poisson process and then select their purchase quantities randomly. For example, Poisson arrivals with a purchase quantity selected from a logarithmic distribution result in a negative binomial distribution for total demand during any fixed period. Smith and Agrawal (2000) used the negative binomial distribution and found that a linear approximation to the newsvendor objective function also worked well in that case. Other papers on assortment optimization (Van Ryzin and Mahajan 1999; Mahjan and van Ryzin 2001, Guar and Honhon 2006) have used a normal approximation for demand to obtain a newsvendor expected profit function.

When there are time based holding costs, it may be advantageous for retailers to restock more frequently than once per season. This feature can be added to the newsvendor model (7), provided that the assortment does not change in midseason. If there is an additional cost h = unit holding cost for one restocking period, a cost term of the form $0.5h\left[s + |s - D_j(y)|^+\right]$ is subtracted from the

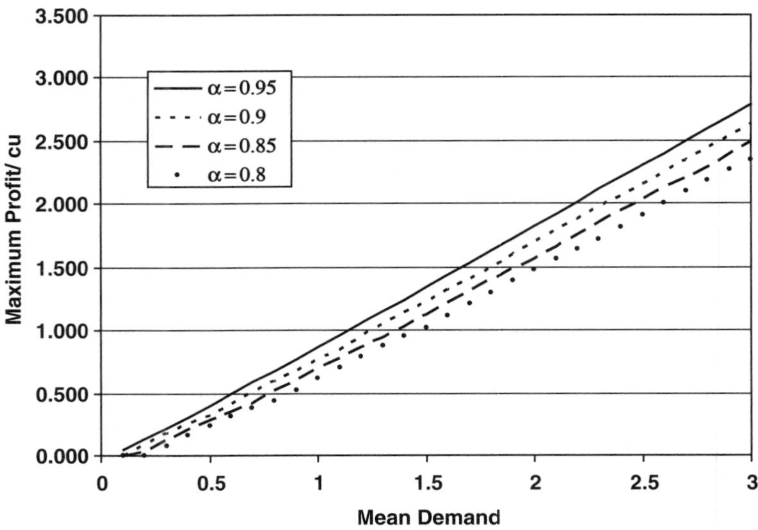

Figure 8-1. Maximum Profit vs. Mean Demand =cu, co=(1/α-1)

objective function. The critical ratio stock level formula still holds, where c_{oj} is replaced by $c_{oj} + h$ and c_{uj} is replaced by $c_{uj} - h/2$. The costs c_{oj} and c_{uj} may also be allowed to vary by time period.

A Linear Approximation for the Objective Function

It can be verified by numerical calculation that for common ratios of profit margin to overstock and understock costs, the newsvendor expected profit function (7) is approximately linear in the expected demand $\mu_j(y)$ for the Poisson distribution. That is, when the various costs are held fixed and expected demand increases, the target service level remains constant and the safety stock increases in such a way that the sum of the terms in (7) increases approximately linearly in the mean.

For the Poisson demand distribution, this approximation is illustrated for a range of parameter ratios in Figure 8-1. To simplify the graph, all $F_j = 0$ and all profits have been divided by c_u. That is, when all cost parameters are expressed as multiples of c_u, the graphs can be expressed as (expected profit)/c_u, which implies that the only required variables are the service level α and the mean demand. Using linear regression, the R^2 values for all the linear fits to the points in this figure are at least 0.998.

The linear approximation implies that are constants π_j and b_j derived from the slope and intercept of the regression line for product j such that the expected profit can be approximated as follows

$$E[\Pi_j(D_j(y))] \approx \pi_j\mu_j(y) - y_j(b_j + F_j).$$

In general, it appears that the quality of the fit improves as the mean increases and as the service level α increases. When $F_j = 0$, Figure 8-1 shows that the b_j values are positive. This is because the expected profit becomes negative for low enough mean demand, but in these cases $y_j = 0$ will be optimal.

Using $C_j = b_j + F_j$ to combine the constants b_j with the fixed costs F_j, and recalling that $\mu_j(y) = \sum_i \lambda_i q_{ij}(y)$, the retailer's approximate objective function can therefore be written as

$$\Pi*(y) = \sum_j \pi_j \sum_i \lambda_i q_{ij}(y) - \sum_j C_j y_j. \quad (8)$$

This objective can be maximized with respect to y, subject to various constraints such as display space or brand representation in the assortment.

2.3 Properties of the optimal assortment

When customers' utilities are Gumbel distributed with homogeneous means, Van Ryzin and Mahajan (1999) showed that the optimal assortments form nested sets. This case corresponds to $U_{ij} = U_j$ for all i in this paper's notation. That is, if S^K is the best assortment of size K, then $S^K \subseteq S^{K+1}$ for all K. With general means U_{ij}, however, this property no longer holds, as demonstrated by the following counterexample. Let $\lambda_i = 1$ and $\exp(a_i) = 10$ for all i and consider the following matrix of $\exp(U_{ij})$ values

		Products		
		1	*2*	*3*
	1	1000	2	1000
Customers	*2*	1000	1000	2
	3	2	1000	2
	4	2	2	1000

Let the unit profits for the three products be 10, 9, 9 respectively. Clearly, the best single product is Product 1. But it can be seen from the table of expected profits below that the best two products are 2 and 3.

	y		Expected Profit
0	1	1	35.6
1	1	0	31.0
1	0	1	31.0

Thus, although Product 1 is the best single product, it is not part of the best set of two products.

Nested set properties do hold for two limiting cases, however. First, let us consider the case in which $a_i = a$ for all i and a is very large. Then rewrite $\Pi^*(y)$ as

$$\Pi*(y) = e^{-a}\sum_{i,j}\lambda_i\pi_j\left(\frac{y_jX_{ij}e^{U_{ij}}}{1+e^{-a}P_i(y)}\right) - \sum_j C_jy_j. \tag{9}$$

As a becomes sufficiently large, the term in parenthesis approaches $y_{ij}X_{ij}e^{Uij}$. Thus, if the products are ordered so that

$$\pi_1\sum_i\lambda_iX_{i1}e^{U_{i1}} - C_1 \geq \pi_2\sum_i\lambda_iX_{i2}e^{U_{i2}} - C_2 \geq ..., \tag{10}$$

then the optimal assortments will be $\{1\}$, $\{1, 2\}$, ... for a sufficiently large. This implies that there is an optimal product ordering for the assortment, if the retailer's competition is sufficiently strong, even when consumer preferences are heterogeneous. In microeconomic terms, this might be called the "perfectly competitive" case.

A second special case arises when $\exp(a_i)$ approaches 0 for all i. In this case, the retailer is effectively a monopolist, since any consumer who purchases will choose this retailer. For the case in which $X_{ij} = 1$ for all i, j, every product in the retailer's assortment is in every customer i's choice set. Thus, the optimal strategy for a monopoly retailer is to rank products in order of profitability, based on ranking the expected profits as follows

$$\pi_1\sum_i\lambda_i - C_1 \geq \pi_2\sum_i\lambda_i - C_2 \geq .... . \tag{11}$$

But if some $X_{ij} = 0$, this property may not hold, because some customers may not consider the retailer's most profitable product and thus would not choose it. Thus, with considerations sets, there may be no specific nested set property when $\exp(a_i)$ approaches 0, for all i.

Sensitivity to the Retailer's Market Strength

To illustrate the difference in the two rankings (10) and (11), let us consider an example with 5 customer types and 20 products, where the utilities U_{ij} were

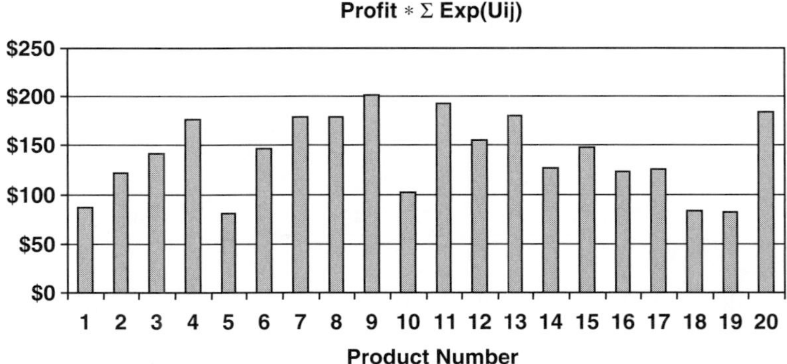

Figure 8-2. Profitability Calculations for 20 Randomly Generated Products

generated by taking samples from a uniform distribution on [0, 2]. Let all $X_{ij} = 1$ and all $\lambda_i = 1$ in this example. The 20 products are assigned gradually decreasing unit profits π_j : \$10.00, \$9.90, \$9.80, ..., \$8.10 and fixed costs $C_j = 0$ for all j. Thus, for the case in which the retailer's competitive position is very strong, the products' ranking is based on (11), which implies that the products would be ranked in order of the unit profits, 1, 2, 3, ... Therefore, for a very dominant retailer, the optimal assortment of K products is $\{1, 2, \ldots, K\}$.

On the other hand, for a retailer in a weak competitive position, the product rankings are based on the rankings in (10). The calculated results for (10) are illustrated in Figure 8-2.

The height of the bars in Figure 8-2 shows that the expected values for this case are quire different from those that would produce the ranking of 1, 2, 3, ... determined by (11). For example, the top 5 products based on ranking the values in Figure 8-2 are: {9, 11, 20, 13, 7}.

2.4 Solving the Optimization Problem

If the total number of products is small, optimal assortments can be obtained by an exhaustive search, but this becomes more difficult for larger numbers of products. Based on the structure of the problem, certain products may be eliminated from the assortment *a priori*, which reduces the problem size. Substituting the definition (6) of $q_{ij}(y)$ into (8), the objective function can be written as

$$\text{Max } \Pi^*(y) = \sum_{j\geq 1} y_j\{\pi_j r_j(y) - C_j \quad \}\text{with } y_j = 0, 1 \text{ for all } j \geq 1,$$

$$\text{where } r_j(y) = \sum_i \lambda_i \left(\frac{y_j X_{ij} e^{U_{ij}}}{e^{a_i} + \sum_{k\geq 1} y_k X_{ij} e^{U_{ij}}} \right) \tag{12}$$

For any y such that $y_k = 0$, define

$$\Delta_k r_j(y) = r_j(y + e_k) - r_j(y), \quad \text{where } e_k = \text{ the unit vector with k}^{\text{th}} \text{ element} = 1.$$

It can be verified that

$$\text{If } y_k = 0, \text{then } \Delta_k y_j[r_j(y) - C_j] \leq 0 \text{ for all } j \neq k.$$

This has the implication that if $\pi_k r_k(e_k) - C_k \leq 0$ for any k, then $y_k = 0$ must hold. That is, $y_k = 1$ cannot be optimal since y_k could be changed to 0 and all terms in the objective function will improve or stay the same. This observation can used to eliminate some products before searching on y. However, it appears that an exhaustive search over the remaining 0,1 variables is required to optimize the assortment.

Retailer imposed constraints, such as the number of products must be at least K, or at least one product of Brand B must be included, can be added as linear constraints on y. For example, if the assortment must include at least one product of Brand B, define the logical inputs

$$I_{Bj} = 1 \text{ if product } j \text{ is of brand } B\text{ , and 0 otherwise.}$$

Then the brand constraint is of the form

$$\sum_j y_j I_{Bj} \geq 1 \text{ for brand } B.$$

We can also include a display space constraint of the form

$$\sum_j d_j y_j \leq D, \text{ where}$$

d_j = the space required for product j
D = total available display space for this category.

These additional constraints also reduce the number of alternatives to be searched.

3. ILLUSTRATIVE APPLICATION FOR A DVD PLAYER DATA BASE

This section illustrates the application of the optimization model to a set of customer utilities derived from conjoint analysis of Internet responses. The preference data were collected through the ***Active Decisions'*** *Active Buyers Guide Sales Assistant* website. [See www.activedecisions.com. This company has recently been acquired by Knova Systems, who plan to offer conjoint utility encoding as a consulting service.] Visitors to *activebuyersguide.com, yahoo.com* and other e-commerce sites completed an interactive survey to elicit their preference tradeoffs for product attributes. These are defined *independently* of the specific products in the market. Product utilities were then derived from additive conjoint analysis of 2213 customer responses for the DVD player category. That is, each customer's net utility for a particular product was calculated as the sum of his or her "part worths" for the attributes of that product, including the price. (See Green and Srinivasan 1978; Wittink and Cattin 1989 for discussions of conjoint analysis. The conjoint analysis of this data was performed by *Active Decisions* and the author is indebted to them for sharing their results.)

The utility values were then normalized by dividing each utility U_{ij} by customer i's maximum utility to obtain

$$S_{ij} = \frac{U_{ij}}{\max\limits_{k \in \Omega} U_{ik}} \text{ for all } i, j.$$

After this normalization, it was assumed that $\xi_i = 1$ for all i. Consideration sets based on utility thresholds can then be defined as a fixed percentage θ of each customer's maximum utility over all products. That is,

$$u_i = \theta \max_{j \in \Omega} U_{ij}, \text{ where } \Omega = \text{the set of all products in the market.}$$

Thus, $X_{ij} = 1$ if and only if $S_{ij} > \theta$.

Assortment optimization for this example was done for the case of "large" a_i, i.e., the retailer's competitive position is very weak. Thus, the optimal assortments will form nested sets according to (10), as discussed previously. Because of the highly competitive nature of the DVD player market and because this retailer was not a dominant player in consumer electronics, this assumption seemed appropriate. However, the database had no data available on retailer preference so this assumption could not be tested.

3.1 Comparing the Model's Predictions to a Retailer's Sales Data

In order to test the predictive accuracy of utilities in the data base and the MNL choice models, we obtained data on the observed selling proportions for an assortment of 30 DVD payers offered by a major retail chain. These selling proportions were compared to those predicted by the MNL choice model fitted to the product attribute utilities in the DVD Player data base. The actual selling proportions of the products ranged from 0.2% to 16%. [There were 117 different DVD player products at the time the preference data set was collected, and the retailer data was obtained for the same time period.] A variety of θ values were tested to obtain the correlations and the R-square values shown below in the table below.

Actual vs. Predicted Selling Proportions for 30 Products

θ	Correlation	R-square
0	69%	47%
0.9	78%	60%
0.95	79%	62%
1.0	72%	53%

This table indicates that the fit is reasonably good for all θ values, but the accuracy improves somewhat when customers are assumed to use moderately restrictive consideration sets. Further investigation also revealed that most of the error in these predictions resulted from over-predictions for three products, which the retailer reported were unavailable in some stores. This test supports the use of the utilities in the data base, and also suggests a fairly high θ value such as $\theta = 0.9$ or 0.95 for this data set.

3.2 Comparing the expected revenue of the retailer's assortment vs. the optimal assortment

The objective function in (12) was then applied to the set of 117 DVD player products available at that point in time to determine the optimal assortment of 30 products. For the optimization, it was assumed that each of the 2213 respondents to the online survey represents a customer segment of equal size, *i.e.*, the λ_i were assumed to be equal for all i. The fixed costs C_j were set to zero and the product prices from the DVD Player data base were used to compute the expected revenue from a given assortment. Since the revenue comparisons will be done on a percentage basis, it is not necessary to know the actual number of buyers in a segment. For percentage calculations with $\lambda_i = \lambda$ for all i, the λ will cancel out of the profit comparisons. Therefore, for the case of "very large" a_i, the objective function in (12) can be maximized by substituting a linear objective function that is similar to the ranking calculation in (10),

$$\text{Max } \Pi_0(y) = \sum_{j\geq 1} y_j \pi_j X_{ij} e^{U_{ij}},$$
$$\text{subject to } y_j = 0, 1 \text{ for all } j \geq 1 \text{ and } \sum_{j=1}^{n} y_j = 30. \tag{13}$$

The optimal assortments were then determined for various values of $\theta = 0.9$, 0.95 and 1.0, which are captured by changes in the X_{ij}. The table below compares the percentage improvements achieved by the optimal assortment over the retailer's current assortment, for the various θ choices.

θ	Revenue Improvement	Common Products
0.9	169%	11 (37%)
0.95	185%	9 (30%)
1.0	208%	7 (23%)

The revenue improvements in this table are, of course, optimistic because they assume that each customer i's buying behavior exactly matches the MNL model. However, even recognizing this, it appears that using the MNL-based optimal assortment with consideration sets has substantial potential to improve this retailer's revenues.

3.3 The impact of customer preference structure

The analysis above is based on both the use of consideration sets and heterogeneous customer market segments. To test the impact of these structural assumptions, we focus on three sensitivity questions:

1. *What is the impact of including customer preference heterogeneity in determining optimal assortments?*
2. *How does customers' use of consideration sets impact the optimal assortments and expected profits?*
3. *How does the expected profit increase with assortment size, i.e., how does the optimal assortment size depend on the fixed costs of offering additional products?*

Customer Heterogeneity

To examine the role of customer preference heterogeneity in developing the optimal assortment, optimal assortments for homogeneous preferences were generated by replacing the S_{ij} with "average" values S_j, which equal the average S_{ij} value over all customer types i. The expected profits for these optimal assortments were then compared to the profits for the optimal assortment with heterogeneous preferences S_{ij} in Figure 8-3.

The potential revenues of the two optimal assortments converge when essentially all positive utility products are carried by the retailer. However, for assortment sizes 10 – 30 that are relevant to most retailers, the optimal assortments for heterogeneous preferences result in profits almost twice as large. Examining the contents of the assortments produced by the two methods found only about 5% common items in the assortments of sizes 5 to 30. Thus, for this data set, ignoring customer heterogeneity has significant financial consequences and major impacts on the optimal assortment.

The Use of Consideration Sets

To analyze the impact of consideration sets, the optimal assortments for $\theta = 0, 0.9$ and 1.0 are compared in Figure 8-4, where $\theta = 0$ is interpreted as "no consideration sets." Figure 8-4 shows that when customers use consideration sets and the retailer uses this information correctly in developing the optimal

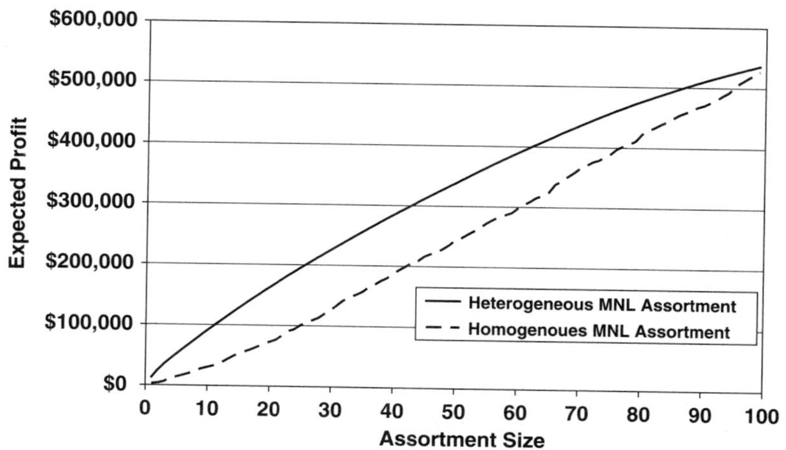

Figure 8-3. Including Preferences Heterogcneity in assortment Optimization

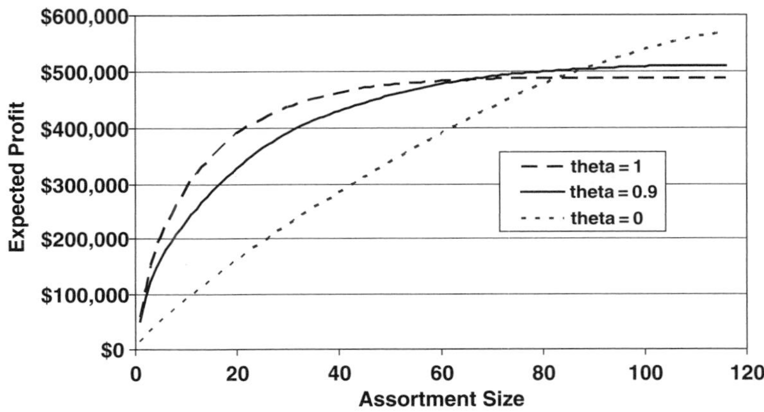

Figure 8-4. The Effect of Consideration Sets on Expected Profit

assortments, a substantial increase in expected profit results for typical assortment sizes. For assortments in the 5 to 10 item range, the $\theta = 0.9$ or 1.0 cases yielded two to three times the profit of the optimal assortment without consideration sets.

Consideration sets allow the retailer to use a more focused assortment. When customers use consideration sets, the retailer can achieve 80% to 90% of the maximum possible profit with assortment size of only about 30 items, while these assortment sizes can achieve only about 50% of the maximum without consideration sets. For $\theta = 1$, all customers can receive their first choice product with an assortment size of 66, but for $\theta = 0$ additional products always increase expected sales.

The shape of the curves in Figure 8-4 also determines the impact of the fixed costs C_j on the optimal assortment size. For an assortment of size 30, for example, the slopes of the lines are approximately, $3500, $5000 and $6000, respectively, which correspond to the marginal benefits of an additional product. [These dollar figures correspond to one purchase decision by each of the 2213 active shoppers in the category. This level of sales would correspond to an aggregate across multiple stores.] Thus, consideration sets allow high fixed costs to be justified for small numbers of products, but tend to limit the optimal assortment size as the number of products increases.

It was assumed in Figure 8-4 that the optimal assortment was determined for the correct θ value in each case. But since customers' behavior with regard to consideration sets may be difficult to predict, it is interesting to consider the impact of incorrect assumptions about consideration sets. This calculation is illustrated in Figure 8-5, where the optimal assortment for $\theta = 0$ was used when the correct value was $\theta = 0.9$, and vice versa.

This shows that if customers form consideration sets based on $\theta = 0.9$, the optimal assortment for $\theta = 0$ results in a reduction in expected profit of 12% to 50% for assortments in the range of 10 to 30. On the other hand, if customer do

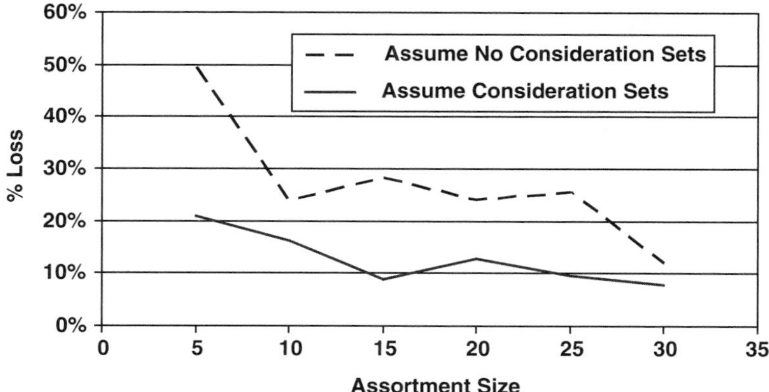

Figure 8-5. Profit Loss for Incorrect Consideration Set Assumption with $\theta = 0, 0.9$

not use consideration sets to prescreen the products, i.e., $\theta = 0$ is correct, the optimal assortment for $\theta = 0.9$ results in a 10% to 20% reduction in expected profit. Thus, for this data set, the less risky alternative is to assume that customers do use considerations sets.

4. SUMMARY AND CONCLUSION

This paper has developed an operational model for assortment optimization, based a multinomial logit choice model with general heterogeneous customer preferences. The structure of the model allows the required input parameters for product choice and retailer choice to be estimated separately from product sales and retailer market shares. These estimates are based on observations from previous assortments, which need not be optimal. The linear approximation of the newsvendor cost function assumes that temporary stockouts result in lost sales, which restricts the model's use to retailers or categories of products with relatively high service levels. However, this assumption leads to a closed form objective function that captures the impact of the assortment on both retailer choice and product choice. While the optimal assortments may no longer form nested sets for heterogeneous preferences, it is shown that the special cases of perfect competition and retailer monopoly do lead to nested sets, and it is illustrated how the optimal assortment transitions between these two extremes as the retailer's market share increases.

The optimization model can accommodate a variety of additional retailer constraints. For example, it may be important to: (1) require that certain top brands be represented in the assortment; (2) provide some level of assortment stability across time for customers; (3) stay within a given display space constraint; or (4) carry products with the full range of price points to promote the image of a category killer. The analysis of the DVD player data base illustrated the decreasing marginal benefits associated with increasing assortment size and also the sensitivity of the optimal assortment to the input assumptions regarding the customer choice process. Including customer heterogeneity had significant impacts on both the optimal assortments and the expected profits. Consideration sets, which have been studied in the context of modeling customer choice but have not previously been included in assortment optimization, were found to strongly influence the optimal assortment for the DVD player data base. This analysis supports the importance of using a consumer choice model that includes heterogeneous preferences and consideration sets in obtaining optimal assortments. The sensitivity analysis also illustrates the potential profit improvement for additional selling effort designed to influence customers' product choices.

There are a number of promising avenues for future research. Clustering customers into fewer classes can reduce the problem size and lead to shorter computation times for the general competitive case. Analytical methods for choosing the best customer clusters for a given database of utilities could therefore extend the applicability of the optimization model. Clusters based on customers' preferences for product attributes, as opposed to individual product utilities, may lead to clusters that are more stable over time. Better optimization approaches that exploit the specific structure of the assortment problem may also exist. It is hoped that this paper will also lead to additional research on the development of decision support systems for assortment planning that implement this optimization model for choosing assortments, taking into account both product choice and retailer choice.

Acknowledgment The author is grateful to Dale Achabal, Kirthi Kalyanam, Shelby McIntyre and Chris Miller for many valuable discussions and to Active Decisions, Inc. for providing the data base that was used for testing the optimization model. This research was partially supported by the Retail Workbench Research and Education Center at Santa Clara University.

References

Andrews, Rick L. and T. C. Srinivasan (1995), "Studying Consideration Effects in Empirical Choice Models," *Journal of Marketing Research*, **32** (February), 30–41.

Ben Akiva, Moshe and Steven Lerman (1985), *Discrete Choice Analysis*, Cambridge: MIT Press.

Bucklin, Randall and James Lattin (1991) A Two State Model of Purchase Incidence and Brand Choice," *Marketing Science*, **10**, (Winter), 24–39.

Cachon, Gerard, Christian Terwiesch and Yi Xu (2005), "Assortment Planning in the Presence of Consumer Search," *Manufacturing and Service Operations Management*, 7 (4) 330–346.

Cachon, Gerard and A. Gurhan Kok, (2007) "Category Management and Coordination in Retail Assortment Planning in the Presence of Basket Shopping Consumers," Management Science, **53**, (June), 934–951.

Chen, Kyle D. and Warren H. Hausman (2001), "Technical Note - Mathematical Properties of the Optimal Product Line Selection Problem Using Choice-Based Conjoint Analysis," *Management Science*, **46** (2), 327332.

Cintagunta, Pradeep K. (1993) "Investigating Purchase Incidence, Brand Choice, and Purchase Quantity Decisions of Households," *Marketing Science*, **12**, 184–208.

Chong, Juin-Kuan, Teck-Hua Ho and Christopher Tang (2001), "A Modeling Framework for Category Assortment Planning," *Manufacturing and Service Operations Management*, **3** (3), 191–210.

Dobson, Gregory and Shlomo Kalish (1988), "Positioning and Pricing a Product Line," *Marketing Science*, **7** (2), 107–125.

Dobson, Gregory and Shlomo Kalish (1993), "Heuristics for Positioning and Pricing a Product Line Using Conjoint and Cost Data," *Management Science*, **39** (2), 160–175.

Green, Paul E. and Abba M. Krieger (1985), "Models and Heuristics for Product Line Selection," *Marketing Science*, **4** (1), 1–19.

Green, Paul E. and V. Srinivasan (1978), "Conjoint Analysis in Consumer Research: Issues and Outlook," *Journal of Consumer Research*, 5 (2), 103–123.

Guar, Vishal and Dorothy Honhon, (2006) "Assortment Planning and Inventory Decisions Under a Locational Choice Model," *Management Science*, 52 (10), 1528–1543.

Hanson, Ward and Kalyanam, Kirthi (2006) *Internet Marketing and e-Commerce*, South-western College Publishing.

Honhon, Dorothee, Vishaul Guar and Sridhar Seshadri (2006), "Assortment Planning and Inventory Management Under Dynamic Substitution," working paper, Stern School of Business, New York University.

Kahn, Barbara E. and Donald R. Lehmann (1991), "Modeling Choice Among Assortments," *Journal of Retailing*, **67**, (Fall) 274–299.

Kohli, Rajeev and R. Sukumar (1990), "Heuristics for Product-Line Design Using Conjoint Analysis," *Management Science*, **36** (12), 1464–1478.

Kok, A. Gurhan and Marshall Fisher (2007), "Demand Estimation and Assortment Optimization Under Substitution: Methodology and Application," *Operations Research*, **55**, (Nov) 1001–1021.

Kok A Gurhan, Marshall Fisher and Ramnath Vaidyanathan (2008), "Assortment Planning: Review of Literature and Industry Practice," to appear in *Retail Supply Chain Management*, N. Agrawal and S. Smith editors.

Mahajan, Siddarth and Garrett van Ryzin (2001) "Stocking Retail Assortments under Dynamic Substitution," *Operations Research*, **49**, (3) 334–351.

McBride, Richard D. and Fred S. Zufryden (1988), "An Integer Programming Approach to the Optimal Product Line Selection Problem," *Marketing Science*, **7** (2), 126–140.

Roberts, John H. and James M. Lattin (1991), "Development and Testing of a Model of Consideration Set Composition," *Journal of Marketing Research*, **28** (November), 429–440.

Roberts, John H. and James M. Lattin (1997), "Consideration: Review of Research Prospects and Future Insights," *Journal of Marketing Research*, **34** (August), 406–410.

Siddarth, S. Randolph Bucklin, and Donald Morrison (1995), "Making the Cut: Modeling and Analyzing Choice Set Restriction in Scanner Panel Data," *Journal of Marketing Research*, **32**, (August) 255–266.

Smith, Stephen A. and Naren Agrawal (2000), "Management of Multi-item Retail Inventories Systems with Demand Substitution," *Operations Research*, **48**, 50–64.

Van Ryzin, Garrett and Siddarth Mahajan (1999), "On the Relationship Between Inventory Costs and Variety Benefits in Retail Assortments," *Management Science*, **45**, 1496–1509.

Wittink, Dick R., and Philippe Cattin (1989), "Commercial Use of Conjoint Analysis: An Update," *Journal of Marketing*, **53** (3), 91–96.

Chapter 9
MULTI-LOCATION INVENTORY MODELS FOR RETAIL SUPPLY CHAIN MANAGEMENT

A Review of Recent Research

Narendra Agrawal and Stephen A. Smith
Department of Operations and MIS, Leavey School of Business, Santa Clara University, Santa Clara, CA 95053, USA

1. INTRODUCTION

Research on multi-level inventory systems is critical to retail supply chain management. Multi-level systems are commonly observed in most retail environments, where regional distribution centers (warehouses) stock products to replenish inventory at the retail stores. There is a rich and vast literature in the field of operations management that focuses on the design and management of multi-echelon inventory systems, which can be applied to retailing. Even so, a variety of open problems remain, and this continues to be a fruitful area for researchers. While more than two echelons are also observed in practice, most retailers now prefer to move toward the simpler, two-echelon systems. Such structures are common even in pure play "E-tailers," such as Amazon.com. Amazon.com started with the idea of owning no distribution centers at all, and relying on direct shipments of books from publishers to customers for demand fulfillment. However they now manage a small number of distribution centers, and use a combination of direct shipments from vendors and shipments from their warehouses for demand fulfillment. Traditional "bricks and mortar" retailers today also face the problem of designing inventory management systems for items that are purchased through their Internet sales channels, in combination with normal store replenishment.

This review paper covers a subset of the research on this topic. Because of the vastness of the literature on multi-level inventory systems, we felt it was important to limit the scope of our survey in a meaningful way. First, we restrict our attention to papers after 1993, and refer the reader to the reviews in other papers for articles prior to 1993. For example, Axsater (1993a), Federgruen (1993), and

N. Agrawal, S.A. Smith (eds.), *Retail Supply Chain Management*,
DOI: 10.1007/978-0-387-78902-6_9, © Springer Science+Business Media, LLC 2009

Nahmias and Smith (1993) contain excellent reviews of the work up to that point. We discuss some of the earlier articles that provide foundations for results that we are presenting, or were not included in the reviews listed above. Second, we omit papers on certain model formulations that are not typical of retail inventory management. For example, we exclude the literature on serial systems, since they are not representative of typical retail chains, and are a special case of general multi-location multi-echelon systems. Also excluded are papers that assume deterministic demand, since demand uncertainty is a key aspect of most retail systems.

Finally, we focus our attention primarily on periodic review systems. Most retail chains today employ technologies such as point-of-sale (POS) scanner systems that provide real time access to sales and inventory data. Consequently, in principle, continuous review models could be an appropriate construct for these retail systems. However, two issues limit the practical applicability of this assumption. First, due to contracts with vendors and shipping companies, shipments occur primarily on a pre-specified schedule, and often a variety of items are delivered simultaneously. Second, despite the real time access to sales information, the ERP databases and inventory allocation algorithms are typically updated periodically. Thus, strictly speaking, inventory decisions must be made by planners according to predefined cycles. Consequently, periodic review systems are a better representation of the inventory management systems used by most retailers. For the sake of completeness, in the appendix we briefly present the formulation of some continuous review models along with a few key references.

The rest of the paper is organized as follows: We begin by discussing the key modeling issues in Section 2. In Section 3, we present the general formulation for periodic review inventory model, and review the relevant literature. Key conclusions and opportunities for further research are discussed in section 4. The continuous review model is discussed briefly in the Appendix.

2. MODELING ISSUES

2.1 The key decision

The fundamental decision to be made in two-echelon retail inventory systems is the appropriate division of inventory between the central (warehouse) location, and each of the retail stores.[1] Clearly, more inventory at the retail stores provides a higher service level to customer demand, but this also increases costs associated with carrying the inventory. The holding cost

[1] Earlier papers used the term "retailers" to refer to individual retail locations, while more recent papers have used the term "stores." In this paper, we will use the term stores, retail stores, or retailers for the lowest echelon level in the inventory system.

is higher at stores, due to increased shrinkage and because space in retail stores is typically more costly than warehouse space. Higher costs also result from transporting additional items to stores, which increases the product's value. Also, immediate distribution of a large proportion of the inventory to stores makes it difficult to address subsequent inventory imbalances across stores, because lateral shipments between stores are not part of normal replenishment. That is, keeping additional inventory at the warehouse offers the advantage of risk pooling, since inventory can be directed to those stores that need it most. This can potentially reduce overall inventory investments and costs. However, the resulting shipment delays may adversely affect customer service levels. This type of risk pooling has been referred to as the *depot effect*. The other advantage of having a warehouse is the possibility of risk pooling over the length of the replenishment lead time from the external supplier. This is sometimes referred to as the *joint replenishment effect*. In other words, while replenishment orders placed by the warehouse take into account actual demands at the retail stores, the actual decision to allocate this inventory to stores can be delayed until the replenishment order is received. The additional demand information gained during this lead time can be used to make more efficient inventory decisions. Note that this benefit can be realized even if the warehouse holds no inventory.

2.2 Modeling demand

The Poisson distribution is often used to model retail store demand, using a probability function of the form

$$P\{\text{Demand} = k\} = e^{-\lambda}\lambda^k/k! \quad k = 0, 1, 2, \ldots$$

with mean = variance = λ. The Poisson distribution is a particularly attractive assumption for modeling demand in continuous review systems because it requires only a single parameter (λ), and the resulting analysis is more tractable.

When mean demand per period is large, the normal distribution can be used to approximate the Poisson. To model discrete demand, the discrete probabilities can be approximated by

$$\text{P}\{\text{Demand} = k\} \approx \Phi(k + 1/2|\mu, \sigma) - \Phi(k - 1/2|\mu, \sigma) \quad k = 0, 1, 2, \ldots$$

where $\Phi(x|\mu,\sigma)$ = normal cumulative distribution with mean μ and variance σ^2.

Some empirical studies of retail data (e.g., Agrawal and Smith 1996) have found that retail demands are more variable than the Poisson distribution, which has a fixed variance to mean ratio of one. There are some practical reasons why actual demand may have higher variance than would be predicted by a Poisson distribution. Random variations may occur in the underlying Poisson arrival rate due to the weather, competitors' promotions, or special events that are not captured by the inventory system's forecasts. Second, customers whose purchases are Poisson arrivals may introduce additional variability by purchasing multiple items of the same kind. The normal distribution can accommodate more variation, by selecting a larger variance, but the empirical analysis mentioned above found that the normal distribution fit low demand items poorly because it assigns probability to negative values and because it is symmetric about its mean.

This suggests that a compound Poisson distribution or a negative binomial distribution may provide a better choice for modeling retail store demand. In particular, the negative binomial can be generated either from a Poisson distribution whose paratmeter λ has a gamma distribution, or from a compound Poisson with a geometrically distributed purchase quantity. Agrawal and Smith (1996) found that the negative binomial fit the store level demand data better than either the Poisson or normal distributions. The negative binomial distribution with parameters N and p has the following discrete probability function:

$$P(D = k|N, p) = f_k(N, p) = \binom{N + k - 1}{N - 1} p^N (1 - p)^k,$$

$0 < p < 1, \quad N > 0, \quad k = 0, 1, \ldots$
where the cumulative probability distribution is

$$F_k(N, p) = \sum_{j=0}^{k} \binom{N + j - 1}{N - 1} p^N (1 - p)^j.$$

The mean and variance are

$$\mu = N\left(\frac{1}{p} - 1\right), \text{and } \sigma^2 = N\left(\frac{1 - p}{p^2}\right).$$

The ratio of the variance to the mean is $1/p$, which is greater than one and can be arbitrarily large. This makes the negative binomial distribution particularly attractive for retailing applications that have high demand variability.

Other assumptions for modeling retail demand include the Gamma (Bradford and Sugrue 1990), Gumbel (Lariviere and Porteus 1999), and the general exponential family of distributions (Agrawal and Smith 2007).

We also note that the majority of papers assume that demand at different locations is independently distributed. There are a few exceptions that allow correlations across stores or across time, which are described later in this chapter.

Finally, in any store level model, it is important to specify assumptions regarding the treatment of excess demand at the stores. Primarily for analytical tractability, most papers assume that unmet demand is backordered, not lost. While backordering is common for some classes of expensive retail items, excess demands for most department store and grocery items result in lost sales to another retailer, or possibly substitution of another item in the store. Backordering can serve as a good approximation to the lost sales case, provided that the inventory service level at the store is sufficiently high.

A few researchers have assumed lost sales for unmet store demands. Because of the complexity of modeling lost sales, these papers generally assume that the latest store demand information is available with zero delay prior to store replenishment. This zero delay assumption is generally correct in today's retail environment, since electronic data interchange (EDI) can provide essentially continuous communication of demand information across locations, and stores are typically replenished after hours, when no sales are occurring. But the lost sales case is significantly more complex analytically than the backorder case. With lost sales, the inventory level at any time t depends on all the individual demands and replenishments that have occurred previously, while in the backorder case, computing the inventory level requires knowledge of only the total demand over the previous periods. That is, in the backorder case, the inventory level at time t ($IL(t)$) follows from the well known relationship between inventory position ($IP(t)$) and total demand during the lead time ($D(t\text{-}L,t)$), i.e., $IL(t) = IP(t\text{-}L) - D(t\text{-}L,t)$. Therefore, knowledge of the actual demand or order placed in every period is not needed to determine the inventory level in a given period. This does not hold for lost sales, adding significant complexity to the analysis.

2.3 Lead times

Two types of lead times are relevant in such systems. The first is the replenishment lead time at the warehouse for orders placed with external suppliers. Since most researchers assume no capacity constraints on the supplier, these lead times may be assumed to be constant. Exceptions are papers that explicitly model production capacity constraints. We briefly mentioned this literature later. The second lead time is for orders placed by retail stores at the warehouse. This consists of two components – the shipment time, which is generally assumed to be constant (but may vary across locations), and the lead time due to shortage delays at the warehouse, which is random. Consequently, the effective lead time at the stores, i.e., the sum of the two components is always stochastic due to the possibility of stockouts at the warehouse. It is also a function of the specific allocation rules at the warehouse when shortage occurs. Thus, determining the store lead time distribution is a key analytical challenge.

2.4 Allocation policies used at the warehouse

How the warehouse allocates inventory among competing store demands in shortage situations is a critical determinant of the complexity of multi-location inventory models. It also affects the service level and the cost structure for the retail stores. Conceptually, researchers have considered four different policies for what the warehouse does with the inventory it receives from the external supplier (McGavin *et al.* 1993). The first policy is essentially a "pass-through," where the warehouse holds no stock, but allocates and ships it to the stores as soon as stock is received from the supplier(s). This is similar to the cross-docking policy that is practiced at many retail warehouses today. The second policy, called the equal interval policy, attempts to balance the stores' inventory at regular intervals. The third policy is called a two-interval policy, where the warehouse makes two shipments during the period between consecutive replenishments from the supplier. The final policy is called as the virtual allocation policy, where units of inventory at the warehouse are reserved for specific demands as they occur at the retail stores. This essentially imposes a first come first served discipline on demand fulfillment. We will discuss the modeling implications of each of these policies in the next section.

3. THE GENERAL PERIODIC REVIEW INVENTORY MODEL

Consider a single-item discrete-time, two-echelon system, where the top echelon consists of a depot (also referred to as the warehouse) which supplies a collection of N retail stores, numbered 1,...,N with l_0 and l_i corresponding to the lead times for the depot and the retail outlet i respectively. Random demand occurs in each period at each retail store, with

$D_i(t,t+s)$ = the total demand at location i during periods $t, \ldots, t+$ s, and

$$D_0(t,t+s) = \sum_{i=1}^{N} D_i(t,t+s)$$

is the system wide demand during the same period. We let $D_i^{(l)}$ and $D_0^{(l)}$ be the l-period demand at retailer i and the warehouse with cumulative distribution functions $F_i^{(l)}$ and $F_0^{(l)}$ respectively. Unmet demand is backlogged at the retailer, with a penalty cost of p_i per unit backordered and h_0 and $(h_0 + h_i)$ are the inventory holding costs assessed on ending inventory at the depot and the retailer i, respectively.

In each period, we define the following sequence of events:

1. Current period's ordering and shipment decisions are made.
2. Shipments are received.

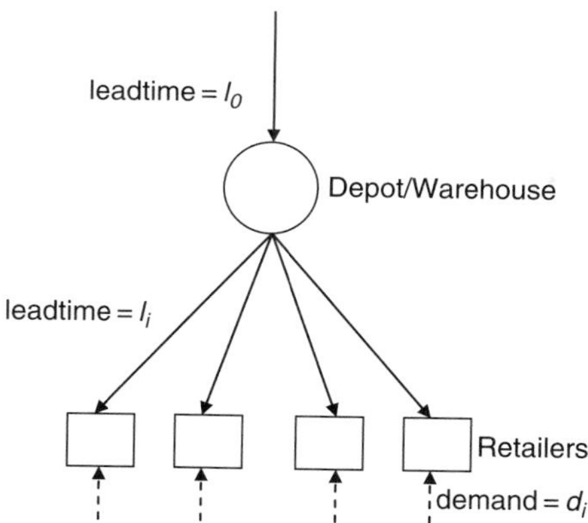

3. Demand occurs.
4. Holding and penalty costs are assessed based on ending inventory levels.

Define $I_i(t)$as the echelon stock (stock on hand plus in transit to and on hand at successor points minus backorders from external customers) at location i at the beginning of any period t just after the receipt of a shipment, and $\hat{I}_i(t)$as the corresponding value at the end of the period t. Define $\hat{I}_i(t) = \hat{I}_i^+(t) - \hat{I}_i^-(t)$. Then $\hat{I}P_i(t)$ and $IP_i(t)$ are the echelon inventory positions just before and after ordering (at the depot) or shipment (if i is a retailer), where echelon inventory position is the echelon stock level plus all orders in transit to that location.

At the end of any period t, the total cost for the whole system, which includes holding and penalty costs, can be expressed as

$$h_0\left(\hat{I}_0(t) - \sum_j \hat{I}_i(t)\right) + \sum_j (h_0 + h_i)\hat{I}_i^+(t) + \sum_j p_i\hat{I}_i^-(t)$$
$$= h_0\hat{I}_0(t) + \sum_i \left(h_i\hat{I}_i(t) + (h_0 + h_i + p_i)\hat{I}_i^-(t)\right).$$

Then, using the notation

$$C_0(t) = h_0\hat{I}_0(t), \text{and} C_i(t) = h_i\hat{I}_i(t) + (h_0 + h_i + p_i)\hat{I}_i^-(t).$$

The total cost is equal to:

$$C_0(t) + \sum_{i=1}^{N} C_i(t).$$

The expected system costs then depend on the *ordering decision* at the warehouse (which raises the inventory position $IP_0(t)$ of the system to, say, y_0), and on how shipment quantities for retail stores are determined, i.e., the *allocation decision*. Let the corresponding inventory positions at the retailers be denoted by $y_1, \ldots, y_N$. The first decision determines the expected cost at a warehouse at the end of period $(t + l_0)$, and limits the amount that the aggregate echelon inventory positions of the retail stores can be raised in period $(t + l_0)$. The later decision is particularly relevant in case of shortage situations. These decisions are not independent, which makes the overall optimization problem challenging. So, the upper limit on the aggregate echelon inventory position of the stores can be specified as

$$\sum_{i=1}^{N} IP_i(t + l_0) \leq y_0 - D_0(t, t + l_0 - 1).$$

Obviously, these decisions influence the cost at echelon i at the end of period $(t + l_0 + l_i)$. Therefore, the effect of decisions made in period t, $C(t)$, is

$$C(t) = C_0(t + l_0) + \sum_{i=1}^{N} C_i(t + l_0 + l_i).$$

Thus, for any given ordering policy, the expected long-run average cost is given as

$$\lim_{T\to\infty} \frac{1}{T} E\left[\sum_{t=0}^{T-1}\sum_{i=0}^{N} C_i(t)\right] = \lim_{T\to\infty} \frac{1}{T}\sum_{t=0}^{T-1} E[C(t)].$$

Minimization of the long run average expected value of this function is the overall objective in the two echelon system.

3.1 Solution methodologies

Determination of optimal strategies for general two echelon systems remains difficult. Consequently, most papers use approximations. While some papers make use of relaxation techniques to obtain bounds on the true costs or profits, others impose specific restrictions on the class of inventory policies and then determine the optimal policy within that class. In all cases, the issue of inventory allocation must be addressed carefully.

The form of the optimal solution can be characterized in special cases. One way of rationing, called the *myopic allocation method*, allocates the echelon stock of the warehouse at the beginning of period $(t + l_0)$ such that the sum of the expected costs at the stores in period $(t + l_0 + l_i)$ is minimized, without regard to later periods. A relaxation of this problem allows the quantities allocated to stores to be negative (by ignoring the constraint that the retail stores' inventory positions must be greater than at the beginning of period $t + l_0$. This is called as the *balance assumption*. The key advantage of the balance assumption is that the echelon stock (sum of the total inventory in the system) suffices to determine the warehouse ordering decision. Further, it also makes the myopic allocation policy optimal. The drawback is that this approach gives up the risk pooling advantage associated with holding stock back at the warehouse. In any case, the balance assumption underestimates the total costs since it is a relaxation. However, absent these assumptions, it turns out that base stock policies are not optimal for such systems (Clark and Scarf 1960). Van Donselaar and Wijngaard (1987), Eppen and Schrage (1981) and Federgruen and Zipkin (1984a) discuss the consequences of making this assumption in detail. These early papers consider special cases of the problem: for example, Eppen and Schrage (1981) consider a 2 echelon model with identical retailers and a depot that doesn't carry any stock. Jackson (1988) extends the Eppen and Schrage model to allow the warehouse to carry stock, while Jackson and Muckstadt (1989) allow non-identical retailers, but with identical cost parameters. Federgruen and Zipkin extend the Eppen and Schrage model to include non-identical retailers, non-stationary demand, and (s,S) ordering at the warehouse, but they determine their allocation policies under the assumption that the warehouse is stock-less. Jonsson and Silver (1987) also assume that the warehouse is stock-less, but extend the Eppen and Schrage model to include the possibility of a single, complete redistribution of inventory between the retailers in the period before the end of any review cycle for the warehouse. Erkip *et al.* (1990) consider a model like Eppen and Schrage (1981) but allow demand correlation across retailers as well as time. Chen and Zheng (1994) develop lower bounds for costs, based on a cost allocation mechanism, for serial, assembly and distribution systems. Our system is an example of their distribution system.

McGavin *et al.* (1993) model a system with identical retailers, zero lead times for shipments from the warehouse to each retailer, centralized control and periodic replenishment at the warehouse. The overall stock allocation consists of four decisions: the number of withdrawals from the warehouse stock (which is an opportunity to allocate inventory to retailers), the time between these withdrawals, the quantity withdrawn, and the division of the withdrawn stock to each retailer. The first three decisions are set when the warehouse is replenished and the last one depends on retailer inventories. In particular, they model two opportunities for allocating stock from the warehouse to the retailers, which need not be equally spaced between warehouse replenishments. They seek to determine the effective timing of these two instances and the allocated quantities, so as to minimize lost sales per retailer. This assumption of lost sales makes this paper's contribution a significant departure from the majority of the literature in this stream of work. However, as noted before, this requires the retailer lead time to be zero. They show that the best allocation policy is one that balances retailer inventories (i.e., maximizes the minimum retailer inventory). Heuristic policies are developed assuming that the number of retailers is infinitely large, and are numerically tested in the finite retailer case. In particular, they test the 50/25 heuristic, where the first interval is 50% of the replenishment cycle and the second withdrawal quantity is 25% of the replenishment cycle's mean demand. The resulting analysis suggests the insight that the choice of the withdrawal quantity and division of inventory may matter more than the number of withdrawals.

Ahire and Schmidt (1996) consider a mixed continuous and periodic review system with one warehouse and multiple, non-identical retailers. While the retailers follow a continuous review (r,Q) policy, the warehouse follows a periodic review policy (with review period T). At the warehouse, the review period is divided into equally spaced intervals, where at each such point, a group of identical retailers (say, within a geographic zone) are reviewed. Each such zone, however, is reviewed only once per review cycle. The implication of this setup is that the retailer system is equivalent to a (nQ, r, T) system. The lead time consists of a deterministic component, the shipping lead time from the warehouse, and a stochastic component, due to possible shortages at the warehouse (however, order splitting is not allowed), and due to the fact that their orders are only reviewed periodically. Thus, an order may have to wait for anywhere from 1 to T periods before it is even reviewed by the warehouse. Results from Little's Law are used to approximate the shortage delays. Retailer demand is assumed to be Poisson, while the warehouse demand is approximated by a normal distribution, whose parameters are computed. The resulting approximations for financial and operational performance metrics compare well to those obtained through simulation.

Graves (1996) considers a general distribution network following a periodic review, order up to policy at each location. Under the assumption that each location orders at pre-set and known times, he specifies a *virtual allocation* policy where a unit at the supply location is committed/reserved for each unit demanded at the time of the occurrence of the demand. This assumes that the warehouse has real time visibility into the retail demands. Shipments, however, occur only at the next appropriate time after order receipt. The committed units cannot be used to satisfy any other order. Unmet demand at the warehouse is backordered and satisfied in a first-come-first-served manner. Independent demand occurs at each retail location following a Poisson process, and excess demand is backordered. Since the order interval is present and excess demand is backordered, each location orders an amount that equals the total demand since the last order. The analysis requires the characterization of the run-out time, the time at which the warehouse runs out of inventory to allocate to the retail sites. The demand at the warehouse is approximated with a Negative Binomial distribution, whose moments can be determined. Various performance metrics can then be quantified using this approximation. Diks and de Kok (1998) model a general N-echelon divergent system where every location can hold stock, and determine policies that minimize long run average costs.

This idea of pre-set, staggered schedules for ordering is also considered in Chen and Samroengraja (2000). In a one-warehouse, multi-retailer model, where retailers are identical, and face i.i.d. demands, they assume that the warehouse follows a periodic review (s, S) policy to receive shipments from a source of unlimited supply with lead time L. The warehouse orders are based on its local inventory position. Between consecutive warehouse ordering epochs, the retailers, whose ordering points are pre-set and equally staggered (there can be groups of retailers ordering at each such epoch), place orders, following base-stock policies (with a common order up to level). Two different allocation policies are evaluated. The first, called past priority allocation (PPA) backlogs the unmet demand from a retailer, and fills it in a first-come-first-served manner from the inventory at the warehouse. However, actual shipment occurs only at the next epoch when the retailer places an order with shipment lead time l. The second policy, called current priority allocation (CPA) gives priority to the current order and backorders for the retailer designated to order in a given period. Thus, under PPA, the warehouse may carry inventory earmarked for a retailer while it denies inventory to orders from other retailers. In the second case, some retailers may be backlogged for several consecutive periods while others get replenished. On the other hand, the PPA model lends itself better to exact analysis. Solutions for this formulation are obtained through an approximation procedure. The CPA model is harder to evaluate exactly, but simulation studies indicate that the optimal policies are close to those under the PPA regime. Unlike in the Graves (1996) paper where inventory at the warehouse is committed to demands as they occur, here, the allocation decision is delayed until the retailer actually places an order. Their derivation of the exact cost function in the PPA case is based on a different accounting scheme. Warehouse holding costs occurring in period $(t + L)$ are charged to period t. For retailers, in period t, they charge the total holding and backorder cost over the next N periods (N is the number of ordering epochs within each warehouse cycle) for the retailer designated to order in that period. The exact calculation under the CPA method is difficult since the distribution of a retailer's inventory position at any time depends not only on the inventory position L periods ago, but also on the exact pattern of deliveries from the outside supplier.

Continuing in the spirit of generalization, Axsater *et al.* (2002) allow the retailers to be non-identical. The warehouse holds stock and orders from an external supplier in multiples of a given batch size, receiving shipments after a fixed lead time. Lead times for shipping to retailers is constant, but can vary by retailer. Instead of the balance assumption, they consider the virtual assignment rule, where the inventory ordering decision at the warehouse accounts for all retailer inventory positions and assigns inventory to retailers as soon as orders are placed. The final inventory allocation, however, is made only upon the arrival of the replenishment. This is a more restrictive policy that overstates costs. Instead of the myopic allocation policy, they consider a two-step allocation policy, which allows some inventory to be retained at the warehouse. Essentially, at the beginning of each period, the remaining time until the next ordering opportunity is assumed to consist of two intervals, the second one being a single period, at which point reallocation can be done again. An optimization methodology is developed under these assumptions and the results are found to compare very favorably with the case of balance assumption and myopic allocation.

Under the balance assumption, Dogru *et al.* (2004) establish the convexity of the cost function for the infinite horizon case and discrete demand case, which implies the existence of optimal policies that are base stock policies. They also characterize newsvendor inequalities that must be satisfied by the optimal solutions. For example, for the special case of identical holding and penalty costs at the retailers, and under the myopic allocation and balance assumptions, the well known *critical fractile* solution yields the optimal stocking policy for each location.

3.2 Batch ordering

The use of batch ordering policies imposes additional complexities on the model since the demand at the warehouse is no longer a simple convolution of the retailers' individual demands. Further, if the retailers follow a periodic review policy, a retailer's order consisting of multiple batches may have to be split across multiple shipments. Of course, the issue of allocation of scarce warehouse inventory remains. Analytically, the key challenge is to determine the distribution of the retailers' replenishment lead time, which consists of both the shipping time (constant) and additional delays due to shortages at the warehouse. Two approaches have been used in the literature for this purpose. One is to evaluate when a batch is ordered by the retailer relative to when the warehouse orders it (as in Svoronos and Zipkin 1988). The second is to evaluate when a batch is ordered by the warehouse relative to when the retailer orders it. In cases with a single warehouse, the later approach is more tractable. This is the approach used in the following two papers.

Cachon (1999) considers a 1 warehouse N (non-identical) retailer model where the retailers as well as warehouse follow (R, nQ) policies. Retailers follow a periodic review policy with period T, but the ordering process is balanced in the sense that a fixed number N/T of retailers order every period. Unmet demand is backordered, and partial fulfillment is allowed. Retailer orders are randomly shuffled upon receipt, and fulfilled in a first-come-first-serve manner. Exact expressions are derived for costs, as well as demand variability at the warehouse. The key result is that the warehouse demand variability decreases due to balancing (rather than synchronizing retailer orders, where all retailers order simultaneously). Further, under a balanced system, increasing the length of the review period T and decreasing the order batch size also helps lower the supplier's demand variability. However, these strategies may not necessarily decrease total supply chain costs, since they might actually increase the retailers' ordering or inventory costs.

Cachon (2001) considers a similar model but with identical retailers, and where each location reviews and orders in each period. All locations follow a batch ordering policy. Demand is stochastic and discrete. Average inventory and backorder levels and fill rates are evaluated exactly at each location. Safety stock requirements are determined exactly at the retailers, but approximately at the warehouse.

3.3 Lost sales

All papers described thus far assume that unmet demand is backordered, McGavin *et al.* 1993. Another exception is Nahmias and Smith (1994), which focuses on a one warehouse multi retailer system, and assumes that a given fraction of unmet retailer demand is lost. Order up-to policies are used at the retailers, and the replenishment lead time from the warehouse is assumed to be zero. The warehouse also uses an order up to policy with zero lead times. The length of the review period at the warehouse is a multiple of the retailer's review period, and the stock levels are such that shortages only occur in the mth period within any cycle. This assumption, along with that of zero lead times, is necessary to lend tractability to the model.

In contrast to most other papers, they assume that the demand at the retailers follows a negative binomial distribution, which has been shown to fit retail data well (Agrawal and Smith 1996) because the variance to mean ratio is often larger than one. Since the warehouse supports many stores, the warehouse demand can be approximated by a normal distribution. Exact expressions are derived for the average inventory level and lost sales at stores and the warehouse. Representative retail data is used to illustrate the results and generate managerial insights. For example, they show that the benefits of holding stock at the warehouse depend upon item characteristics – items with low optimal service levels at stores derive the most benefit by holding the majority of the stock at the warehouse. Increasing the frequency of store delivery can also reduce costs, especially for items that require high optimal service levels at stores.

Anupindi and Bassok (1999) quantify the benefit of centralizing stocks in a single warehouse, two-retailer setting, where a fixed fraction, $1 - \alpha$, of unmet demand at the retailers is lost. The remaining customers look for the product at the other retailer. They too assume zero lead times for shipments to retailers. Each retailer faces an independent demand (with known distribution), buys from the warehouse at a unit cost w and sells it to their customers for a price p. Since they consider a stationary, infinite horizon model, the problem boils down to a single period newsvendor-type problem. In the simplest case where $\alpha = 0$, i.e., all unmet demand is lost, they show that centralization does not necessarily increase sales. This depends upon the nature of the demand distribution, as well as the value of the critical fractile. For example, for demand with a normal or exponential distribution, centralization leads to higher sales, while for a uniform distribution, this happens only if the critical fractile has a value less than 0.77.

In the general case when $\alpha > 0$, the solution corresponds to a Nash equilibrium. They find that the expected total profits for the retailers are greater when stocks are centralized. However, the total sales are greater in the centralized case only if α is smaller than a certain threshold. The manufacturer/warehouse will prefer the centralized case only if α is smaller than a threshold (one interpretation for α in their model is the fraction of customers that, when unsatisfied at a local retailer due to stockouts, search for the goods at other retailers). Interestingly, even the total supply chain profit may decrease due to centralization in some cases. This happens when α is larger than some threshold value, which in turn is a function of the wholesale price w. These insights apply even when coordinating contracts are used. Thus, the main insight from this analysis is that while conventional wisdom dictates that costs decrease (and profits increase) under centralized systems due to risk pooling benefits, this benefit may not result for all parties in the supply chain.

3.4 Decentralized environments (quantifying the value of information sharing)

The discussion thus far assumed that the entire supply chain was under central control, and information about all locations was available to the central decision maker. This assumption is not appropriate when the entities at the different echelons operate independently. When decisions are made so as to optimize local incentives, the overall supply chain performance may not be optimal. The consequences of the resulting actions by the supply chain participants include the well known bull whip effect, as discussed in Lee *et al.* (1997a) and (1997b).

In an early paper, Eppen (1979) showed that in a multi-location model with normal and correlated demand, the total holding and penalty costs are lower in a centralized system than in a decentralized system. This result was later generalized for other distributions in Chen and Lin (1989) and Stulman (1987), and to include inter-node transportation costs in Chang and Lin (1991).

Recently, however, spurred by the advances in information technology and software solutions, explicitly quantifying the potential value of information sharing in supply chains has been the subject of a number of papers. For example, Cachon and Fisher (2000) quantify this value in the case of a single warehouse multi-retailer environment. The retailers are identical, and use periodic review batch ordering policies. Retailers order periodically, in batches of a given size Q, and receive shipments after a fixed lead time. The warehouse also orders in multiples of Q, and receives its orders from an external supplier after a constant lead time. Inventory is allocated using a batch priority rule, where each batch ordered is assigned a priority, and shipments are done in the order of priority. By comparing the total supply chain costs with and without information sharing, they conclude that the value of information sharing is rather limited, 2.2% on average. However, the benefit from shorter lead times and smaller batch sizes was nearly 20% each. The explanation they offer is that demand information only matters when the retailer inventory levels are very low, since otherwise, they don't need to place orders. However, this is precisely when retailers actually place orders, so essentially, the demand data is already captured in the order information.

Lee *et al.* (2000) quantify the value of information sharing, albeit in a one warehouse one retailer supply chain. In contrast to the earlier papers which assume the demand is independent and identical across time, they assume that demand at the retailer is auto-correlated [AR(1)], such that

$$D_t = d + \rho D_{t-1} + \varepsilon_t,$$

where $d > 0$, $-1 < \rho < 1$, and ε_t is normally distributed with mean zero and standard deviation of σ. Both locations order every period in a periodic review system, with fixed lead times for shipments to each location. Unmet demand at the retailer is backordered, while at the warehouse excess demand is met with a special order placed at an external supplier at an additional cost. They assume that the manufacturer bears the full cost of guaranteeing supply to the retailer. They characterize the retailer's ordering process, which becomes the demand process for the manufacturer. In the case of no information sharing, the manufacturer only receives the retailer's orders. In the case of information sharing, the manufacturer also receives information about actual demand, which allows him to obtain the value of the error term ε_t , thereby lower demand variability. Since the manufacturer bears the full cost of assuring supply, the retailer's inventory costs remain unchanged with information sharing. However, information sharing leads to lower inventory levels as well as costs for the manufacturer. Further, they show that the benefit of information sharing is greater when the auto-correlation or demand variance is high. This analysis is complicated by the fact that when demand is auto-correlated, exact expressions for average inventory levels cannot be derived. Consequently, they make use of approximations for the retailer's and manufacturer's inventory levels.

Chen (1998) also quantifies value of information, but in a serial system with continuous review policies. They report cost benefits in the range of 2–9%. Gavirneni *et al.* (1999) also consider a serial system (1 warehouse, 1 retailer), but extend the model to the case where the manufacturer's capacity is limited. By comparing the base case to one in which the manufacturer obtains information about the retailers' demand distribution and inventory policy parameters, they are able to quantify the value of information. They find that the value of information is more compelling when end item demand is not very variable, when the retailer's $(S - s)$ is not very large or very small, or, when supplier's capacity is large. Aviv and Federgruen (1998) also consider the benefits resulting from sharing demand forecasts, also with limited supplier capacity.

3.5 Lateral pooling

There is a large body of research that focuses on the issue of lateral pooling, also referred to as transshipments. In practice, this is rarely done for low-ticket items, since the cost and time involved in repackaging leftover inventory, shipping it to another location, and unpacking it again can easily wipe out the margins. However, for bigger ticket items, like electronics, expensive jackets and suits, and automobiles, this practice is common. Obviously, the presence of an information technology solution that provides information about inventory levels is a prerequisite for this system. One stream of research on transshipments addresses the problem in the context of repairable items. In the interest of staying focused on the retail environment, we will not review this literature, but direct the interested reader to Cohen *et al.* (2006), Muckstadt (2004), Axsater (1990) and Lee (1987), and the references contained therein.

Since the other locations serve as a backup location from which to meet unmet demand, albeit at some cost, this alters the penalty incurred due to shortages. Similarly, since there is the possibility of selling excess inventory to other locations, it alters the salvage value. Depending upon the cost of transshipment and the terms of the exchange, a retail location may, in some conditions, find it profitable to transfer its inventory to another location even when it has its own demand to meet. Clearly, each location will need to determine rules for when is it appropriate to give up its inventory. In any case, the inventory stocking policy must be modified. A second factor to consider is whether the stocking decisions are made centrally, or in a decentralized manner. In the later case, a game theoretic formulation is necessary to determine the optimal inventory ordering and allocation rules to appropriately model the incentives for each party. This results from the externality created due to decentralized decision making – larger inventory carried by one location could lower the stock out cost for others. Similarly, lower inventory levels at one location make it more economical for another location to dispose off its excess inventory. An important source of distinction between papers on this topic is whether the redistribution of stock occurs *after* or *before* demand is realized.

We begin with the former category first. Early works on this topic include Krishnan and Rao (1965) and Karmarkar and Patel (1977). Both assume identical costs at retailers, an assumption later generalized by Tagaras (1989). Robinson (1990) formulates the problem for an arbitrary number of non-identical retailers, and shows the optimality of order up to policies. However, analytical solutions can be determined only for the case of identical retailers, or when there are only two retailers. Consequently, Monte Carlo simulation has been used to solve the general case. All these papers assume zero replenishment and shipment lead times. This assumption leads to the result of "complete pooling" (Tagaras 1989), which implies that if transshipment is economically viable, then it is optimal for each location to make its excess inventory available for lateral shipments, *i.e.,* there is no reason for holding inventory back at any location. This logic, *a priori*, may not hold if the replenishment lead times are non-zero. This factor is the focus of Tagaras and Cohen (1992), which we discuss next due to its generality.

Tagaras and Cohen (1992) model a multi-period, one-warehouse, two-retailer locations system, where demands occur independently at the retail locations. Shipments from the warehouse to retailer i arrive after l_i periods. Order-up-to policies are followed by each retailer, who faces a unit holding cost c_{hi} as well as shortage cost c_{pi} on the ending inventory OH_i and backorders BO_i, respectively. Additionally, there is a unit lateral shipment cost c_{ij} incurred for the X_{ij} units shipped from i to j. The transshipment policy is determined by whether the inventory level (or inventory position) at the shipping location i is above a threshold level r_i, and target inventory level t_j, (or inventory position) at the receiving location j, which must not be exceeded after transshipment. Four transshipment policies are thus generated. The first two involve on-hand inventory level as the criteria. In the first case, transshipment occurs only if a location faces a shortage (i.e., $t_i = t_j = 0$). Under the second policy, transshipment can take place even if there are no shortages (i.e., $t_i = r_i = 0$, $i = 1,2$). Obviously, $r_i = r_j = 0$ implies complete pooling in this case. The third and fourth policies are similar to these two, except that the triggers are inventory positions. The objective is to determine order quantities Q_i that minimize total expected costs, as given by:

$$E(C) = \sum_{1}^{2} \left\{ c_i E(Q_i) + c_{hi} E(OH_i) + c_{pi} E(BO_i) + \sum_{j=1, j \neq i}^{2} c_{ij} E(X_{ij}) \right\}.$$

Exact analysis of this formulation is mathematically intractable. Consequently, search procedures are used to determine optimal solutions. They also derive heuristics based on the assumption of zero lead times. The key finding is that the complete pooling policy always dominates, as was the case when lead times were zero. In other words, hedging, by holding back inventory, or transshipping in anticipation of shortages is not optimal. Also, the heuristics were found to be near-optimal. These results are extended to the case where the transshipment lead times are non-zero in Tagaras and Vlachos (2002).

Archibald *et al.* (1997) also consider a two-location model, but assume that unmet demand at a location can be met either through transshipment from the other location, or through an emergency shipment from the supplier (no warehouse is assumed). The demand distribution is assumed to be Poisson. A Markov chain formulation is developed to characterize the optimal policies, which are shown to be of the order up to type. The model is then extended to the case of multiple items with constraints on the amount of inventory that can be carried at any location.

Herer *et al.* (2006) generalize Robinson (1990) to include more general cost structures, and develop an optimization approach that is guaranteed to converge, as compared to Robinson's heuristic, which does not provide such a guarantee. They too assume zero lead times, show optimality of order up to policies, which are computed using Infinitesimal Perturbation Analysis. The transshipment quantities are determined by solving a linear programming formulation.

Bertrand and Bookbinder (1998), on the other hand, consider a general, periodic review model for the case where the redistribution decision is made *before* demand realization. They consider a model with multiple non-identical retailers that are supplied by a warehouse. The warehouse does not carry any stock, but allocates it to stores on the basis of their inventory levels so as to minimize total costs. In the period immediately before the end of the cycle (after which the warehouse orders again), inventory can be redistributed so as to minimize shortage in the last period. The assumption is that shortages primarily occur in the last period in any cycle. The redistribution decision is determined using a greedy heuristic. The optimal policies, and the corresponding costs and service level are determined using simulation, since any analytical treatment is intractable. Similar assumptions were made earlier in Jonsson and Silver (1987), but the objective was to minimize the total number of stockouts.

Anupindi and Bassok (1999), discussed earlier, model interactions between retailers when transshipments are possible. Similarly, Rudi *et al.* (2001) consider interactions between retailers in a game-theoretic setting, although their work is based on ideas contained in earlier papers by Parlar (1988) and Lippman and McCardle (1997). In the later two papers, in case of stockouts, it is the customer demand that is directed to the other location. This is different from the currently assumed scenario more relevant to us where products are transferred (albeit at a cost). Nonetheless, the modeling mechanics are similar. Rudi *et al.* (2001) consider the interactions between two firms, each modeled as a newsvendor within a single period framework. They assume that transshipment occurs *after* demand is realized, and the number of units exchanged from location i to location j is

$$T_{ij} = \min\{(D_j - Q_j)^+, (Q_i - D_i)^+\}.$$

A unit cost is incurred for each unit shipped, and a unit price is charged that varies by shipping location. The resulting profit functions follow in a straightforward manner following the newsvendor methodology. They characterize the optimal decision in the centralized as well as the decentralized cases by solving for the Nash equilibrium. The pricing decision is also evaluated. Extending this approach to the case of more than two locations is obviously complicated by the specific construction of the schedule of transshipment prices and costs.

Anupindi *et al.* (2001) develop a more generalized framework for the analysis of decentralized distribution systems. They assume N retailers who face stochastic demands and hold stocks locally and/ or at one or more central locations. An exogenously specified fraction of any unsatisfied demand at a retailer could be satisfied using excess stocks at other retailers and/ or stocks held at a central location. The operational decisions of ordering inventory and allocation of stocks and the financial decision of allocation of revenues/costs must be made in a way consistent with the individual incentives of the various independent retailers. They develop a "coopetitive" framework for the sequential inventory and allocation decisions. They define *claims* that allow them to separate the ownership and the location of inventories in the system. For the cooperative shipping and allocation decision, they develop sufficient conditions for the existence of the core of the game. For the inventory decision, they develop conditions for the existence of a pure strategy Nash Equilibrium.

They show that there exists an allocation mechanism that achieves the first-best solution for inventory deployment and allocation, and develop conditions under which the first best equilibrium will be unique.

Dong and Rudi (2004) include the consequences of lateral shipments between retailers on the warehouse/manufacturer in their study. However, they do so in a single period setting with identical retailers. Recall that Anupindi and Bassok (1999) solved only the 2 retailer case. They analyze the case where the manufacturer is a price taker as well as one where he is a price setter (i.e., a Stackelberg leader). Following an analysis in a newsvendor type setting, they find that the benefit of transshipment is no longer guaranteed, rather it depends upon the parameters of the problem.

In an interesting paper, Zhao *et al.* (2005) formulate the problem faced by a network of decentralized retailers who stock inventory of a common item (they consider this problem in the context of a spare parts dealer network). Each location follow an (S, K)-type policy. S denotes the order up to level while K denotes a threshold rationing level such that inventory will be shared with the other dealer only if the inventory level exceeds the threshold. Higher values of K imply that smaller portions of inventory are available for sharing. While demand occurs independently at each location, this possibility of inventory sharing changes the cost structure. Thus, each location needs an incentive to share inventory. Otherwise, it might be profitable to retain inventory to satisfy future demand (understandably, the complete pooling result does not always hold in the decentralized setting). This manufacturer can either provide incentives for sharing, or subsidize the cost of sharing the inventory. The consequences differ. The first incentive induces the locations to lower their threshold rationing levels instead of increasing their stocking levels. The second induces them to lower their stocking levels, which results in lower service levels. Thus, from the manufacturer's point of view, a combination of such incentives may be best.

3.6 Fashion products

The majority of the papers discussed thus far model environments in which the product being managed is a basic, replenishable item. In contrast, there is a significantly smaller literature that explores issues relevant to the management of fashion products in large, multi-echelon retail chains. Fashion products tend to have very short selling seasons, with replenishment lead times that may be substantially longer than the length of the selling season. Consequently, these environments differ in that the retailer may have a very limited number of opportunities (often one or two) to place inventory in stores, and demand uncertainty tends to be large. At the same time, for many fashion forward retailers, sales from such products form the bulk of revenues.

For single retail location environments, the problem can be modeled in a straightforward manner using the well known newsvendor formulation.

Extensions to the case of multiple locations, but only a single opportunity to position retail inventory, are fairly straightforward too. However, the problem is more complicated when there are multiple locations, limited inventory on hand, and more than one opportunity to stock stores. Multiple stocking opportunities also offer the possibility of forecast updates based on observed sales.

Fisher and Rajaram (2000) consider a demand model, with different store types. They consider the problem of determining the optimal set of test stores to stock prior to the beginning of the selling season. Using sales histories of comparable products from a prior season, they cluster the stores in the chain deterministically using a store similarity measure and then choose one test store from each cluster. Then, in the test period, inventory is placed in the test stores so that demand can be observed, from which, regression is used to estimate sales for the season. They use linear regression to estimate forecasts for season sales. Test stores are obtained deterministically by considering only the prior season sales.

Agrawal and Smith (2008) develop a two period inventory decision model for seasonal items at a retail chain with nonidentical stores. As is typical in such scenarios, they assume that store demands can be correlated across the chain, and across the two time periods. At the beginning of the second period, demand forecasts and inventory policies can be revised, based on the observed demands in the first period. They develop a generalized Bayesian inference model assuming that the store demand distributions share a common unknown parameter. They also develop a two stage optimization methodology to determine the total order quantity, as well as the initial and revised store stocking policies for the two periods, taking into account the fact that store stocking policies in the first period affect the demand information that is collected. If many stores are stocked in the first period, better information about demand may be possible, but fixed costs associated with stocking stores, especially at low-volume ones, can lower profits. Additionally, ordering and inventory allocation decisions made in the first period also affect the amount of inventory that will be available for stores in the second period. To reduce the state space of this problem, they develop a normal approximation for the excess inventory left over at the end of the first period, which greatly simplifies the analysis.

By comparing the performance of the system under different supply chain flexibility arrangements, they develop insights regarding the magnitude of benefits resulting from 1) using updated demand information to modify store inventory levels and the set of stores that are stocked in mid-season, and 2) flexible supply arrangements that allow the total replenishment quantity to be adjusted in mid-season.

3.7 **Transportation issues**

A closely related problem in multi-location systems is that of determining optimal policies and routes for scheduling vehicles to deliver products to the various retailers in the network. The well known joint replenishment problem is also a part of this stream of work. This area represents a substantial body of research, and we will not review it in this paper. However, we will briefly point to some of the papers, and encourage the interested readers to follow the references therein.

Papers that focus on the joint replenishment problem when demand is deterministic include Jackson *et al.* (1985), Anily and Federgruen (1991), Federgruen and Zheng (1992), Vishwanathan and Mathur (1997), Speranza and Ukovich (1994) and Bramel and Simchi-Levi (1995). Papers that consider stochastic demands include Balintfy (1964) (can order, must order, order up to levels in a continuous review setting); Silver (1981) and Federgruen *et al.* (1984) (determining can-order policies); Atkins and Iyogun (1988) (periodic review policies for coordinated replenishments); Pantumsinchai (1992) (heuristics for *QS* policies for multiple items); Vishwanathan and Mathur (1997) ((*T*,*s*,*S*) policies); Pryor *et al.* (1999) (single item with transportation set up costs), and Cachon (2001) (single store but multiple items, capacitated vehicles).

There are also many papers that consider vehicle routing along with inventory costs, but the few among these that allow for stochastic demand include Federgruen and Zipkin (1984b), McGavin *et al.* (1993), Adelman and Kleywegt (1999) and Reinman *et al.* (1999).

3.8 Additional issues

While the focus of the papers discussed thus far was primarily on cost minimization, another approach to system design may be driven by service level targets. For this type of problem, de Kok (1990) assumes that the depot does not carry any stock and imposes a service level target at the retail locations. This model is extended in Verrijdt and de Kok (1995) for more general N-echelon networks, and in de Kok *et al.* (1994) to allow the depot to hold stock as well. Diks and de Kok (1998) derive newvendor equalities for such systems under continuous demand.

In an interesting paper, Erkip *et al.* (1990) consider a multi-echelon model with multiple retail outlets whose demands may be correlated with each other and also across time, but do not consider forecast revision as demand data becomes available. They model demand at retailer j in period t as

$$d_{jt} = R_j \hat{D}_t L_t + \varepsilon_{jt},$$

where R_j is the average fraction of chain-wide demand at store j, $\hat{D}_t$ is the forecasted chain-wide demand, L_t is the normally distributed (with unit mean) *index variable* for period t, and ε_{jt} is the normally distributed (with zero mean) random forecast error at store j. The index variable parameter, common to all stores, is assumed to be an autoregressive process of order one. This is what induces correlation across stores and time. To lend tractability to their analysis, they need to assume that the coefficient of variation of demand at each store is equal. This assumption, along with the allocation assumption at the warehouse allows them to derive newsvendor type cost minimizing solutions for the problem.

While allocation policies are clearly important in the papers discussed above, this issue is also the subject of other papers that have been written in the context of assembly/production systems. In this case, when multiple products require the same common component, the available stock of components needs to be allocated in shortage situations. Similarly, in single location problems where there are multiple "classes" of demand, some allocation mechanism must be designed. Comparing these settings to distribution systems, it is clear that in both these cases, the inventory dynamics at the retail locations are not relevant, but the problem of inventory allocation is similar to that faced by the warehouse in our model. Without reviewing in detail, we list some of the papers in this category for the sake of completeness: Collier (1982), Baker *et al.* (1986), Gerchak and Henig (1986), Gerchak *et al.* (1988), Ha (1997) and Agrawal and Cohen (2001).

Finally, for versions of our problem that include capacity constraints, i.e., capacitated production/distribution systems, see Glasserman and Tayur (1994) and Rappold and Muckstadt (2000), and the references therein.

4. CONCLUSIONS

The review presented in this paper, as well as earlier ones, clearly show that much has been accomplished in the area of designing and managing multi-location retail supply chain structures. However, our collaboration with a number of prominent retail chains has identified a number of practical issues that have yet to be examined in any detail. The brief description of these issues that follows here is by no means an exhaustive list, and the interested reader should append this list to the other open questions discussed in many of the papers that we have reviewed here.

The trend towards micro-merchandising presents the first set of opportunities. Since local consumer preferences vary by location, retailers are attempting to customize their product assortments and model stocks to such local needs. However, this requires investing in mechanisms and methodologies that can allow retailers to determine what such differences are, and how best to let inventory policies be influenced by such information. Correlations between demand across stores and across time add additional complexity to such decisions in general. Agrawal and Smith (2007) present one approach for addressing this problem. This work can be generalized to include multiple products, multiple planning periods, and the potential to use pricing as yet another instrument for supply chain flexibility.

As we move from planning of one product to multiple products that form an assortment, practical considerations relating to product packaging become important. Products often move in supply chains in the form of pre-packs. For example, for an apparel retailer, a pre-pack might consist of one red, 2 black, and 1 grey t-shirt. Such pre-packs may also contain products corresponding to different sizes. Designing such pre-packs is critical to supply chain efficiency. Obviously, smaller pre-packs maximize the ability of stores to match supply and demand cost effectively. However, larger pre-packs minimize packaging and material handling costs throughout the supply chain. They also result in the possibility of shipping more units than are really needed at stores. When retail stores vary greatly in their sales rates, the problem of pre-pack design assumes even greater complexity.

While the mathematical models described in this paper have the ability to make unique inventory decisions at the store level, in practice, for large chains with thousands of stores, managing such a large number of policies is prohibitive. Consequently, stores are often grouped into a manageable number of categories (4–10), such that the same policy can be implemented within a category. While mathematically suboptimal, the practical advantages are substantial. However, this presents us with the interesting question of how best to specify such categories, particularly considering store differences across geographies and product categories.

Pricing and markdown strategies in retail chains are yet another rich area of research. The majority of papers we have discussed here ignore the pricing decision. Most pricing papers that we are aware of are single location models. How best to determine pricing and inventory policies simultaneously across chains is an important research topic for retailing.

Finally, no discussion of the retail industry can be complete without recognizing the tremendous opportunities afforded by multi-channel formats, where retailers attempt to access customers using the traditional store, plus the Internet and catalog channels. Retailers vary greatly in their capabilities to deliver their products and services in this manner, and few appear to have realized any potential supply chain synergies from jointly optimizing such formats. This, we hope, will be a topic that researchers in the area of supply chain management will explore in the coming years.

5. APPENDIX: CONTINUOUS REVIEW INVENTORY SYSTEMS

Many of the results in this research area, particularly for centrally controlled continuous review systems, grew out of the METRIC approximation derived in the seminal work done by Sherbrooke (1968). Consider a one-warehouse multi-retailer system where inventory is managed using a one-for-one $(S\text{-}1,S)$ inventory policy. Further, let the demand distribution at each retailer i be independent and Poisson (λ_i). Then, it follows that the demand faced by the warehouse is Poisson $(\lambda_0 = \Sigma_{i=1..N} \lambda_i)$. Using Palm's theorem, it then follows that the number of outstanding orders at the warehouse has a Poisson distribution with mean $\lambda_0 L_0$, where L_0 is the replenishment lead time at the warehouse. Then, for a given order up to level S_0, expressions for expected backorders (B_0), waiting time (W_0) as well as inventory levels (I_0) can be derived as follows:

$$E(B_0) = \sum_{j=S_0+1}^{\infty} (j - S_0) \frac{(\lambda_0 L_0)^j}{j!} \exp(-\lambda_0 L_0),$$

$$E(W_0) = E(B_0)/\lambda_0,$$

$$E(I_0) = \sum_{j=0}^{S_0-1} (S_0 - j) \frac{(\lambda_0 L_0)^j}{j!} \exp(-\lambda_0 L_0).$$

While the actual lead time is random, the average lead time for retailer orders now equals the shipping lead time plus the average delay time due to shortages at the warehouse. The problem is that the random replenishment lead times for retailers are not independent, since they all depend upon the inventory situation at the warehouse. The METRIC approximation ignores this correlation, and replaces the random lead time with its expected value. This allows results similar to the ones for the warehouse to be derived for the retailers as well. Thus, cost expressions can be derived and optimized.

Exact expressions can be obtained by characterizing the steady state distributions of inventory levels. While the previous papers focused on characterizing the distribution of the retailer lead times, an alternate approach was taken by Axsater (1990) to develop an exact evaluation methodology for the costs directly. In particular, he observed that any unit ordered by facility i will be used to fill the S_i-th unit of demand at this facility following that particular order, where S_i is the order up to level. Therefore, the distribution of the time elapsed between an order and the occurrence of the unit of demand that it will satisfy will have an Erlang (λ_i, S_i) distribution, with the following density function:

$$g_i^{S_i}(t) = \frac{(\lambda_i^{S_i} t^{S_i - 1})^j}{(S_i - 1)!} \exp(-\lambda_i t).$$

Now, conditioning on the delay at the warehouse (which also has an Erlang distribution similar to the one above), cost expressions for that unit can be derived (consisting of holding and backordering costs). Axsater derived a recursive procedure for evaluating the resulting costs. Thus, this method primarily focuses on keeping track of costs associated with arbitrary supply units.

Such procedures and results become ineffective when we consider general systems where one-for-one policies are replaced by batch ordering policies (R, Q) due to fixed ordering costs. In this case, the demand arising from retailers is no longer Poisson, but Erlang instead. Consequently, the demand process at the warehouse is the sum of N Erlang processes, which is more complicated to analyze.

This generalization is considered in Axsater (1993b), where the authors consider a one warehouse multi-retailer inventory system, with *N identical* retailers facing independent Poisson demand. However, all locations are allowed to order in batches using a (R, Q) policy, and the policies at the warehouse are defined in terms of retailer batches. Lead times are assumed to be constant. Unmet demand is assumed to be backordered, and costs include proportional holding as well as backordering costs. The basic idea stems from a similar observation in Axsater (1990). In this case, a sub-batch ordered at the warehouse will fill the $(R_w + 1)^{th}$ subsequent order for a retailer batch at the warehouse. Of course, this will happen after a random number of system demands. The costs are then derived by conditioning on which subsequent demand triggers an order. Exact as well as approximate evaluation procedures are derived.

Following a similar logic, in Axsater (1997), the results are further generalized to a two-level inventory system with one warehouse N retailers and constant lead times (transportation times), but where the retailers face *different compound Poisson* demand processes. All facilities apply continuous review ***echelon*** stock (R, Q) policies and backorder unmet demands. They provide a method for exact evaluation. Note however that echelon stock based policies may not always dominate installation stock based policies.

The third approach to solving such problems is based on characterizing the steady state distribution of inventory levels. For example, Graves (1985) fitted a two parameter Negative Binomial distribution to the number of outstanding orders for the basic METRIC model. In a similar manner, Chen and Zheng (1997) consider a one warehouse N retailer system where the retailers face different but independent compound Poisson demands, lead times are fixed, and orders are restricted to be batches of some specified lot size. They too assume *installation* stock based replenishment policies. For the case of simple Poisson demands, exact results are possible. The inventory level at the warehouse can be determined easily, since its echelon inventory position has a uniform distribution. The distribution of the inventory level at the retailer locations is more complicated, for which the authors determine an exact procedure. For the case of compound Poisson demand, approximate evaluation methods are derived.

REFERENCES

Adelman, D., A. Kleywegt. 1999. Price directed inventory routing. Working paper, Univ. of Chicago, Chicago, IL.

Agrawal, N., S.A. Smith. 1996. Estimating negative Binomial demand for retail inventory management with unobservable lost sales. *Naval Research Logistics*. 43, 839–861.

Agrawal, N., S.A. Smith. 2008. A Bayesian framework for seasonal inventory management in a retail chain with non-identical stores. Working paper. Santa Clara University, CA.

Agrawal, N., M.A. Cohen. 2001. Optimal material control in an assembly system with component commonality. *Naval Research Logistics*. 48, 409–429.

Ahire, S.L., C.P. Schmidt. 1996. A model for a mixed continuous-periodic review one warehouse, N retailer inventory system. *European Journal of Operational Research*. 92, 69–82.

Anily, S., A. Federgruen. 1991. Capacitated two-stage multi-item production/inventory model with joint set up costs. *Operations Research*. 39 (3), 443–455.

Anupindi, R., Y. Bassok. 1999. Centralization of stocks: Retailers vs. manufacturer. *Management Science*. 45 (1) 178–191.

Anupindi, R., Y. Bassok. E. Zemel. 2001. A general framework for the study of decentralized distribution systems. *Manufacturing and Service Operations Management*. 3 (4), 349–368.

Archibald, T.W., S.A.E. Sassesn, L.C. Thomas. 1997. An optimal policy for a two depot inventory problem with stock transfer. *Management Science*. 43(2), 173–183.

Atkins, D., P. Iyogun. 1988. Periodic versus "can-order" policies for coordinated multi-item inventory systems. *Management Science*. 34 (6) 791–796.

Aviv, Y., A. Federgruen. 1998. The operational benefits of information sharing and vendor managed inventory (VMI) programs. Working paper, Washington University, St. Louis, MO.

Axsater, S. 1990. Modeling emergency lateral transshipments in inventory systems. *Management Science*. 36, 1329–1338.

Axsater, S. 1993a. Continuous review policies for multi-level inventory systems with stochastic demand. In Handbooks in Operations Research and Management Science, (Eds.) S.C. Graves, A.H.G. Rinnooy Kan and P.H. Zipkin, Volume 4 (Logistics of Production and Inventory), Elsevier Science Publishing Company B.V., Amsterdam, The Netherlands. 175–197.

Axsater, S. 1993b. Exact and approximate evaluation of batch ordering policies for two-level inventory systems. *Operations Research*. 41 (4), 777–785.

Axsater, S. 1997. Simple evaluation of echelon stock (R,Q) policies for two-level inventory systems. *IIE Transactions*. 29, 661–669.

Axsater, S., J. Marklund, E.A. Silver. 2002. Heuristic methods for centralized control of one-warehouse, N-retailer inventory systems. *Manufacturing and Service Operations Management*. 4 (1), 75–97.

Baker, K. R., M. J. Magazine, H. L. Nuttle. 1986. The effect of commonality on safety stock in a simple inventory model. *Management Science*. 32, 982–988.

Balintfy, J. 1964. On a basic class of multi-item inventory problems. *Management Science*. 10, 287–297.

Bertrand, L.P., J.H. Bookbinder. 1998. Stock redistribution in two-echelon logistics systems. *Journal of the Operations Research Society*. 49, 966–975.

Bradford, J.W., P.K. Sugrue. 1990. A Bayesian approach to the two-period style-goods inventory problem with single replenishment and heterogeneous Poisson demands. *The Journal of the Operational Research Society*. 41 (3), 211–218.

Bramel, J. D. Simchi-Levi. 1995. A location based heuristic for general routing problems. *Operations Research*. 43, 649–660.

Cachon, G.P. 1999. Managing supply chain demand variability with scheduled ordering policies. *Management Science*. 45 (3), 843–856.

Cachon, G. 2001. Managing a retailer's shelf space, inventory and transportation. *Manufacturing and Service Operations*. 3 (3), 211–229.

Cachon, G.P. 2001. Exact evaluation of batch-ordering inventory policies in two-echelon supply chains with periodic review. *Operations Research*. 49 (1), 79–98.

Cachon, G.P., M.L. Fisher. 2000. Supply chain inventory management and the value of shared information. *Management Science*. 46 (8), 1032–1048.

Chang, P.L.., C.T. Lin. 1991. On the effects of centralization on expected costs in a milti-location newsboy problem. *Journal of Operational Research Society*. 42, 1025–1030.

Chen, M.S., C.T. Lin. 1989. Effects of centralization on expected costs in a milti-location newsboy problem. *Journal of Operational Research Society*. 40, 597–602.

Chen, F., 1998. Echelon reorder points, installation reorder points, and the value of centralized demand information. *Management Science*. 44 (12), S221-S234.

Chen, F., Y.S. Zheng. 1994. Lower bounds for multi-echelon stochastic inventory systems. *Management Science*. 40 (11), 1462–1443.

Chen, F., Y.S. Zheng. 1997. One-warehouse multiretailer systems with centralized stock information. *Operations Research*. 45 (2), 275–287.

Chen, F., R. Samroengraja. 2000. A staggered ordering policy for one-warehouse multiretailer systems. *Operations Research*. 48 (2), 281–293.

Clark, A.J., and H. Scarf. 1960. Optimal policies for a multi-echelon inventory problem. *Management Science*, 40, 1426–1443.

Cohen, M.A., N. Agrawal, V. Agrawal. 2006. Achieving breakthrough service delivery through dynamic asset deployment strategies. *Interfaces*. 36 (3), 259–271.

Collier, D. A. 1982. Aggregate safety stock levels and component part commonality. *Management Science*. 28, 1296–1303.

de Kok, A.G. 1990. Hierarchical production planning for consumer goods. *European Journal of Operations Research*, 45, 55–69.

de Kok, A.G., A.G. Lagodimos, H.P. Siedel. 1994. Stock allocation in a two-echelon distribution network under service constraints. Working Paper. Department of Industrial Engineering and Management Science, Eindhoven University of Technology, EUT 94-03.

Diks, E.B., A.G. de Kok. 1998. Optimal control of a divergent multi-echelon inventory system. *European Journal of Operations Research*. 111, 75–97.

Dong, L., N. Rudi. 2004. Who benefits from transshipment? Exogenous vs. endogenous wholesale prices. *Management Science*. 50 (5), 645–657.

Dogru, M.K., A.G. de Kok, G.J. van Houtum. 2004. Optimal control of one-warehouse multi-retailer systems with discrete demand. *Working paper*, Department of Technology Management, Technische Universiteit Eindhoven, Eindhoven, Netherlands.

Eppen, G.D. 1979. Effect of centralization on expected costs in a milti-location newsboy problem. *Management Science*. 25, 498–501.

Eppen, G. and L. Schrage. 1981. Centralized ordering policies in a multi-warehouse system with lead times and random demands. In L.B. Schwarz (ed.) *Multi-level production/inventory control systems: Theory and practice*, North-Holland, Amsterdam, 51–67.

Erkip, N., W. Hausman, S. Nahmias. 1990. Optimal centralized ordering policies in multi-echelon inventory systems with correlated demands. *Management Science*. 36 (3), 381–392.

Federgruen, A., 1993. Centralized planning models for multi-echelon inventory systems under uncertainty. In Graves, S.C., Rinnooy Kan, A,H.G., Zipkin, P.H. (Eds.), *Logistics of production and inventory. Handbooks in Operations Research and Management Science*, vol. 4, ch. 3. Elsevier, Amsterdam, pp. 133–173.

Federgruen, A., H. Groenevelt, H. Tijms. 1984. Coordinated replenishments in a multi-item inventory system with compound Poisson demands and constant lead times. *Management Science*. 30, 344–357.

Federgruen, A., P.H. Zipkin. 1984a. Approximation of dynamic, multi-location production and inventory problems. *Management Science*. 30, 69–84.

Federgruen, A., P. Zipkin. 1984b. A combined vehicle routing and inventory allocation problem. *Operations Research*. 32 (5), 1019–1037.

Federgruen, A., Y.S. Zheng. 1992. The join replenishment problem with general joint cost structures. *Operations Research*. 40, 384–403.

Fisher, M.L., K. Rajaram. 2000. Accurate retail testing of fashion merchandise: Methodology and application. *Marketing Science*, 19 (3), 266–278.

Gavirneni, S., R. Kapuscinski, S. Tayur. 1999. Value of information in capacitated supply chains. *Management Science*. 45 (1), 16–24.

Gerchak, Y., M. Henig. 1986. An inventory model with component commonality. *Operations Research Letters*. 5, 157–160.

Gerchak, Y., M. J. Magazine, A. B. Gamble. 1988. Component commonality with service level requirements. *Management Science*. 34, 753–760.

Glasserman, P., S. Tayur. 1994. The stability of a capacitated, multi-echelon production-inventory system under a base-stock policy. *Operations Research*. 42 (5), 913–925.

Graves, S.C. 1985. A multi-echelon inventory model for a repairable item with one-for-one replenishment. *Management Science*. 31 (10), 1247–1256.

Graves, S. C. 1996. A multi echelon inventory model with fixed replenishment intervals. *Management Science*. 42 (1), 1–18.

Ha, A. 1997. Inventory rationing in a make-to-stock production system with several demand classes and lost sales. *Management Science*. 43 (8), 1093–1103.

Herer, Y.T., M. Tzur, E. Yucesan. 2006. The multilocation transshipment problem. *IIE Transactions*. 38, 185–200.

Jackson, P.L. 1988. Stock allocation in a two-echelon inventory system or what to do until your ship comes in. *Management Science*. 34, 880–895.

Jackson, P., W. Maxwell, J. Muckstadt. 1985. The joint replenishment problem with power-of-two intervals. *IIE Transactions*. 17, 25–32.

Jackson, P.L., J.A. Muckstadt. 1989. Risk pooling in a two-period multi-echelon inventory stocking and allocation problem. *Naval Research Logistics*. 36, 1–26.

Jonsson, H., E.A. Silver. 1987. Analysis of a two-echelon inventory control system with complete redistribution. *Management Science*. 33 (2), 215–227.

Karmarkar, U.S., N.R. Patel. 1977. The one-period, N-location distribution problem. *Naval Research Logistics Quarterly*. 24, 559–575.

Krishnan, K.S., V.R.K. Rao. 1965. Inventory control in N warehouse. *Journal of Industrial Engineering*. 16, 212–215.

Lariviere, M.A., E. L. Porteus. 1999. Stalking information: Bayesian inventory management with unobserved lost sales. *Management Science*. 45 (3), 346–363.

Lee, H.L. 1987. A multi-echelon inventory model for repairable items with emergency lateral transshipments. *Management Science*. 45(5), 633–640.

Lee, H.L., P. Padmanabhan, S. Whang. 1997a. Information distortion in a supply chain: The bullwhip effect. *Management Science*. 43, 546–558.

Lee, H.L., P. Padmanabhan, S. Whang. 1997b. Bullwhip effect in a supply chain. *Sloan Management Review*. 38, 93–102.

Lee, H.L., C. So, C.S. Tang. 2000. The value of information sharing in a two-level supply chain. *Management Science*. 46 (5), 626–643.

Lippman, S.A., K. F. McCardle. 1997. The Competitive Newsboy. *Operations Research*. 45 (1), 54–65.

McGavin, E.J., L.B. Schwarz, J.E. Ward. 1993. Two-interval inventory allocation policies in a one-warehouse N-identical retailer distribution system. *Management Science*. 39 (9), 1092–1107.

Muckstadt, J. A.. 2004. Analysis and algorithms for service parts supply chains. Springer Series in Operations Research and Financial Engineering. Springer Verlag.

Nahmias, S., S.A. Smith. 1993. Mathematical models of retailer inventory systems: A review. In *Perspectives in Operations Management*, R. K. Sarin Editor, Kluwer Academic Publishers. MA. 249–278.

Nahmias, S. and S.A. Smith. 1994. Optimizing inventory levels in a two-echelon retailer system with partial lost sales. *Management Science*. 40.(5),582–596.

Pantumsinchai, P. 1992. A comparison of three joint ordering inventory policies. *Decision Science*. 23 111–127.

Parlar, M.. 1988. Game theoretic analysis of the substitutable product inventory problem with random demands. *Naval Research Logistics*. 35, 397–409.

Pryor, K., R. Kapuscinski, C. White. 1999. A single item inventory problem with multiple setup costs assigned to delivery vehicles. Working paper, University of Michigan, Ann Arbor.

Rappold, J.A., J.A. Muckstadt. 2000. A computationally efficient approach for determining inventory levels in a capacitated multi-echelon production-distribution system. *Naval Research Logistics*. 47, 377–398.

Reinman, M., R. Rubio, L. Wein. 1999. Heavy traffic analysis of the dynamic stochastic inventory-routing problem. *Transportation Science*. 33 (4), 361–380.

Robinson, L.W. 1990. Optimal and approximate policies in multi-period, multi-location inventory models with transshipments. *Operations Management*. 38 (2), 278–295.

Rudi, N., S. Kapur, D.F. Pyke. 2001. A two location inventory model with lateral shipment and local decision making. *Management Science*. 47 (12), 1668–1680.

Sherbrooke, S.C. 1968. Metric: A multi-echelon technique for recoverable item control. *Operations Research*. 16 (1), 122–141.

Silver, E. 1981. Establishing reorder points in the (S, c, s) coordinated control system under compound Poisson demand. *International Journal of Production Research*. 19, 743–750.

Speranza, M.G. W. Ukovich. 1994. Minimizing transportation and inventory costs for several products on a single link. *Operations Research*. 42 (5), 879–896.

Stulman, A. 1987. Benefits of centralized stocking for the multi-center newsboy problem with first-come-first-serve allocation, *Journal of Operational Research Society*. 38, 827–832.

Svoronos, A., P. Zipkin. 1988. Estimating the performance of multi-level inventory systems. *Operations Research*. 36 (1), 57–72.

Tagaras, G. 1989. Effects of pooling on the optimization and service levels of two-location inventory systems. *IIE Transactions*. 21 (3), 250–257.

Tagaras, G., M.A. Cohen. 1992. Pooling in two-location inventory systems with non-negligible replenishment lead times. *Management Science*. 38 (8), 1067–1083.

Tagaras, G., D. Vlachos. 2002. Effectiveness of stock transshipment under various demand distributions and nonneglibible transshipment times. *Production and Operations Management*. 11 (2), 183–198.

Van Donselaar, K. and J. Wijngaard. 1987. Commonality and safety stocks. *Engineering Costs and Production Economics*. 12, 197–204.

Verrijdt, J.H.C.M., de Kok, A.G. 1995. Distribution planning for a divergent N-echelon network without intermediate stock under service restrictions. *International Journal of Production Economics*, 38, 225–243.

Viswanathan, S. 1997. Periodic review (s,S) policies for joint replenishment inventory systems. *Management Science*. 43 (10), 1447–1453.

Vishwanathan, S., K. Mathur. 1997. Integrating routing and inventory decisions in a one-warehouse multi-retailer multi-product distribution system. *Management Science*. 43 (3), 294–312.

Zhao, H., V. Deshpande, J.K. Ryan. 2005. Inventory sharing and rationing in decentralized dealer networks. *Management Science*. 51 (4), 531–547.

Chapter 10
MANUFACTURER-TO-RETAILER VERSUS MANUFACTURER-TO-CONSUMER REBATES IN A SUPPLY CHAIN

Goker Aydin[1] **and Evan L. Porteus**[2]
[1]*Department of Industrial and Operations Engineering, The University of Michigan, Ann Arbor, MI, USA*
[2]*Graduate School of Business, Stanford University, Stanford, CA, USA*

Abstract Starting with a newsvendor model (single-product, single-period, stochastic demand), we build a single-retailer, single-manufacturer supply chain with endogenous manufacturer rebates and retail pricing. The demand uncertainty is multiplicative, and the expected demand depends on the effective (retail) price of the product. A retailer rebate goes from the manufacturer to the retailer for each unit it sells. A consumer rebate goes from the manufacturer to the consumers for each unit they buy. Each consumer's response to consumer rebates is characterized by two exogenous parameters: α, the effective fraction of the consumer rebate that the consumer values, leading to the lower effective retail price perceived by the consumer, and β, the probability that a consumer rebate will be redeemed. The type(s) of rebate(s) allowed and the unit wholesale price are given exogenously. Simultaneously, the manufacturer sets the size of the rebate(s) and the retailer sets the retail price. The retailer then decides how many units of the product to stock and the manufacturer delivers that amount by the beginning of the selling season. Compared to no rebates, an equilibrium retailer rebate leads to a lower effective price (hence, higher sales volume) and higher profits for both the supply chain and the retailer. An equilibrium consumer rebate also leads to a lower effective price and higher profits for the retailer, but not necessarily for the chain. Under our assumptions, such a consumer rebate (with or without a retailer rebate) allocates a fixed fraction of the (expected) supply chain profits to each player: The retailer gets $\alpha/(\alpha + \beta)$ and the manufacturer gets the rest, leading to interesting consequences. However, both firms prefer that α be higher and β lower: Even though the manufacturer gets a smaller share of the chain profits, the total amount received is higher. Neither the retailer nor the manufacturer always prefers one particular kind of rebate to the other. In addition, contrary to popular belief, it is possible for both firms to prefer consumer rebates even when all such rebates are redeemed.

N. Agrawal, S.A. Smith (eds.), *Retail Supply Chain Management*,
DOI: 10.1007/978-0-387-78902-6_10,

Keywords Supply Chain Management · Inventory and Pricing Decisions · Consumer Rebates · Channel Rebates

1. INTRODUCTION

Rebates are widely used as promotional tools. In this paper we investigate the effects of two kinds of rebates (from the manufacturer) on supply chains: retailer rebates and consumer rebates. Retailer rebates, also known as channel rebates, are payments from the manufacturer to the retailer based on the sales performance of the retailer. Taylor (2002) cites several examples of the use of retailer rebates, in industries that range from software to printers, from network hardware switching to automotive. Consumer rebates, which are no less widespread than retailer rebates, are payments from the manufacturer to the consumer upon the consumer's purchase of the manufacturer's product. Most everybody is familiar through personal experience with the use of consumer rebates in consumer electronics, automotive and food products industries. The magnitude of rebate offers can reach surprisingly large numbers: A *New York Times* article reports that $10 billion worth of consumer rebates were offered in 2002 (Millman, 2003).[1] Although some consumers do not claim their rebates (especially when the rebate size is small), the number of claims for consumer rebates is not negligible either: In 1998 Young America Inc. was reported to mail out 30 million rebate checks a year on behalf of companies like PepsiCo Inc., Nestle SA and OfficeMax (Bulkeley, 1998).

For both retailer and consumer rebates, there do exist different implementations. Retailer rebates can be paid for each unit the retailer sells to the end customer or only for units sold in excess of a target number (Taylor, 2002). Here we focus on the former type. In our model, the manufacturer uses consumer rebates for the sole purpose of selling more to the retailer. Thus, we do not address the role they may have early in a product's life cycle to learn more about demand or later to increase demand for unintended excess inventories. Consumer rebates can be in the form of mail-in rebates or coupons. Moreover, there are different kinds of coupons; some can be instantly redeemed at the time of purchase and some can be used only the next time a product is purchased. Of course, the specifics of the rebate offer have an influence on how attractive consumers find the rebate and how many customers will redeem the rebate. Here we use a stylized model of consumer rebates. We assume that (all) consumers treat a rebate of $1 as being equivalent to a price discount of $\$\alpha$ and will redeem their rebates with probability β, where $0<\alpha \leq 1$ and $0<\beta \leq 1$. Thus, if a consumer rebate of x is offered on a product with price p, then the *effective retail price* is $p - \alpha x$ and if y customers buy the product, then the expected

[1] In some cases, retailers themselves offer rebates to consumers. It is possible that the amount $10 billion quoted in the article includes the rebates offered by the retailers themselves.

number of claims will be βy. Note that consumers are homogeneous in regard to the parameter α and we do not explicitly model a customer's decision of whether to claim a rebate or not. We shall see that modeling consumer rebates at this aggregate level allows us to identify the roles of the *claim rate* β and the *effective fraction* α in splitting the supply chain profit between the retailer and the manufacturer.

The values of both α and β are likely to differ from one product category to another. For example, according to a survey of AC Nielsen's Homescan Consumer Panel, 27.7% of households that reported buying computer products said mail-in rebates were very important when they bought PCs, monitors, printers and peripherals; 35.7% said they were somewhat influenced by rebates (Ricadela and Koenig, 1998). The same article reports, however, that consumers are less influenced by rebates when purchasing software. This example suggests that the value of α is likely to differ from one product category to another. The claim rate, on the other hand, is likely to depend on the size of the rebate itself. For example, an educational software vendor reports that 8 to 10% of its customers claim \$10 rebates, and the claim rate increases to 20% for \$20 rebates (Bulkeley, 1998). Nevertheless, the rebate sizes tend to be similar within a product category and, hence, the product category seems to be a more important determinant of the claim rate than the size of the rebate. For example, in contrast to the software vendor who faced claim rates in the 10% to 20% range, eMachines' mail-in rebate program had a 70 to 90% claim rate prior to its cancellation (Olenick, 2002). In the case of new automotive purchases, where the rebates are even larger, the usual practice is for the rebate to be instantaneously redeemable at the time of purchase, which suggests that $\alpha = \beta = 1$. In summary, while consumer response to rebate offers may vary in the size of the rebate, much of this variation may be accounted for by the product category.

In order to compare and contrast the effects of the two rebate types on the supply chain, we consider a single-retailer, single-manufacturer supply chain selling a single product, and we analyze the equilibrium outcome under each rebate policy. (The decision of what rebate type to use is not endogenous to our model; instead, we analyze and compare the equilibria under each rebate type.) We assume that the wholesale price for the product is exogenously fixed. This assumption is mainly for tractability, but it is also an approximation of an environment where rebate offers constitute a further stage of decision making in a supply chain with a well-established wholesale price. The consumer demand for the product is stochastic and depends on the effective retail price. In the case of a retailer rebate, the effective retail price is simply the retail price, whereas in the case of a consumer rebate, the effective retail price is the retail price minus the effective fraction of the consumer rebate. We assume that the expected demand for the product is a function of the effective retail price, and the realized demand is a multiplicative random perturbation of that expected demand. The assumption of a multiplicative model is not without consequence; it implies that the coefficient of variation of demand is constant with respect to price.

Under either rebate policy, before the start of the single-period selling season, the retailer must determine the retail price, and the manufacturer needs to choose the size of the rebate (or rebates, if both rebate types are used in the supply chain) simultaneously. This simultaneous determination of the rebates and the retail price can be seen as approximating a negotiation process between the manufacturer and the retailer in setting the terms of a rebate offer. Once the price and rebate(s) are announced, the retailer decides how many units of the product to purchase. The manufacturer builds that amount and delivers it to the retailer by the beginning of the selling season. At the end of the selling season, all unmet demands become lost sales, and leftover inventory is salvaged. This model would be particularly applicable to high-tech products where the short life cycle of the product can be modeled as covering a single season with a single ordering and pricing opportunity. The more replenishments take place during the life cycle of the product and the more price adjustments made, the more approximate our model becomes.

Of course, both retailer and consumer rebates provide the retailer with an incentive to stock more. However, the two rebates differ in how they achieve this result: Retailer rebates do so by increasing the retailer's margin on every unit sold, whereas consumer rebates do so by boosting the demand for the product. We find that, as expected (in equilibrium), when retailer rebates are present, the retailer will reduce the retail price (by an amount less than the rebate itself) to increase the sales volume of an item and collect a larger sum from the manufacturer in rebates, thereby passing on to the consumer some of the benefits it receives. On the other hand, a consumer rebate will induce the retailer to increase the retail price (by an amount less than the effective rebate) to take advantage of the boost in demand that arises from a consumer rebate, thereby sharing in some of the benefits offered to consumers. We show that the total supply chain profit always improves under retailer rebates, compared to no rebates. The same is true for consumer rebates, provided that the effective fraction (α) is larger than the claim rate (β). However, if $\alpha < \beta$, then total supply chain profit may suffer. We provide numerical examples to demonstrate that neither the retailer nor the manufacturer always prefers one particular kind of rebate to the other. In addition, our numerical examples suggest that, contrary to popular belief, it is possible for both firms to prefer consumer rebates even when all such rebates are redeemed.

In comparing the two rebate types, we find that the split of supply chain profits under consumer rebates depends critically on α and β. In particular, we obtain the following results:

- Under the consumer rebate equilibrium, the retailer's share of the supply chain profit will be $\frac{\alpha}{\alpha+\beta}$, and the manufacturer's $\frac{\beta}{\alpha+\beta}$. In other words, the profit will be divided so that the ratio of the retailer profit to the manufacturer profit will be α/β.
- The higher α is with respect to β (i.e., the higher consumers value the rebate relative to the rate at which consumers redeem them), the more attractive the

consumer rebate becomes from the overall supply chain's perspective. Therefore, one can conclude that, everything else being equal, the more attractive the consumer rebate from the overall supply chain's perspective, the larger the retailer's share of the supply chain profit will be in equilibrium.

- Note that the retailer's share is increasing in α and decreasing in β, and the opposite is true for the manufacturer. Nevertheless, as we demonstrate through a numerical example, this does not mean that the retailer and the manufacturer are at odds in terms of what α and β they prefer. It turns out that, under a consumer rebate equilibrium, both firms can prefer α to be larger and β to be smaller; even though the manufacturer's share of supply chain profits is smaller, the manufacturer gets more, because the increase in the supply chain profits more than compensates for the decrease in the share it gets.

In the next section, we review the related literature and compare our model to those in earlier research. Section 3 describes our model and discusses our results for the case where both rebate types are used simultaneously. In Sections 4 and 5, we discuss our results when retailer rebates and consumer rebates are used in isolation. We provide a number of numerical examples in Section 6 to demonstrate some interesting equilibrium outcomes. We conclude in Section 7. All proofs are provided in the appendix.

2. LITERATURE REVIEW

The marketing and economics literature has investigated the use of consumer rebates. For example, Gerstner and Hess (1991, 1995) use a demand model where the consumer population consists of two segments; the size and reservation price of each segment is deterministic and known. The higher-end segment has a cost associated with redeeming a consumer rebate, reflecting the higher disutility price-insensitive customers have for claiming rebates. The supply chain is assumed to be serving only the higher-end segment in status quo. They examine how retailer rebates (called push price promotions) and consumer rebates (called pull price promotions) can be used to induce the retailer to serve the lower-end segment as well as the higher-end one, and how such promotions affect manufacturer and supply chain profits. Narasimhan (1984) offers a price discrimination argument to explain the use of consumer rebates. He considers a model where the firm offering the rebate is selling directly to the end consumer. In his model, a consumer need not redeem a rebate every time she purchases a product. He models the consumer's decision of how many rebates to use as a utility maximization problem, and shows that the more price-sensitive a customer, the more she engages in consumer rebates. Therefore, rebates result in the firm selling at a lower price to consumers who are more price sensitive. In this sense, the consumer rebate acts as a price discrimination device. Our model is less general than this stream of research because we do not model how individual consumers respond differently to rebate offers.

Instead, we model the effect of rebates at the aggregate demand level, through the effective fraction parameter α and the claim rate parameter β. Our model is more general in the sense that we incorporate demand uncertainty and retail stock level decisions.

There is also a stream of research in marketing that considers the use of trade promotions; i.e., a discount in wholesale price offered by the manufacturer in order to induce the retailer to lower the retail price. Since the typical assumption of this research stream is that all demand is met (i.e., sales equals demand), such a discount in wholesale price is equivalent to a retailer rebate. Most of the model-based work in this research stream involves multiple competing manufacturers, and the emergence of trade promotions is explained through the equilibrium of the game among these multiple manufacturers. In this setting, the manufacturer is assumed to be selling directly to the end consumers, and the role of the retailer is ignored. See, for example, Raju, Srinivasan and Lal (1990), Lal (1990) and Rao (1991). Our model has only a single manufacturer, but we add explicit consideration of a retailer, demand uncertainty, and the retailer's decision of the stock level.

There is another marketing research stream on trade promotions that considers manufacturers selling through a retailer. For example, Lal, Little and Villas-Boas (1996) consider an infinite horizon model where two identical manufacturers sell through a single retailer. Their customer population consists of three customers: one switcher and two loyals. In this model, trade promotions exist because the manufacturers compete for the switcher. Dreze and Bell (2003) consider a single-retailer, single-manufacturer setting where customer demand is a deterministic function of price. They compare the effects of two different contractual arrangements for trade promotions: off-invoice deals that correspond to a wholesale price discount and scan-back deals that correspond to retailer rebates. In this model, even though demand is deterministic, the retailer may choose to carry inventories to take advantage of a temporary promotional offer from the manufacturer. In our model, the reason a retailer chooses to carry inventories is due to demand uncertainty. We also emphasize how the rebates affect supply chain profits and the shares that the two firms get.

There is some recent work in the operations management literature that considers the role played by retailer rebates in the presence of operational concerns like inventory costs. Taylor (2002) considers retailer rebates in a model where demand is stochastic, but the retail price is exogenously given. He shows that retailer rebates paid for units sold beyond a target level can be used to achieve supply chain coordination. He also analyzes a model where the retailer can exert sales effort to influence demand. In this case, retailer rebates can still achieve coordination, but a returns policy should also be implemented. Using a more general model, Krishnan, Kapuscinski and Butz (2004) focus on the use of retailer rebates in the presence of retailer efforts. Their main focus is finding coordinating contracts. Unlike these two, we do not model the retailer's sales effort; however, we consider a model with price-dependent stochastic demand, and retail price is endogenous to our model in that the retailer decides what price to charge. We do not seek to establish channel-coordinating mechanisms, but we do show that retailer rebates improve supply chain profits. We also compare the supply chain profit under retailer rebates with that under consumer rebates.

More recently, Chen, Li, Rhee and Simchi-Levi (2007) consider the question of consumer rebates from an operations management perspective. As in our model, they consider a single-retailer, single-manufacturer supply chain where one-shot inventory and pricing decisions are made to satisfy price-dependent uncertain customer demand. Their consumer rebate is an exogenously fixed fraction of the wholesale price and the decision making is sequential: the manufacturer chooses the wholesale price first, and the retailer chooses the retail price second. Our wholesale price is exogenous but our consumer rebate is a decision variable. We add consideration of retailer rebates and our assumptions allow us to show how the claim rate and the effective fraction parameters affect the split of supply chain profits between the retailer and the manufacturer.

There is an extensive operations management literature on the price setting newsvendor problem, in which a retailer faces a single-period inventory and pricing problem with stochastic, price-dependent demand. See, for example, Petruzzi and Dada (1999) for a review with extensions. Our analysis benefits from Petruzzi and Dada (1999); in particular, Lemma 4(a) in the appendix is due to them. In their multiplicative model, they assume that demand is given by $ap^{-b}\epsilon$, where ϵ is a random variable. In this demand model, the price elasticity of expected demand is constant. Our assumptions do not cover this specific model, but we do allow the (absolute) price elasticity of expected demand to be increasing in price, thereby complementing some of the existing structural results on the price setting newsvendor problem. Kalyanam (1996) finds empirical support for both constant and increasing price elasticity of demand.

In this chapter, we use an inverse demand representation to write the retailer's and manufacturer's expected profit functions, which facilitates our analysis. (See the next section.) Aydin and Porteus (2006) study an inventory and pricing problem where a retailer sets the prices and inventory levels for an assortment of substitutable products, and they take advantage of a similar representation.

3. CONSUMER AND RETAILER REBATES TOGETHER

In this section, we describe our model when the manufacturer uses both retailer and consumer rebates, and we derive some preliminary results. The use of both rebates at the same time is quite common in the automotive industry, where retailer rebates are usually called dealer incentives and the consumer rebates are offered in the form of cashback allowances. In the following sections, we will focus on the cases where each rebate type is used in isolation, and the results developed in this section will apply to those special cases. Let r_{R} denote the retailer rebate and r_{C} the consumer rebate, each paid to their respective recipients for every unit the customer buys. Also, let p be the retail price of the product.

Let us first describe the demand model. First, the higher the consumer rebate the larger the stochastic demand will be. Therefore, the demand should be a function of r_C as well as p. Let $D(p, r_C)$ denote the stochastic demand for the product. We assume that consumers treat a \$1 rebate as the equivalent of an \$$\alpha$ price discount; i.e., consumers act as if the unit retail price they are paying is $p - \alpha r_C$. We will impose the following assumptions on the demand model:

(A1) $D(p, r_C) = f(p - \alpha r_C)\epsilon$,
(A2) ϵ is a strictly positive random variable with a strictly increasing failure rate (IFR),
(A3) $f(\cdot)$ is strictly decreasing, and $f(x) \to 0$ as $x \to \infty$, and
(A4) $\frac{f'(\cdot)}{f(\cdot)}$ is non-increasing.

The first assumption implies that the expected demand is a function of the retail price minus the effective consumer rebate (i.e., the price after rebate). (*A*1) and (*A*2) implicitly assume that ϵ is independent of price and any rebate. Thus, (*A*1) implies that the coefficient of variation of demand for the product does not change with price. The requirement in (*A*2) that ϵ be IFR is not very restrictive as many probability density functions, including the normal and Weibull with shape parameter greater than one, satisfy this assumption. (For more on IFR distributions, see Barlow and Proschan, 1965.) (*A*3) is a natural assumption that means the expected demand is decreasing in price. This assumption is violated only for very few luxury items. (*A*4) implies that the magnitude of the expected demand's elasticity to price is increasing in p; i.e., as price gets larger the percentage change in demand in response to a percentage change in price gets larger. (*A*4) is satisfied by many commonly used forms of price dependency. For example, it is easy to check that (*A*4) will be satisfied when expected demand is exponentially decreasing in price; i.e., $f(x) = e^{-ap}$, or when expected demand is linearly decreasing in price; i.e., $f(x) = a - bx$, or when expected demand is given by the logit demand model; i.e., $f(x) = \frac{exp(u_1 - x)}{exp(u_0) + exp(u_1 - x)}$.

We define the following notation:

w : unit wholesale price charged by the manufacturer

c : unit production cost

v : unit salvage value

$\Phi(x, p - \alpha r_C)$: cumulative distribution function (cdf) of $D(p, r_C)$

$\phi(x, p - \alpha r_C)$: probability density function (pdf) of $D(p, r_C)$

$\Phi_\epsilon(\cdot)$: cdf of ϵ

$\phi_\epsilon(\cdot)$: pdf of ϵ

We assume that $\Phi(x, p - \alpha r_C)$ is twice-continuously differentiable in both its arguments. Throughout the remainder of the paper, given a function g of vector

x, we use $\nabla_i g(\tilde{x})$ to denote the partial derivative of $g(x)$ with respect to the ith component of x evaluated at $x = \tilde{x}$. Similarly, $\nabla^2_{ij} g(\tilde{x})$ and $\nabla^2_{ii} g(\tilde{x})$ denote the cross-partial and second partial of $g(x)$ at $\tilde{x}$, respectively.

Before the selling season starts, the retailer determines p, and, simultaneously, the manufacturer chooses $r_{\mathbf{R}}$ and $r_{\mathbf{C}}$. The assumption of simultaneous decision making implies that one party in the supply chain is not particularly more powerful than the other, so one party cannot impose its respective decision on the other. We assume that all the parameters and distributions are known by both the retailer and the manufacturer.

Once the price and rebates are announced, the retailer chooses the stock level and the manufacturer then builds that amount, which is delivered to the retailer by the beginning of the selling season. After the selling season is over, the retailer will salvage the leftover inventory at unit salvage value of v. We assume that $w > c > v$ for the problem to make economic sense. In the presence of retailer rebates, there is the possibility that the retailer could misreport the amount of sales to collect larger rebates from the manufacturer. For example, the retailer could dump all the leftover inventory and claim that it had been sold. While the existence of a salvage value alleviates this moral hazard problem, a complete avoidance of such misreporting of sales requires some form of possibly costly monitoring of retail sales. We return to this issue in Section 7.

The retailer's profit function is given by

$$\begin{aligned}\Pi_{\mathbf{R}}(p, y, r_{\mathbf{C}}, r_{\mathbf{R}}) = (p + r_{\mathbf{R}})\left[\int_0^y x\phi(x, p - \alpha r_{\mathbf{C}})dx + y(1 - \Phi(y, p - \alpha r_{\mathbf{C}}))\right] \\ + v\int_0^y (y - x)\phi(x, p - \alpha r_{\mathbf{C}})dx - wy.\end{aligned} \tag{1}$$

Note that the optimal stock level for the product, $y^*(p, r_{\mathbf{C}}, r_{\mathbf{R}})$, is given for each given retail price p, wholesale price w and rebates $r_{\mathbf{R}}$ and $r_{\mathbf{C}}$ as the critical fractile solution:

$$\Phi(y^*(p, r_{\mathbf{C}}, r_{\mathbf{R}}), p - \alpha r_{\mathbf{C}}) = \frac{p + r_{\mathbf{R}} - w}{p + r_{\mathbf{R}} - v} \tag{2}$$

It is important to note how the two different kinds of rebates affect $y^*(p, r_{\mathbf{C}}, r_{\mathbf{R}})$: the stock level chosen depends on the retailer rebate since the critical fractile itself is a function of the retailer rebate, whereas the consumer rebate affects the stock level through its impact on the demand distribution.

The retailer's profit function can be rewritten as the following induced profit function, obtained by substituting for $\Phi(y^*(p, r_{\mathbf{C}}, r_{\mathbf{R}}), p - \alpha r_{\mathbf{C}})$ in (1) (see, for example, Porteus, 2002):

$$\Pi_{\mathbf{R}}(p, r_{\mathbf{C}}, r_{\mathbf{R}}) = (p + r_{\mathbf{R}} - v)\int_0^{y^*(p, r_{\mathbf{C}}, r_{\mathbf{R}})} x\phi(x, p - \alpha r_{\mathbf{C}})dx, \tag{3}$$

where $y^*(p, r_C, r_R)$ is as defined by (2). Define the inverse demand function $z(p, r_C, \xi)$ as

$$\Phi(z(p, r_C, \xi), p - \alpha r_C) = \xi. \tag{4}$$

With this definition, $z(p, r_C, \xi)$ is the demand that corresponds to the ξ fractile of Φ, given the retail price p and consumer rebate r_C. We use this representation as it provides a more convenient way of dealing with the pricing problems to be solved. Using the inverse demand function, we can rewrite (2) as $y^*(p, r_C, r_R) = z(p, r_C, \frac{p+r_R-w}{p+r_R-v})$. Also, we can rewrite the retailer's induced profit function in (3) as

$$\Pi_R(p, r_C, r_R) = (p + r_R - v) \int_0^{\frac{p+r_R-w}{p+r_R-v}} z(p, r_C, \xi) d\xi. \tag{5}$$

The following proposition states our structural result on $\Pi_R(p, r_C, r_R)$.

Proposition 1 *Suppose (A1) through (A4) hold. Then, given r_C and r_R, there is a unique $p > w - r_R$ that optimizes the retailer's profit, and this unique* p *satisfies the first order condition (FOC) for $\Pi_R(p, r_C, r_R)$.*

The manufacturer's profit function is given by

$$\begin{aligned} \Pi_M(p, r_C, r_R) =& (w - c) y^*(p, r_C, r_R) \\ & - (\beta r_C + r_R) \left[\int_0^{\frac{p+r_R-w}{p+r_R-v}} z(p, r_R, \xi) d\xi + \frac{w - v}{p + r_R - v} y^*(p, r_C, r_R) \right]. \end{aligned} \tag{6}$$

The first term in (6) is the profit margin of the manufacturer multiplied by the number of units ordered by the retailer. The term in brackets is the expected sales. Note that the rebate the manufacturer pays per unit sold is the retailer rebate r_R, plus a fraction β of the consumer rebate r_C (since a fraction β of consumers claim their rebate). Therefore, the expected total rebate payment made by the manufacturer is $\beta r_C + r_R$ multiplied by the expected sales. The following proposition states some structural results on $\Pi_M(p, r_C, r_R)$:

Proposition 2 *Suppose (A1) through (A4) hold. Then, given p:*

(a) *Suppose r_R is fixed so that $p + r_R > v$. Then, either the manufacturer's profit is optimized at $r_C = 0$, or there exists a unique r_C that satisfies the FOC for $\Pi_M(p, r_C, r_R)$ and such r_C optimizes the manufacturer's profit.*
(b) *Suppose r_C and p are fixed. Then, either the manufacturer's profit is optimized at $r_R = 0$, or there exists a unique r_R that satisfies the FOC for $\Pi_M(p, r_C, r_R)$ and such r_R optimizes the manufacturer's profit.*
(c) *At any r_C and r_R such that $\nabla_2 \Pi_M(p, r_C, r_R) = \nabla_3 \Pi_M(p, r_C, r_R) = 0$, we have $\nabla_2 y^*(p, r_C, r_R)/\beta > \nabla_3 y^*(p, r_C, r_R)$.*

Parts (a) and (b) of the proposition establish that the manufacturer's profit is well-behaved in the rebates. We cannot rule out the possibility that the manufacturer's profit will be decreasing in the retailer or the consumer rebate. Therefore, the manufacturer's optimal solution may involve a zero rebate. To understand part (c), note that increasing the retailer rebate by \$1 costs the manufacturer \$1 for every unit sold to consumers, while increasing the consumer rebate by \$1 costs only \$$\beta$ for every unit sold to consumers. Thus, part (c) says that, given a pair of rebates that is a candidate for the manufacturer's optimal solution, the marginal increase in units sold to the retailer, per manufacturer's effective (at-risk) cost of a rebate-dollar, is higher for consumer rebates than retailer rebates.

Let $\Pi_{\mathrm{SC}}(p, r_{\mathrm{C}}, r_{\mathrm{R}})$ be the profit of the supply chain for a given retail price p, consumer rebate r_{C} and retailer rebate r_{R}. Note that $\Pi_{\mathrm{SC}}(p, r_{\mathrm{C}}, r_{\mathrm{R}}) = \Pi_{\mathrm{R}}(p, r_{\mathrm{C}}, r_{\mathrm{R}}) + \Pi_{\mathrm{M}}(p, r_{\mathrm{C}}, r_{\mathrm{R}})$ where $\Pi_{\mathrm{R}}(p, r_{\mathrm{C}}, r_{\mathrm{R}})$ and $\Pi_{\mathrm{M}}(p, r_{\mathrm{C}}, r_{\mathrm{R}})$ are as defined by (5) and (6), respectively. The following proposition states how the supply chain profit will be split between the two parties under an equilibrium solution.

Proposition 3 *Suppose (A1) through (A4) hold. Furthermore, suppose that a pure-strategy Nash equilibrium exists for the game between the retailer and the manufacturer. Let $\tilde{p}$ be an equilibrium retail price, and $\tilde{r}_{\mathrm{R}}$ and $\tilde{r}_{\mathrm{C}}$ the corresponding equilibrium rebates. The stock level that arises under this equilibrium is given by $y^*(\tilde{p}, \tilde{r}_{\mathrm{C}}, \tilde{r}_{\mathrm{R}})$ where y^* is given in (2). Under this equilibrium, if $\tilde{r}_{\mathrm{C}} > 0$, then $\frac{\Pi_{\mathrm{R}}(\tilde{p}, \tilde{r}_{\mathrm{C}}, \tilde{r}_{\mathrm{R}})}{\Pi_{\mathrm{M}}(\tilde{p}, \tilde{r}_{\mathrm{C}}, \tilde{r}_{\mathrm{R}})} = \frac{\alpha}{\beta}$.*

As we will see later on, this particular division of the supply chain profit under an equilibrium solution is due to the use of the consumer rebate, and, as stated in the proposition, will be true whenever the equilibrium consumer rebate is (strictly) positive. The key assumption that leads to this interesting result is that the demand uncertainty is multiplicative. We will discuss the rationale behind this result in detail when we discuss the use of consumer rebates in isolation. Also, this constant-split property allows us to conclude that, even when multiple Nash equilibria (with strictly positive consumer rebates) exist, there is one equilibrium that is preferred by both parties to all other equilibria, and the equilibrium preferred by both parties is the one under which the supply chain profit is at its highest among all other equilibria. If one could argue that our model captured the first order issues addressed in the automotive industry, where it is plausible to assume that both α and β are equal to one (due to the large sums involved in cashback allowances), one could say that the rebates would lead to dividing the channel profits evenly between the manufacturers and the dealers.

In the next two sections, we will consider the cases that arise when either only retailer rebates or only consumer rebates are used.

4. RETAILER REBATE ONLY

The *retailer rebate game* is the game between the retailer and the manufacturer in the previous section with the restriction that $r_{\mathrm{C}} = 0$. We will continue to use the same notation as before, replacing r_{C} with zero where necessary. The structural results on the profit functions of the manufacturer and the retailer (adapted for $r_C = 0$) will carry over directly from the previous section. In addition, the following proposition states how the optimal decision of one player changes with the decision of the other one.

Proposition 4 *Suppose (A1) through (A4) hold and* $r_C = 0$. *Let* $p^*(r_R)$ *be the optimal price chosen by the retailer as a response to a given* r_R *and* $r_R^*(p)$ *the optimal retailer rebate chosen by the manufacturer as a response to a given* p. *Then:*

(a) $-1 \leq \frac{dp^*(r_R)}{dr_R} < 0$.

(b) $-1 < \frac{dr_R^*(p)}{dp} \leq 0$.

(c) *There exists a unique Nash equilibrium for the retailer rebate game.*

The first part of the proposition above implies that when the manufacturer offers an additional \$1 rebate to the retailer for every unit sold, the retailer will decrease the selling price of the product, but the price discount will be less than \$1. Therefore, the retailer rebate results in some savings being passed on to the customer. Likewise, when the retailer reduces the price of the product by \$1, the manufacturer will increase the rebate paid to the retailer, but by less than \$1. The following proposition summarizes our results in this setting.

Proposition 5 *Suppose (A1) through (A4) hold and* $r_C = 0$. *Let* p_o *be the retail price and* y_o *the stock level chosen by the retailer when* $r_R = 0$. *Let* $\tilde{p}$ *be the equilibrium retail price, and* $\tilde{r}_R$ *the equilibrium rebate that will arise under the retailer rebate game. The stock level that arises under this equilibrium is given by* $y^*(\tilde{p}, 0, \tilde{r}_R)$ *where* y^* *is as defined by (2). Then:*

(a) $p_o - \tilde{r}_R \leq \tilde{p} \leq p_o$,

(b) $y_o \leq y^*(\tilde{p}, 0, \tilde{r}_R)$ *and*

(c) *If* $0 < \tilde{r}_R \leq w - c$, *then* $\Pi_{SC}(\tilde{p}, 0, \tilde{r}_R) > \Pi_{SC}(p_o, 0, 0)$.

The first two parts of the proposition state that, as expected, the retail price will decrease and the stock level will increase when retailer rebates are used. We should note that, in parts (a) and (b) of the proposition, the inequalities are not strict, since the equilibrium may turn out to be the no-rebate case; i.e., $\tilde{r}_R$ may be zero. It is interesting to note here how the role played by retailer rebates under endogenous retail pricing differs from that under an exogenously-fixed retail price. When the retail price is exogenous, the rebate helps the manufacturer by increasing the retailer's margin on every unit sold, thereby increasing the quantity ordered by the retailer. On the other hand, when the retail price is endogenous, the rebate serves a dual purpose for the manufacturer: As before, the rebate increases the order quantity of the retailer by increasing the retailer's margin on every unit sold, but, in addition, the rebate causes a decrease in the retail price (as stated in part (a) of the proposition), thereby increasing the customer demand, which causes a further increase in retailer's order quantity.

The last part of the proposition states that if the equilibrium rebate is (strictly) positive and below the manufacturer's unit profit margin (which would be expected to be the case in practice), then the supply chain will be strictly better off as a result of the use of the retailer rebate. This result is not surprising. Intuitively speaking, the higher the retailer rebate, the closer the supply chain becomes to one that is owned by a single decision maker, since increasing the retailer rebate brings the retailer's underage cost closer to the integrated supply chain's underage cost. Therefore, the higher the retailer rebate, the closer the performance of the supply chain becomes to that of the integrated one. We should note that the constant-split property does not hold when only retailer rebates are used.

Next, we discuss the case in which only consumer rebates are used.

5. CONSUMER REBATE ONLY

The *consumer rebate game* is the game between the retailer and the manufacturer in Section 3 with the restriction that $r_{\mathrm{R}} = 0$. The structural results on the retailer's and manufacturer's profit functions stated in Section 3 (adapted for $r_{\mathrm{R}} = 0$) carry over. The following proposition states how the optimal price chosen by the retailer responds to a change in the consumer rebate.

Proposition 6 *Suppose (A1) through (A4) hold and $r_{\mathrm{R}} = 0$. Let $p^*(r_{\mathrm{C}})$ be the optimal price chosen by the retailer as a response to a given r_{C}. Then:*

(a) $0 < \frac{dp^*(r_{\mathrm{C}})}{dr_{\mathrm{C}}} < \alpha$.

(b) *There exists a Nash equilibrium for the consumer rebate game.*

The proposition states that when the manufacturer offers an additional \$1 rebate to the consumer, the retailer will take advantage of this offer, and will increase the retail price, but the increase will be less than α. This means that, as is commonly thought, a consumer rebate will bring about a price increase, however the effective retail price paid by the consumer will still be less than the price that would be paid if the rebate did not exist. Unfortunately, a result on how the optimal consumer rebate responds to price eludes us. In the absence of such a result, we are not able to claim that the Nash equilibrium under the consumer rebate game will be unique. The following proposition summarizes our results for this game.

Proposition 7 *Suppose (A1) through (A4) hold and $r_{\mathrm{R}} = 0$. Let p_o denote the price and y_o the stock level chosen by the retailer when $r_{\mathrm{C}} = 0$. Let $\tilde{p}$ be an equilibrium retail price under the consumer rebate game, and $\tilde{r}_{\mathrm{C}}$ the corresponding equilibrium rebate. Suppose that $\tilde{r}_{\mathrm{C}} > 0$. The stock level that arises under this equilibrium is given by $y^*(\tilde{p}, \tilde{r}_{\mathrm{C}}, 0)$ where y^* is as defined by (2). Then:*

(a) $p_o \leq \tilde{p} \leq p_o + \alpha\tilde{r}_{\mathrm{C}}$,

(b) $y_o \leq y^*(\tilde{p}, \tilde{r}_{\mathrm{C}}, 0)$,

(c) *If* $\alpha \geq \beta$ *and* $\tilde{r}_C \leq w - c$, *then* $\Pi_{SC}(\tilde{p}, \tilde{r}_C, 0) \geq \Pi_{SC}(p_o, 0, 0)$ *and*
(d) $\frac{\Pi_R(\tilde{p},\tilde{r}_C,0)}{\Pi_M(\tilde{p},\tilde{r}_C,0)} = \frac{\alpha}{\beta}$

The first two parts of the proposition state the intuitive results that the retail price and the stock level will increase when consumer rebates are used. However, the increase in retail price will not be larger than the effective fraction of the consumer rebate, so consumers are still better off as a result of the rebate. The third part of the proposition states that, if α is larger than β, the supply chain profit will improve as a result of the consumer rebate (provided that the rebate is less than the manufacturer's profit margin, which we would expect to be the case). This result is expected: Essentially, when the cost of a \$1 rebate, modeled by β, is less than the effective fraction of the \$1 rebate, modeled by α, the supply chain is able to achieve the demand impact of an α-dollar price discount at a cost of $\beta<\alpha$ dollars. Also, we see from the last part of the proposition that the constant-split property of supply chain profit continues to hold when consumer rebates are used in isolation. Due to this constant-split property, we conclude that, even when multiple Nash equilibria exist, the equilibrium under which the supply chain profit is at its highest (among all other equilibria) is the one preferred by both parties. Furthermore, the constant-split property shows that, in an equilibrium solution, neither party is able to extract the entire supply chain profits. (Unless α or β is zero, which are not likely to be the case. Here, we assume that both α and β are strictly positive, and we do not cover the cases that arise when one or the other is zero.)

An interesting consequence of the constant-split property is that if the retailer's share of the supply chain profit under consumer rebates is larger than the manufacturer's, then it must be that $\alpha>\beta$ for the product in question, and, hence, by Proposition 7(c), the use of consumer rebates must have improved total supply chain profits.

From part (d) of Proposition 7, we observe that the manufacturer's share of the supply chain profit under consumer rebate equilibrium is $\frac{\beta}{\alpha+\beta}$. However, this observation does not imply that the manufacturer would necessarily like to design rebates so that β is high or α is low. In fact, in many numerical examples, we observed the opposite to be true. One such example is depicted in Figure 10-1. In this example, with β fixed at 0.9, the manufacturer prefers a large α to a small one, since the manufacturer prefers getting a smaller

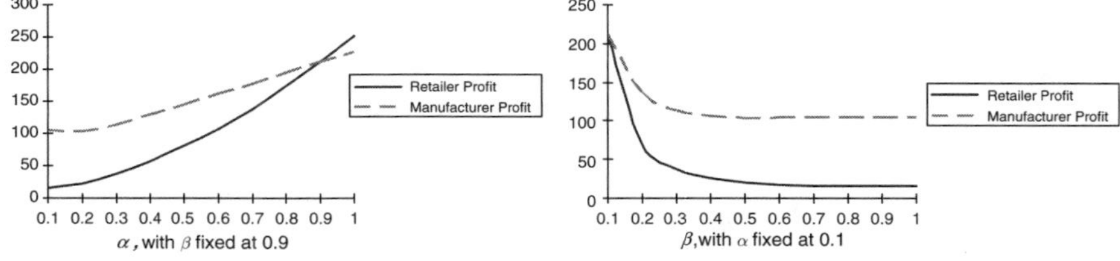

Figure 10-1. Equilibrium Retailer and Manufacturer Profits as a Function of α (*left*) and β (*right*)

share of the large supply chain profit achieved under a large α value. Likewise, with α fixed at 0.1, the manufacturer prefers a small β to a large one. Note that the manufacturer's profit is not necessarily monotonic in α or β, which can be confirmed with careful scrutiny of the graphs. Another conclusion that applies to this example is that there is no conflict between the retailer and the manufacturer in terms of the attributes of a rebate: To the extent possible, both parties would like a rebate with a high customer valuation α and a small redemption rate β. We observed this to be the case in many other numerical examples. We return to this point in Section 7.

It is worthwhile to discuss the rationale behind the constant-split property. We will do so through a marginal analysis discussion. For the sake of the following discussion, define $\gamma(p) := -\frac{f'(p)}{f(p)}$; i.e., $\gamma(p)$ is a positive number representing the fractional decrease in expected demand in response to a marginal increase in price p. Under the consumer rebate equilibrium, the retail price must satisfy the FOC for the retailer. Hence, by part (a) of Lemma 6, the retail price p must satisfy

$$\int_0^{\frac{p+r_{\mathbf{R}}-w}{p+r_{\mathbf{R}}-v}} z(p, r_{\mathbf{C}}, \xi)d\xi + \frac{w-v}{p+r_{\mathbf{R}}-v} y^*(p, r_{\mathbf{C}}, r_{\mathbf{R}}) = \gamma(p + r_{\mathbf{R}} - v) \int_0^{\frac{p+r_{\mathbf{R}}-w}{p+r_{\mathbf{R}}-v}} z(p, r_{\mathbf{C}}, \xi)d\xi \tag{7}$$

The left-hand side of (7) is the expected sales of the product; a \$1 price increase means the retailer will make \$1 more on every unit sold, so the retailer's profit will increase by an amount equal to the expected sales. The right-hand side of (7) is γ times the (expected) profit of the retailer; a \$1 price increase will lead to a demand reduction, which will cause the retailer to lose some profit, and this loss turns out to be equal to γ times the profit of the retailer. (This is a consequence of the multiplicative demand model.) Therefore, as the FOC given by (7) implies, the optimal price chosen by the retailer must set the expected sales volume equal to γ times the retailer's profit.

Likewise, under the consumer rebate equilibrium, the consumer rebate must satisfy the FOC for the manufacturer. Hence, by part (b) of Lemma 6, the consumer rebate $r_{\mathbf{C}}$ must satisfy

$$\begin{aligned}\beta\left(\frac{w-v}{p+r_{\mathbf{R}}-v} y^*(p, r_{\mathbf{C}}, r_{\mathbf{R}}) + \int_0^{\frac{p+r_{\mathbf{R}}-w}{p+r_{\mathbf{R}}-v}} z(p, r_{\mathbf{C}}, \xi)d\xi\right) &= \gamma\alpha\Bigg[(w-c)y^*(p, r_{\mathbf{C}}, r_{\mathbf{R}}) - (\beta r_{\mathbf{C}} + r_{\mathbf{R}}) \\ &\quad \left(\frac{w-v}{p+r_{\mathbf{R}}-v} y^*(p, r_{\mathbf{C}}, r_{\mathbf{R}}) + \int_0^{\frac{p+r_{\mathbf{R}}-w}{p+r_{\mathbf{R}}-v}} z(p, r_{\mathbf{C}}, \xi)d\xi\right)\Bigg].\end{aligned} \tag{8}$$

The left-hand side of (8) is β times the expected sales of the product; a \$1 rebate increase means the manufacturer will pay β dollars more per each unit sold, so the manufacturer's profit will decrease by an amount equal to β times the expected sales. The right-hand side of (8) is $\gamma\alpha$ times the profit of the manufacturer; a \$1 rebate increase will lead to a demand increase, which will cause the manufacturer to gain some profit, and this gain turns out to be equal to $\gamma\alpha$ times the profit of the manufacturer. (Once again, this is a consequence of the multiplicative demand model.) Therefore, the optimal price chosen by the manufacturer must set β times the expected sales volume equal to $\gamma\alpha$ times the manufacturer's profit.

In summary, both parties are using the (expected) sales volume as a benchmark; one is trying to set its profit equal to the sales volume multiplied by $\frac{1}{\gamma}$, and the other is trying to set its profit equal to $\frac{\beta}{\gamma\alpha}$ times the sales volume. Since both parties will be seeing the same sales volume in equilibrium, the last part of the proposition follows.

In the next section, we provide some numerical examples to compare the effects of the retailer and consumer rebates on the profits of the supply chain partners.

6. NUMERICAL EXAMPLES

One natural question to ask is which rebate type each player in the supply chain prefers. Unfortunately, there is no clear-cut answer to this question. In particular, as one would expect, the values of α and β have a significant impact on the equilibrium that arises under consumer rebates, and, therefore, whether a party prefers consumer rebates to retailer rebates depends very much on the values of α and β. Consider the equilibrium results depicted in Table 10-1. These equilibria are obtained under the assumption that $f(\cdot)$ is given by the logit demand function; i.e., $f(x) = \frac{exp(u_1 - x)}{exp(u_0) + exp(u_1 - x)}$, and ϵ is distributed uniformly between 50 and 250. The other parameter values were as follows: $w = 18.55, c = 4.08, v = 0, u_1 = 22.91, u_0 = 2.70$. (This is one of many randomly-generated numerical examples we tested.) For the three combinations of parameters considered for consumer rebates, there was only a single equilibrium to the "both rebate types" game and it specified zero retailer rebate.[2] Thus, the prices and profits are the same as those given in the table under consumer rebates only. When $\alpha = 1$ and $\beta = 0.8$, both the retailer and the manufacturer prefer consumer rebates to retailer rebates. However, if β increases to 1 while keeping α fixed at 1, the manufacturer will now suffer from the increased claim rate of rebates, and, therefore, will now prefer retailer rebates to consumer rebates, while the retailer's preference is not affected by the change in β. On the other hand, if α decreases to 0.4 while keeping β fixed at 1, consumer rebates will now have a smaller impact on consumer demand, and, hence, the retailer will now prefer retailer rebates to consumer rebates. Therefore, neither party always prefers one rebate type to another.

---

[2] Under consumer rebate equilibria, the manufacturer expected profit to retailer expected profit ratios are not precisely $\beta : \alpha$, since our searches were over fine grids that were nevertheless discrete.

Table 10-1. Equilibria Under Different Rebate Scenarios

Rebate Type	α	β	Retailer Rebate	Consumer Rebate	Price	Manufacturer Profit	Retailer Profit	Expected Demand
No Rebate	–	–	–	–	20.55	417.95	49.67	62.37
Retailer	–	–	3.44	–	19.32	689.79	204.72	106.33
Consumer	1	0.8	–	11.58	29.74	694.90	868.46	132.89
Consumer	1	1	–	8.94	27.37	621.60	621.17	128.35
Consumer	0.4	1	–	7.18	22.33	430.05	172.01	101.94
Consumer + Retailer	1	0.8	0	11.58	29.74	694.90	868.46	132.89
Consumer + Retailer	1	1	0	8.94	27.37	621.60	621.17	128.35
Consumer + Retailer	0.4	1	0	7.18	22.33	430.05	172.01	101.94

A form of prisoner's dilemma in choosing what rebate(s) to offer: Note from Table 1 that for $\alpha = 0.4$ and $\beta = 1$, a supply chain in which both rebate types are allowed will settle in the same equilibrium as a supply chain in which only consumer rebates are allowed. Notice that there is a form of prisoners' dilemma here: The retailer rebate game equilibrium, even though it is preferred by both parties, is not an equilibrium in this game with both types allowed. This leads to an interesting observation: When the supply chain plays the game where both types of rebates are allowed, the supply chain ends up using only consumer rebates in equilibrium, an outcome that hurts both parties when compared to what they could achieve if only retailer rebates are allowed. The policy implication of this observation is that there are environments in which both the retailer and the manufacturer will agree in advance, before prices and rebates are set, to not allow the use of consumer rebates.

Both parties may prefer consumer rebates even when all consumers claim them: There exist cases where both parties prefer to use the consumer rebates to stimulate customer demand. For example, when $w = 10, c = 4, v = 0$, $u_1 = 30, u_0 = 20$, and $\alpha = \beta = 1$, both parties prefer consumer rebates. (Under retailer rebates only, the equilibrium profits are 122.45 for the manufacturer and 32.92 for the retailer. Under consumer rebates only, the equilibrium profits are 151.70 for both the manufacturer and the retailer.) Under the consumer rebate equilibrium, the retail price is 13.28 and the consumer rebate is 3.77, which yield an effective price of 9.51, less than the wholesale price of 10. Note that this is an environment where all consumers claim their rebates, i.e., β is one; nevertheless, both parties prefer consumer rebates to retailer rebates. Moreover, in this supply chain, even when both types of rebates are allowed, it turns out that retailer rebates are not offered in equilibrium. The policy implication is that, contrary to popular belief, there exist environments in which supply chains prefer consumer rebates even when all consumers claim them.

A variant of this result can be seen in Table 10-1, where the supply chain profits are higher under consumer rebates than retailer rebates when $\alpha = \beta = 1$. In this case, because the wholesale price is fixed so much higher than cost, the retail price and consumer rebate are both high, leading to an effective price lower than under retailer rebates, but with a much higher margin to the retailer on units sold with still a good margin to the manufacturer on an increased level of sales. The manufacturer gets slightly lower profits but the retailer gets dramatically more.

Retailer may choose to sell at a loss to make money on rebates: Rebates can play an interesting role in the supply chain when the exogenously-fixed wholesale price is high. For example, if $w = 20, c = 5, v = 0, u_1 = 40, u_0 = 20$, and $\alpha = \beta = 1$, then the retailer rebate game equilibrium has a retail price of 19.24, which is lower than the wholesale price, and the retailer rebate is 4.53. Thus, in this example, the wholesale price is so high that the retailer sells the product at a loss to stimulate customer demand, and makes money only on rebates collected from the manufacturer rather than directly from consumers.

The effect of wholesale price: To further examine the effect of wholesale price on the equilibrium, consider the case where only consumer rebates are allowed.

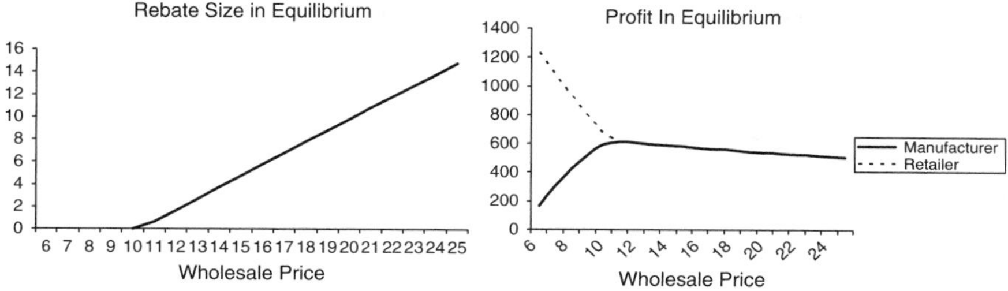

Figure 10-2. Equilibrium Rebate Size (*left*) and Profits (*right*) as a Function of the Wholesale Price When Only Consumer Rebates are Allowed

Figure 10-2 shows the effect of w on the rebate size in equilibrium as well as on the manufacturer's and retailer's profits. In this example, $w = 20, c = 5, v = 0, u_1 = 40, u_0 = 20$, and $\alpha = \beta = 1$.

Observe from the figure that there is a threshold for the wholesale price such that only if the wholesale price exceeds this threshold will the manufacturer offer a strictly positive consumer rebate. This is intuitive: As the wholesale price gets larger, the manufacturer's profit margin per unit gets larger as well, and the manufacturer becomes more willing to pay a rebate to drive the retailer's stock level up. In addition, the figure suggests that the manufacturer's profit is at its highest at the threshold wholesale price. Therefore, if the manufacturer were to choose the wholesale price first, followed by a game where the consumer rebate and retail price are chosen simultaneously, then it would be optimal for the manufacturer to set the wholesale price equal to its threshold value, which would lead to a zero rebate in equilibrium. We have observed the same behavior in a number of numerical examples, but further analysis is needed to determine if this result is true in general.

7. CONCLUSION

We considered a supply chain where the retailer faces stochastic, effective-price-dependent demand and the manufacturer builds to order. We established some properties of the equilibrium that would arise when the manufacturer offers retailer and/or consumer rebates. We showed that supply chain profits are improved by the use of retailer rebates. On the other hand, consumer rebates may reduce the supply chain profit, but they will lead to an improvement whenever the effective fraction, α, is larger than the fraction of customers who claim their rebate, β. Furthermore, we showed that these two parameters have further significance: Under the equilibrium of the consumer rebate game, the ratio of (expected) retailer profits to (expected) manufacturer profits equals the ratio α/β. We discussed some interesting consequences of this property. We provided numerical examples to demonstrate that neither the retailer nor the manufacturer always prefers one particular kind of rebate to the other. In addition, our numerical examples suggest that, contrary to popular belief, it is possible for both firms to prefer consumer rebates even when all such rebates are redeemed.

In our model, we examined how the two rebate types differ from each other through their effects on the pricing and inventory decisions for a product. When the product's price is fixed, but the retailer is able to exert some type of hidden effort to sell the product; e.g., putting up in-store displays or advertising in local media, the effects of retailer and consumer rebates are likely to differ again and are worthy of study. Another extension worthy of study is to address the moral hazard problem of misreporting retailer sales. One approach is to add buy-backs to the model (the manufacturer buys back unsold inventory at the end of the season at a set price), which could reduce the retailer's incentive to misreport sales. It would also be interesting to add a verification cost (of sold units) to the model.

We give a partial answer to the question of why consumer rebates are offered. Our numerical examples illustrate the existence of cases where the manufacturer will prefer offering consumer rebates to offering a retailer rebate. Consumer rebates help the manufacturer by increasing the stock level at the retailer, and our results suggest that they may be useful even when all customers claim them. Bulkeley (1998) cites some alternative explanations for the use of consumer rebates. For example, consumer rebates may be seen as temporary price reductions, used in order to learn more about the customer population's price elasticity. Alternatively, in high-tech products, consumer rebates can be used to offer price discounts to consumers on older-generation products, which would eliminate the need for offering price protection to the retailer. Analysis of such uses for consumer rebates is left for future research. Hopefully, some of the structural results in this paper could prove useful for researchers who would like to further analyze the question of why consumer rebates are used. Another line of extension for this research is using more elaborate models for the redemption of consumer rebates, such as having heterogeneous consumer types, with differing values of α and β. A utility-based model that describes the customer's attitude towards redeeming a rebate would contribute to our understanding of the use of consumer rebates.

It is possible that some retailers will force manufacturers to move away from mail-in consumer rebates in the future. For example, BestBuy announced that it will no longer stock products tied to mail-in rebates, a policy that will be implemented in the span of a few years (Menzies, 2005). BestBuy's stated reason is that mail-in rebates are cumbersome for the consumers. To the extent that our model captures the BestBuy environment (the major violation is likely to be that the wholesale price is not exogenous), it may be that BestBuy prefers the retailer rebate regime, although in our numerical examples where that happens, the manufacturer also prefers the retailer rebate regime, so would not resist dropping consumer rebates and instituting retailer rebates. Another explanation is that BestBuy is lobbying for having the consumer rebates instantaneously redeemable at the time of consumer purchase, as is done in the automotive industry. This might have the effect of increasing both the customer valuation α

and redemption rate β to 1, which would make rebates the equivalent to price discounts offered directly by the manufacturer to consumers. Depending on what the values of α and β are in status quo, this might improve the total supply chain profit as well as be appreciated by consumers. There are other explanations for BestBuy's position that are not covered by our model, such as that it helps BestBuy in its competition with other retailers. In any event, BestBuy could be acting in its self interest, while claiming that its motivation is as a consumer advocate.

Acknowledgments The authors would like to thank the editors and an anonymous referee for their comments that helped improve the paper.

APPENDIX

For the purposes of the appendix, let $h(\cdot) = \frac{\phi_\epsilon(\cdot)}{1-\Phi_\epsilon(\cdot)}$ denote the failure rate of Φ_ϵ. Throughout the appendix, we will use the following short-hand notation by dropping the functional arguments: $f = f(p - \alpha r_{\mathrm{C}})$, $z = z(p, r_{\mathrm{C}}, \xi)$, $y^* = y^*(p, r_{\mathrm{C}}, r_{\mathrm{R}})$ and $h = h\left(\frac{y^*(p,r_{\mathrm{C}},r_{\mathrm{R}})}{f(p-\alpha r_{\mathrm{C}})}\right)$. In addition, define $\gamma := -\frac{f'}{f}$ and $\theta := \frac{f''}{f}$. Hence, by (A3), $\gamma > 0$, and, by (A4), $\gamma' = \gamma^2 - \theta \geq 0$. We first state and prove some lemmas that will be useful in the proofs of the propositions.

Lemma 1 *Suppose (A1) holds. For $z(p, r_{\mathrm{C}}, \xi)$ implicitly defined by (4), we have:*

(a) $\nabla_1 z = -\gamma z$,
(b) $\nabla_2 z = \alpha\gamma z$,
(c) $\nabla^2_{11} z = \theta z$,
(d) $\nabla^2_{22} z = \alpha^2 \theta z$,
(e) $\nabla^2_{12} z = -\alpha\theta z$.

Proof of Lemma 1 By virtue of ($A1$), we can rewrite (4) as $\Phi_\epsilon\left(\frac{z}{f}\right) = \xi$. Now, implicit differentiation of this identity with respect to p yields the following:

$$\nabla_1 z f - z f' = 0.$$

The first part of the lemma follows from the above equality recalling the definition of $\gamma := -\frac{f'}{f}$. The proof of the second part follows the same logic. The third part can be obtained directly by partial differentiation of the expression for $\nabla_1 z$. Likewise, the fourth and fifth parts are obtained by partial differentiation of the expression for $\nabla_2 z$.

Lemma 2 *Suppose (A1) through (A4) hold. For $y^*(p, r_{\mathrm{C}}, r_{\mathrm{R}})$ implicitly defined by (2), we have:*

(a) $\nabla_1 y^* = -\gamma y^* + \frac{f}{(p+r_{\mathrm{R}}-v)h}$,
(b) $\nabla_2 y^* = \alpha\gamma y^* > 0$,

(c) $\nabla_3 y^* = \frac{f}{(p+r_{\mathbf{R}}-v)h} > 0$,
(d) $\nabla_1 y^* = -\frac{1}{\alpha}\nabla_2 y^* + \nabla_3 y^*$,
(e) $\nabla^2_{22} y^* = \alpha^2 \theta y^*$,
(f) $\nabla^2_{33} y^* = -\frac{f}{(p+r_{\mathbf{R}}-v)^2 h} - \frac{fh'}{(p+r_{\mathbf{R}}-v)^2 h^3} < 0$,
(g) $\nabla^2_{23} y^* = \alpha\gamma \frac{f}{(p+r_{\mathbf{R}}-v)h} > 0$,
(h) $\nabla^2_{13} y^* = -\frac{1}{\alpha}\nabla^2_{23} y^* + \nabla^2_{33} y^* < 0$.

Proof of Lemma 2
Proofs of (a) through (d) Due to ($A1$), we can rewrite (2) as

$$\Phi_\epsilon\left(\frac{y^*}{f}\right) = \frac{p + r_{\mathbf{R}} - w}{p + r_{\mathbf{R}} - v}$$

Now, implicit differentiation of (9) with respect to p yields

$$\frac{\nabla_1 y^* f - f' y^*}{f^2}\phi_\epsilon\left(\frac{y^*}{f}\right) = \frac{w - v}{(p + r_{\mathbf{R}} - v)^2}.$$

Recalling the definition of $h(\cdot) = \frac{\phi_\epsilon(\cdot)}{1-\Phi_\epsilon(\cdot)}$ and noting that $1 - \Phi_\epsilon\left(\frac{y^*}{f}\right) = \frac{w-v}{p+r_{\mathbf{R}}-v}$ (this follows from (9)), we can leave $\nabla_1 y^*$ alone in the above expression to obtain part (a) of the lemma. The proofs of parts (b) and (c) follow the same line of argument. Part (d) of the lemma follows directly from parts (a) through (c).
Proofs of (e) through (g) These follow from partial differentiation of the expressions obtained in parts (a) through (c). To see why $\nabla^2_{33} y^* < 0$, recall that $h(\cdot)$ is the failure rate and it is an increasing function by ($A2$). To see why $\nabla^2_{23} y^* > 0$, recall that $\gamma > 0$ by ($A3$).
Proof of (h) This follows from part (d) of the lemma.

Lemma 3 *Given* $r_{\mathbf{C}}$ *and* $r_{\mathbf{R}}$, *if* $\tilde{p}$ *satisfies* $\nabla_1 \Pi_{\mathbf{R}}(\tilde{p}, r_{\mathbf{C}}, r_{\mathbf{R}}) = 0$, *then* $\nabla_1 y^* \nabla_1 y^*(\tilde{p}, r_{\mathbf{C}}, r_{\mathbf{R}}) < 0$.
Proof of Lemma 3 Omitted. See Aydin and Porteus (2006) for the proof of the same result under more general conditions.

Lemma 4 *Let $f(x)$ be a twice-continuously-differentiable function of a single real variable defined on $[a, \infty)$. Suppose that $f''(x) < 0$ at any $x \geq a$ that satisfies $f'(x) = 0$. Then:*

(*a*) (*Petruzzi and Dada, 1999*) *If $f'(a) > 0$* and *$f(x)$ is strictly decreasing in* x *as* x *tends to infinity, then there exists a unique x^* that satisfies $f'(x) = 0$*, and *x^* maximizes $f(x)$.*
(*b*) *If $f'(a) \leq 0$* then *$f(x)$ is non-increasing for all $x \geq a$, and $x^* = a$ maximizes $f(x)$.*

Proof of Lemma 4 Omitted. Lemma 4 (a) is due to Petruzzi and Dada (1999). See Aydin and Porteus (2006) for a detailed proof. The proof of part (b) is very similar.

Lemma 5 *In a two-player game, let $g_i(x_1, x_2)$ be the payoff function of player $i = 1, 2$ when the strategies chosen by players 1 and 2 are x_1 and x_2, respectively. The strategy space for player i is $X_i := \{x : \underline{x}_i \leq x \leq \overline{x}_i\}$. Suppose that g_i is continuous and quasi-concave with respect to x_i, $i = 1, 2$. Let $x_i^*(x_j)$ be the best response of player i when player j chooses strategy x_j; i.e., $x_i^*(x_j) = \arg\max_{x_i}(g_i(x_1, x_2))$. Then:*
(a) There exists at least one pure strategy Nash equilibrium.
(b) If $\frac{dx_1^*(x_2)}{dx_2}\frac{dx_2^*(x_1)}{dx_1} < 1$, then there exists a unique pure strategy Nash equilibrium.

Proof of Lemma 5 Omitted. See Cachon and Netessine (2003) for a summary of standard results in game theory.

Lemma 6 *Suppose (A1) through (A4) hold. Let $\Pi_R(p, r_C, r_R)$ and $\Pi_M(p, r_C, r_R)$ be as defined by (5) and (6), respectively. Then:*
(a)

$$\nabla_1 \Pi_R(p, r_C, r_R) = \int_0^{\frac{p+r_R-w}{p+r_R-v}} z d\xi + \frac{w-v}{p+r_R-v} y^* - \gamma(p + r_R - v) \int_0^{\frac{p+r_R-w}{p+r_R-v}} z d\xi$$

(b)

$$\nabla_2 \Pi_M(p, r_C, r_R) \quad = \quad (w-c)\alpha\gamma y^* - \quad [\beta + \alpha\gamma(r_R + \beta r_C)]\left(\int_0^{\frac{p+r_R-w}{p+r_R-v}} z d\xi + \frac{w-v}{p+r_R-v} y^*\right),$$

(c)

$$\nabla_3 \Pi_M(p, r_C, r_R) = \left[(w-c) - (r_R + \beta r_C)\frac{w-v}{p+r_R-v}\right]\frac{f}{(p+r_R-v)h} - \int_0^{\frac{p+r_R-w}{p+r_R-v}} z d\xi - \frac{w-v}{p+r_R-v} y^*$$

(d) For $p + r_R > v$:

$$\nabla_{11}^2 \Pi_R(p, r_C, r_R)|_{\nabla_1 \Pi_R = 0} \quad = \quad -\gamma \int_0^{\frac{p+r_R-w}{p+r_R-v}} z d\xi + \frac{w-v}{p+r_R-v} \nabla_1 y^* - (p + r_R - v)\gamma' \int_0^{\frac{p+r_R-w}{p+r_R-v}} z < 0$$

(e) For $p + r_R > v$:

$$\nabla_{22}^2 \Pi_M(p, r_C, r_R)|_{\nabla_2 \Pi_M = 0} = -\alpha\beta\gamma\left(\int_0^{\frac{p+r_R-w}{p+r_R-v}} z d\xi + \frac{w-v}{p+r_R-v} y^*\right) + \left[(w-c)y^* - (r_R + \beta r_C)\left(\int_0^{\frac{p+r_R-w}{p+r_R-v}} z d\xi + \frac{w-v}{p+r_R-v} y^*\right)\right] (\alpha^2\theta - \alpha^2\gamma^2)$$

(*f*) For $p + r_{\mathbf{R}} > v$:

$$\nabla^2_{33}\Pi_{\mathbf{M}}(p, r_{\mathbf{C}}, r_{\mathbf{R}})|_{\nabla_3\Pi_{\mathbf{M}}=0} = -\frac{1}{p + r_{\mathbf{R}} - v}\left[\int_0^{\frac{p+r_{\mathbf{R}}-w}{p+r_{\mathbf{R}}-v}} z d\xi + \frac{w - v}{p + r_{\mathbf{R}} - v} y^*\right]$$
$$- \left[(w - c) - (r_{\mathbf{R}} + \beta r_{\mathbf{C}})\frac{w - v}{p + r_{\mathbf{R}} - v}\right]\frac{fh'}{(p + r_{\mathbf{R}} - v)^2 h^3}$$
$$+ \left[-2\frac{w - v}{p + r_{\mathbf{R}} - v} + (r_{\mathbf{R}} + \beta r_{\mathbf{C}})\frac{w - v}{(p + r_{\mathbf{R}} - v)^2}\right]\frac{f}{(p + r_{\mathbf{R}} - v)h} < 0$$

(*g*) For $p + r_{\mathbf{R}} > v$:

$$\nabla^2_{23}\Pi_{\mathbf{M}}(p, r_{\mathbf{C}}, r_{\mathbf{R}})|_{\nabla_2\Pi_{\mathbf{M}}=0} = -\frac{w-v}{p+r_{\mathbf{R}}-v}\frac{\beta f}{(p+r_{\mathbf{R}}-v)h} < 0$$

(*h*) For $p + r_{\mathbf{R}} > v$:

$$\nabla^2_{13}\Pi_{\mathbf{R}}(p, r_{\mathbf{C}}, r_{\mathbf{R}})|_{\nabla_1\Pi_{\mathbf{R}}=0} = \frac{w-v}{p+r_{\mathbf{R}}-v}\nabla_1 y^* - \gamma\int_0^{\frac{p+r_{\mathbf{R}}-w}{p+r_{\mathbf{R}}-v}} z d\xi < 0$$

(*i*) For $p + r_{\mathbf{R}} > v$:

$$\nabla^2_{13}\Pi_{\mathbf{M}}(p, r_{\mathbf{C}}, r_{\mathbf{R}})|_{\nabla_3\Pi_{\mathbf{M}}=0} = -\left[(w - c) - (r_{\mathbf{R}} + \beta r_{\mathbf{C}})\frac{w}{p + r_{\mathbf{R}} - v}\right]\frac{fh'}{(p + r_{\mathbf{R}} - v)^2 h^3}$$
$$- \left[(2w - c) - 2(r_{\mathbf{R}} + \beta r_{\mathbf{C}})\frac{w - v}{p + r_{\mathbf{R}} - v}\right]\frac{f}{(p + r_{\mathbf{R}} - v)^2 h} < 0$$

(*j*) For $p + r_{\mathbf{R}} > v$:

$$\nabla^2_{12}\Pi_{\mathbf{R}}(p, r_{\mathbf{C}}, r_{\mathbf{R}})|_{\nabla_1\Pi_{\mathbf{R}}=0} = \alpha(p + r_{\mathbf{R}} - v)\gamma'\int_0^{\frac{p+r_{\mathbf{R}}-w}{p+r_{\mathbf{R}}-v}} z d\xi > 0$$

Proof of Lemma 6

Proof of (a) The result follows from partial differentiation of $\Pi_{\mathrm{R}}(p, r_{\mathbf{C}}, r_{\mathbf{R}})$ (defined by (5)) with respect to p and substituting for $\nabla_1 z$ using Lemma 1(a).

Proof of (b) The result follows by partial differentiation of $\Pi_{\mathbf{M}}(p, r_{\mathbf{C}}, r_{\mathbf{R}})$ (defined by (6)) with respect to $r_{\mathbf{C}}$ and substituting for $\nabla_2 z$ and $\nabla_2 y^*$ from Lemma 1(b) and from Lemma 2(b).

Proof of (c) The result follows by partial differentiation of $\Pi_{\mathbf{M}}(p, r_{\mathbf{C}}, r_{\mathbf{R}})$ (defined by (6)) with respect to $r_{\mathbf{R}}$ and substituting for $\nabla_3 y^*$ from Lemma 2(c).

Proof of (d) The second partial of $\Pi_{\mathrm{R}}(p, r_{\mathbf{C}}, r_{\mathbf{R}})$ with respect to p is given, after substituting for $\nabla_1 z$ and $\nabla^2_{11} z$ using Lemma 1(a) and (c), by

$$\nabla^2_{11}\Pi_{\mathrm{R}}(p, r_{\mathbf{C}}, r_{\mathbf{R}}) = -2\gamma\int_0^{\frac{p+r_{\mathbf{R}}-w}{p+r_{\mathbf{R}}-v}} z d\xi + \frac{w - v}{p + r_{\mathbf{R}} - v}\nabla_1 y^* - \frac{w - v}{p + r_{\mathbf{R}} - v}\gamma y^*$$
$$+ (p + r_{\mathbf{R}} - v)\theta\int_0^{\frac{p+r_{\mathbf{R}}-w}{p+r_{\mathbf{R}}-v}} z d\xi$$

Thus, when $\nabla_1 \Pi_{\mathbf{R}} = 0$, using part (a) of the lemma, we have

$$\nabla^2_{11}\Pi_{\mathbf{R}}(p, r_{\mathbf{C}}, r_{\mathbf{R}}) = -\gamma \int_0^{\frac{p+r_{\mathbf{R}}-w}{p+r_{\mathbf{R}}-v}} z d\xi + \frac{w-v}{p+r_{\mathbf{R}}-v}\nabla_1 y^* + (p + r_{\mathbf{R}} - v)(\theta - \gamma^2) \int_0^{\frac{p+r_{\mathbf{R}}-w}{p+r_{\mathbf{R}}-v}} z d\xi,$$

which is strictly negative, by Lemma 3 and since $\gamma > 0$ (by $(A3)$) and $\gamma' = \gamma^2 - \theta \geq 0$ (by $(A4)$).

Proof of (e) The second partial of $\Pi_{\mathbf{M}}(p, r_{\mathbf{C}}, r_{\mathbf{R}})$ with respect to $r_{\mathbf{C}}$ is given, after substituting for $\nabla_2 z$, $\nabla^2_{22} z$, $\nabla_2 y^*$ and $\nabla^2_{22} y^*$ from Lemma 1(b) and (d), and from Lemma 2(b) and (e), by

$$\nabla^2_{22}\Pi_{\mathbf{M}}(p, r_{\mathbf{C}}, r_{\mathbf{R}}) = (w - c)\alpha^2\theta y^* - \left[2\alpha\beta\gamma + (r_{\mathbf{R}} + \beta r_{\mathbf{C}})\alpha^2\theta\right]\left(\int_0^{\frac{p+r_{\mathbf{R}}-w}{p+r_{\mathbf{R}}-v}} z d\xi + \frac{w-v}{p+r_{\mathbf{R}}-v} y^*\right)$$

Thus, when $\nabla_2 \Pi_{\mathbf{M}} = 0$, using part (b) of the lemma, we have

$$\nabla^2_{22}\Pi_{\mathbf{M}}(p, r_{\mathbf{C}}, r_{\mathbf{R}}) = -\alpha\beta\gamma\left(\int_0^{\frac{p+r_{\mathbf{R}}-w}{p+r_{\mathbf{R}}-v}} z d\xi + \frac{w-v}{p+r_{\mathbf{R}}-v} y^*\right) + \left[(w-c)y^* - (r_{\mathbf{R}} + \beta r_{\mathbf{C}})\left(\int_0^{\frac{p+r_{\mathbf{R}}-w}{p+r_{\mathbf{R}}-v}} z d\xi + \frac{w-v}{p+r_{\mathbf{R}}-v} y^*\right)\right](\alpha^2\theta - \alpha^2\gamma^2),$$

which is strictly negative since $\gamma > 0$ (by $(A3)$), $\gamma' = \gamma^2 - \theta \geq 0$ (by $(A4)$) and the term in brackets is $\Pi_{\mathbf{M}}$ which should be positive when $\nabla_2 \Pi_{\mathbf{M}} = 0$.

Proof of (f) The second partial of $\Pi_{\mathbf{M}}(p, r_{\mathbf{C}}, r_{\mathbf{R}})$ with respect to $r_{\mathbf{R}}$ is given, after substituting for $\nabla_3 y^*$ and $\nabla^2_{33} y^*$ from Lemma 2(c) and (f), by

$$\nabla^2_{33}\Pi_{\mathbf{M}}(p, r_{\mathbf{C}}, r_{\mathbf{R}}) = \left[(w-c) - (r_{\mathbf{R}} + \beta r_{\mathbf{C}})\frac{w-v}{p+r_{\mathbf{R}}-v}\right]\left[-\frac{f}{(p+r_{\mathbf{R}}-v)^2 h} - \frac{fh'}{(p+r_{\mathbf{R}}-v)^2 h^3}\right] + \left[-2\frac{w-v}{p+r_{\mathbf{R}}-v} + (r_{\mathbf{R}} + \beta r_{\mathbf{C}})\frac{w-v}{(p+r_{\mathbf{R}}-v)^2}\right]\frac{f}{(p+r_{\mathbf{R}}-v)h}.$$

Thus, when $\nabla_3 \Pi_{\mathbf{M}} = 0$, using part (c) of the lemma, we have

$$\nabla^2_{33}\Pi_{\mathbf{M}}(p, r_{\mathbf{C}}, r_{\mathbf{R}}) = -\frac{1}{p+r_{\mathbf{R}}-v}\left[\int_0^{\frac{p+r_{\mathbf{R}}-w}{p+r_{\mathbf{R}}-v}} z d\xi + \frac{w-v}{p+r_{\mathbf{R}}-v} y^*\right] - \left[(w-c) - (r_{\mathbf{R}} + \beta r_{\mathbf{C}})\frac{w-v}{p+r_{\mathbf{R}}-v}\right]\frac{fh'}{(p+r_{\mathbf{R}}-v)^2 h^3} + \left[-2\frac{w-v}{p+r_{\mathbf{R}}-v} + (r_{\mathbf{R}} + \beta r_{\mathbf{C}})\frac{w-v}{(p+r_{\mathbf{R}}-v)^2}\right]\frac{f}{(p+r_{\mathbf{R}}-v)h}.$$

In order to show $\nabla^2_{33}\Pi_{\mathrm{M}}(p,r_{\mathrm{C}},r_{\mathrm{R}})|_{\nabla_3\Pi_{\mathrm{M}}=0}<0$, first note that, by Lemma 6 (c), if $\nabla_3\Pi_{\mathrm{M}}(p,r_{\mathrm{C}},r_{\mathrm{R}})=0$, then we must have $(w-c)-(r_{\mathrm{R}}+\beta r_{\mathrm{C}})\frac{w-v}{p+r_{\mathrm{R}}-v}>0$, in which case we will also have $-2\frac{w-v}{p+r_{\mathrm{R}}-v}+(r_{\mathrm{R}}+\beta r_{\mathrm{C}})\frac{w-v}{(p+r_{\mathrm{R}}-v)^2}<0$. (This can be verified through some algebra.) After making these observations, the desired result now follows since $h'>0$ by assumption ($A2$).

Proof of (g) It can be verified that the cross-partial $\nabla_{23}\Pi_{\mathrm{M}}(p,r_{\mathrm{C}},r_{\mathrm{R}})$ is given, after substituting for $\nabla_2 z$ from Lemma 1(b) and for $\nabla_2 y^*, \nabla_3 y^*$ and $\nabla_{23}y^*$ from Lemma 2(b), (c) and (g), by

$$\begin{aligned}\nabla^2_{23}\Pi_{\mathrm{M}}(p,r_{\mathrm{C}},r_{\mathrm{R}}) = &-\left[(w-c)-(r_{\mathrm{R}}+\beta r_{\mathrm{C}})\frac{w-v}{p+r_{\mathrm{R}}-v}\right]\alpha\gamma\frac{f}{(p+r_{\mathrm{R}}-v)h}\\ &-\alpha\gamma\left(\int_0^{\frac{p+r_{\mathrm{R}}-w}{p+r_{\mathrm{R}}-v}} z d\xi+\frac{w-v}{p+r_{\mathrm{R}}-v}y^*\right)-\beta\frac{w-v}{p+r_{\mathrm{R}}-v}\frac{f}{(p+r_{\mathrm{R}}-v)h}\end{aligned}$$

Thus, when $\nabla_3\Pi_{\mathrm{M}}=0$, using part (c) of the lemma, we have

$$\nabla^2_{23}\Pi_{\mathrm{M}}(p,r_{\mathrm{C}},r_{\mathrm{R}}) \;=\; -\beta\frac{w-v}{p+r_{\mathrm{R}}-v}\frac{f}{(p+r_{\mathrm{R}}-v)h}$$

Proof of (h) It can be verified that $\nabla^2_{13}\Pi(p_c,r_{\mathrm{R}})$ is given, after substituting for $\nabla_1 z$ from Lemma 1(a) and $\nabla_3 y^*$ from Lemma 2(c), by

$$\nabla^2_{13}\Pi_{\mathrm{R}}(p,r_{\mathrm{C}},r_{\mathrm{R}}) \;=\; \frac{w-v}{p+r_{\mathrm{R}}-v}\left(-\gamma y^*+\frac{f}{(p+r_{\mathrm{R}}-v)h}\right)-\gamma\int_0^{\frac{p+r_{\mathrm{R}}-w}{p+r_{\mathrm{R}}-v}} z d\xi$$

Now, from part (a) of Lemma 2, we note that $-\gamma y^*+\frac{f}{h(p+r_{\mathrm{R}}-v)}=\nabla_1 y^*$. The desired conclusion on the sign follows from $\gamma>0$ (by ($A3$)) and Lemma 3.

Proof of (i) It can be verified that $\nabla_{13}\Pi_{\mathrm{M}}(p,r_{\mathrm{C}},r_{\mathrm{R}})$ is given, after substituting for $\nabla_1 z$ from Lemma 1(a) and for $\nabla_1 y^*$, $\nabla_3 y^*$, and $\nabla_{13}y^*$ from Lemma 2(a), (c) and (h), by

$$\begin{aligned}\nabla^2_{13}\Pi_{\mathrm{M}}(p,r_{\mathrm{C}},r_{\mathrm{R}}) = &\left[(w-c)-(r_{\mathrm{R}}+\beta r_{\mathrm{C}})\frac{w-v}{p+r_{\mathrm{R}}-v}\right]\left[-\gamma\frac{f}{(p+r_{\mathrm{R}}-v)h}-\frac{f}{(p+r_{\mathrm{R}}-v)^2h}-\frac{fh'}{(p+r_{\mathrm{R}}-v)^2h^3}\right]\\ &+\gamma\int_0^{\frac{p+r_{\mathrm{R}}-w}{p+r_{\mathrm{R}}-v}} z d\xi-\frac{w-v}{p+r_{\mathrm{R}}-v}\left[-\gamma y^*+\frac{f}{(p+r_{\mathrm{R}}-v)h}\right]\\ &+(r_{\mathrm{R}}+\beta r_{\mathrm{C}})\frac{w-v}{(p+r_{\mathrm{R}}-v)^2}\frac{f}{(p+r_{\mathrm{R}}-v)h}\end{aligned}$$

Now, using part (c) of the lemma and the above expression, one can verify through some algebra that the following is true when $\nabla_3\Pi_{\mathrm{M}}=0$:

$$\begin{aligned}\nabla^2_{13}\Pi_M(p, r_C, r_R) &= -\left[(w-c) - (r_R + \beta r_C)\frac{w-v}{p+r_R-v}\right]\frac{fh'}{(p+r_R-v)^2h^3} \\ &- \left[(2w-c) - 2(r_R + \beta r_C)\frac{w-v}{p+r_R-v}\right]\frac{f}{(p+r_R-v)^2h}\end{aligned}$$

In order to show that $\nabla^2_{13}\Pi_M(p, r_C, r_R)|_{\nabla_3\Pi_M=0} < 0$, note that, by part (c) of the lemma, if $\nabla_3\Pi_M(p, r_C, r_R) = 0$, then we must have $(w-c) - (r_R + \beta r_C)\frac{w}{p+r_R-v} > 0$, in which case we will also have $(2w-c) - (r_R + \beta r_C)\frac{w}{p+r_R-v} > 0$. The desired result now follows since $h' > 0$ by assumption ($A2$).

Proof of (j) It can be verified that $\nabla^2_{12}\Pi_R(p, r_C, r_R)$ is given, after substituting for $\nabla_2 z$ and $\nabla^2_{12}z$ from Lemma 1(b) and (e) and for $\nabla_2 y^*$ from Lemma 2(b) by

$$\nabla^2_{12}\Pi_R(p, r_C, r_R) = \alpha\gamma\int_0^{\frac{p+r_R-w}{p+r_R-v}} zd\xi + \frac{w-v}{p+r_R-v}\alpha\gamma y^* - \alpha\theta(p+r_R-v)\int_0^{\frac{p+r_R-w}{p+r_R-v}} zd\xi$$

Now, when $\nabla_1\Pi_R = 0$, the following relationship can be verified through algebra, using part (a) of the lemma and the above expression:

$$\nabla^2_{12}\Pi_R(p, r_C, r_R) = \alpha(\gamma^2 - \theta)(p + r_R - v)\int_0^{\frac{p+r_R-w}{p+r_R-v}} zd\xi,$$

which is strictly positive since $\gamma' = \gamma^2 - \theta > 0$ by virtue of ($A4$).

Proof of Proposition 1 Using Lemma 6(a), one can verify that $\nabla_1\Pi_R(w - r_R, r_C, r_R) > 0$. Again using Lemma 6(a), one can also verify that $\nabla_1\Pi_R(p, r_C, r_R) < 0$ as $p \to \infty$. Given these observations, the result now follows from Lemma 4(a) and Lemma 6(d).

Proof of Proposition 2

Proof of (a) Given p and r_R, using Lemma 6(b), one can verify that $\nabla_2\Pi_M(p, r_C, r_R) < 0$ as $r_C \to \infty$. From Lemma 6(e), we know that $\nabla^2_{22}\Pi_M(p, r_C, r_R)|_{\nabla_2\Pi_M=0} < 0$. The result now follows by applying parts (a) and (b) of Lemma 4.

Proof of (b) We can focus on r_R such that $p + r_R \geq w$ (and, hence, $p + r_R \geq v$), since the retailer would stock zero units otherwise, and the manufacturer would make zero profits. Given p and r_C, using Lemma 6(c), one can verify that $\nabla_3\Pi_M(p, r_C, r_R) < 0$ as $r_R \to \infty$. From Lemma 6(f), we know that $\nabla^2_{33}\Pi_M(p, r_C, r_R)|_{\nabla_3\Pi_M=0} < 0$. The result now follows by applying parts (a) and (b) of Lemma 4.

Proof of (c) When $\nabla_2\Pi_M(p, r_C, r_R) = 0$, we can use Lemma 6(b) to write

$$\begin{aligned}\beta\int_0^{\frac{p+r_R-w}{p+r_R-v}} zd\xi + \beta\frac{w-v}{p+r_R-v}y^* &= \left[(w-c) - (r_R + \beta r_C)\frac{w-v}{p+r_R-v}\right]\alpha\gamma y^* \\ &- (r_R + \beta r_C)\alpha\gamma\int_0^{\frac{p+r_R-w}{p+r_R-v}} zd\xi.\end{aligned}$$

Note that for the above equality to hold, we need to have $(w-c)-(r_{\mathbf{R}}+r_{\mathbf{C}})\frac{w-v}{p+r_{\mathbf{R}}-v}>0$ (since $\gamma>0$ by assumption (A3)). Similarly, when $\nabla_3\Pi_{\mathbf{M}}(p,r_{\mathbf{C}},r_{\mathbf{R}})=0$, we can use Lemma 6 (c) to write

$$\int_0^{\frac{p+r_{\mathbf{R}}-w}{p+r_{\mathbf{R}}-v}} z d\xi+\frac{w-v}{p+r_{\mathbf{R}}-v}y^*=\left[(w-c)-(r_{\mathbf{R}}+\beta r_{\mathbf{C}})\frac{w-v}{p+r_{\mathbf{R}}-v}\right]\frac{f}{(p+r_{\mathbf{R}}-v)h}.$$

Again, note that for the above equality to hold, we need to have $(w-c)-(r_{\mathbf{R}}+r_{\mathbf{C}})\frac{w-v}{p+r_{\mathbf{R}}-v}>0$. By using the last two equalities, we obtain:

$$-(r_{\mathbf{R}}+\beta r_{\mathbf{C}})\alpha\gamma\int_0^{\frac{p+r_{\mathbf{R}}-w}{p+r_{\mathbf{R}}-v}} z d\xi=\left[(w-c)-(r_{\mathbf{R}}+\beta r_{\mathbf{C}})\frac{w-v}{p+r_{\mathbf{R}}-v}\right]\left[\frac{\beta f}{(p+r_{\mathbf{R}}-v)h}-\alpha\gamma y^*\right]$$

Now, note that the second term in brackets on the right-hand side of the equality above is $-\nabla_2 y^*+\beta\nabla_3 y^*$ (from parts (b) and (c) of Lemma 2). Also, as noted above, we must have $(w-c)-(r_{\mathbf{R}}+r_{\mathbf{C}})\frac{w-v}{p+r_{\mathbf{R}}-v}>0$. The term on the left-hand side is negative (since $\gamma>0$ by assumption $(A3)$). The desired result now follows.

Proof of Proposition 3 Under an equilibrium solution $(\tilde{p},\tilde{r}_{\mathbf{C}},\tilde{r}_{\mathbf{R}})$ with $\tilde{r}_{\mathbf{C}}>0$, we need to have $\nabla_2\Pi_{\mathbf{M}}(\tilde{p},\tilde{r}_{\mathbf{C}},\tilde{r}_{\mathbf{R}})=\nabla_1\Pi_{\mathbf{R}}(\tilde{p},\tilde{r}_{\mathbf{C}},\tilde{r}_{\mathbf{R}})=0$ (by Proposition 1 and part (a) of Proposition 2). Since $\nabla_2\Pi_{\mathbf{M}}(\tilde{p},\tilde{r}_{\mathbf{C}},\tilde{r}_{\mathbf{R}})=0$, we know from Lemma 6(b) that

$$\begin{aligned}\beta\frac{w-v}{\tilde{p}+\tilde{r}_{\mathbf{R}}-v}y^*+\beta\int_0^{\frac{\tilde{p}+\tilde{r}_{\mathbf{R}}-w}{\tilde{p}+\tilde{r}_{\mathbf{R}}-v}} z d\xi &= 38;\ 38;\ (w-c)\alpha\gamma y^* \\ &\quad -(\tilde{r}_{\mathbf{R}}+\beta\tilde{r}_{\mathbf{C}})\alpha\gamma\left(\int_0^{\frac{\tilde{p}+\tilde{r}_{\mathbf{R}}-w}{\tilde{p}+\tilde{r}_{\mathbf{R}}-v}} z d\xi+\frac{w-v}{\tilde{p}+\tilde{r}_{\mathbf{R}}-v}y^*\right), \\ &= \alpha\gamma\Pi_{\mathbf{M}}(\tilde{p},\tilde{r}_{\mathbf{C}},\tilde{r}_{\mathbf{R}})\text{ by (6)}\end{aligned} \tag{10}$$

Also, since $\nabla_1\Pi_{\mathbf{R}}(\tilde{p},\tilde{r}_{\mathbf{C}},\tilde{r}_{\mathbf{R}})=0$, we know from Lemma 6 (a) that

$$\begin{aligned}\frac{w-v}{\tilde{p}+\tilde{r}_{\mathbf{R}}-v}y^*+\int_0^{\frac{\tilde{p}+\tilde{r}_{\mathbf{R}}-w}{\tilde{p}+\tilde{r}_{\mathbf{R}}-v}} z d\xi &= (\tilde{p}+\tilde{r}_{\mathbf{R}}-v)\gamma\int_0^{\frac{\tilde{p}+\tilde{r}_{\mathbf{R}}-w}{\tilde{p}+\tilde{r}_{\mathbf{R}}-v}} z d\xi, \\ &= \gamma\Pi_{\mathbf{R}}(\tilde{p},\tilde{r}_{\mathbf{C}},\tilde{r}_{\mathbf{R}})\text{ by (5)}\end{aligned} \tag{11}$$

Now, (10) and (11) together allow us conclude $\frac{\Pi_{\mathbf{M}}(\tilde{p},\tilde{r}_{\mathbf{C}},\tilde{r}_{\mathbf{R}})}{\Pi_{\mathbf{R}}(\tilde{p},\tilde{r}_{\mathbf{C}},\tilde{r}_{\mathbf{R}})}=\frac{\beta}{\alpha}$.

Proof of Proposition 4

Proof of (a) Throughout the proof, recall that $p^*(r_{\mathbf{R}})$ will satisfy $\nabla_1\Pi_{\mathbf{R}}(p^*(r_{\mathbf{R}}),0,r_{\mathbf{R}})=0$ at any given $r_{\mathbf{R}}$ (by Proposition 1). By implicit differentiation of this identity with respect to $r_{\mathbf{R}}$, we obtain $\frac{dp^*(r_{\mathbf{R}})}{dr_{\mathbf{R}}}=-\frac{\nabla^2_{13}\Pi_{\mathbf{R}}(p^*(r_{\mathbf{R}}),0,r_{\mathbf{R}})}{\nabla^2_{11}\Pi_{\mathbf{R}}(p^*(r_{\mathbf{R}}),0,r_{\mathbf{R}})}$. Hence, we will conclude the proof of part (a) if we can show that $\nabla^2_{11}\Pi_{\mathbf{R}}(p^*(r_{\mathbf{R}}),0,r_{\mathbf{R}})\le\nabla^2_{13}\Pi_{\mathbf{R}}(p^*(r_{\mathbf{R}}),0,r_{\mathbf{R}})<0$. From Lemma

6(d) and (h), we know that $\nabla^2_{11}\Pi_{\mathbf{R}}(p^*(r_{\mathbf{R}}),0,r_{\mathbf{R}})<0$ and $\nabla^2_{13}\Pi_{\mathbf{R}}(p^*(r_{\mathbf{R}}),0,r_{\mathbf{R}})<0$. Again, from Lemma 6(d) and (h), note that:

$$\nabla^2_{11}\Pi_{\mathbf{R}}(p^*(r_{\mathbf{R}}),0,r_{\mathbf{R}}) - \nabla^2_{13}\Pi_{\mathbf{R}}(p^*(r_{\mathbf{R}}),0,r_{\mathbf{R}}) = -(p+r_{\mathbf{R}}-v)\gamma' \int_0^{\frac{p+r_{\mathbf{R}}-w}{p+r_{\mathbf{R}}-v}} z \leq 0, \quad (12)$$

where the inequality follows from $\gamma' \geq 0$ (by (*A*4)). Thus, we are able to conclude that

$$\nabla^2_{11}\Pi_{\mathbf{R}}(p^*(r_{\mathbf{R}}),0,r_{\mathbf{R}}) \leq \nabla^2_{13}\Pi_{\mathbf{R}}(p^*(r_{\mathbf{R}}),0,r_{\mathbf{R}})<0,$$

which concludes the proof of part (a).

Proof of (b) Given p, w and $r_{\mathbf{C}}=0$, it follows from Proposition 2(b) that either $r^*_{\mathbf{R}}(p)=0$ or $r^*_{\mathbf{R}}(p)>0$ in which case $r^*_{\mathbf{R}}(p)$ satisfies $\nabla_3\Pi_{\mathbf{M}}(p,0,r^*_{\mathbf{R}}(p))=0$. If $r^*_{\mathbf{R}}(p)=0$ for all $p>0$, then part (b) holds trivially. Suppose now there exists a p at which $r^*_{\mathbf{R}}(p)>0$ and satisfies $\nabla_3\Pi_{\mathbf{M}}(p,0,r^*_{\mathbf{R}}(p))=0$. By implicit differentiation of this identity with respect to p, we obtain $\frac{dr^*_{\mathbf{R}}(p)}{dp} = -\frac{\nabla^2_{13}\Pi_{\mathbf{M}}(p,0,r^*_{\mathbf{R}}(p))}{\nabla^2_{33}\Pi_{\mathbf{M}}(p,0,r^*_{\mathbf{R}}(p))}$. We already know from Lemma 6(f) and (i) that $\nabla^2_{33}\Pi_{\mathbf{M}}(p,0,r^*_{\mathbf{R}}(p))<0$ and $\nabla^2_{13}\Pi_{\mathbf{M}}(p,0,r^*_{\mathbf{R}}(p))<0$. Furthermore, again from Lemma 6(f) and (i), one can verify that

$$\begin{aligned}\nabla^2_{13}\Pi_{\mathbf{M}}(p,r_{\mathbf{C}},r_{\mathbf{R}})\big|_{\nabla_3\Pi_{\mathbf{M}}=0} &= \nabla^2_{33}\Pi_{\mathbf{M}}(p,r_{\mathbf{C}},r_{\mathbf{R}})\big|_{\nabla_3\Pi_{\mathbf{M}}=0} + \frac{w-v}{p+r_{\mathbf{R}}-v}\frac{f}{(p+r_{\mathbf{R}}-v)h} \\ &\quad - \frac{1}{p+r_{\mathbf{R}}-v}\Bigg\{-\int_0^{\frac{p+r_{\mathbf{R}}-w}{p+r_{\mathbf{R}}-v}} z\,d\xi - \frac{w-v}{p+r_{\mathbf{R}}-v}y^* \\ &\quad + \left[(w-c)-(r_{\mathbf{R}}+\beta r_{\mathbf{C}})\frac{w-v}{p+r_{\mathbf{R}}-v}\right]\frac{f}{(p+r_{\mathbf{R}}-v)h}\Bigg\}\end{aligned}$$

From Lemma 6(c), we observe that the term in curly brackets above is in fact $\nabla_3\Pi_{\mathbf{M}}(p,r_{\mathbf{C}},r_{\mathbf{R}})$. Therefore, from the above expression, we obtain:

$$\nabla^2_{13}\Pi_{\mathbf{M}}(p,r_{\mathbf{C}},r_{\mathbf{R}})\big|_{\nabla_3\Pi_{\mathbf{M}}=0} = \nabla^2_{33}\Pi_{\mathbf{M}}(p,r_{\mathbf{C}},r_{\mathbf{R}})\big|_{\nabla_3\Pi_{\mathbf{M}}=0} + \frac{w-v}{p+r_{\mathbf{R}}-v}\frac{f}{(p+r_{\mathbf{R}}-v)h}$$

Hence, from the last equality, we conclude that $\nabla^2_{33}\Pi_{\mathbf{M}}(p,0,r^*_{\mathbf{R}}(p)) <\nabla^2_{13}\Pi_{\mathbf{M}}(p,0,r^*_{\mathbf{R}}(p))$, which, along with $\nabla^2_{33}\Pi_{\mathbf{M}}(p,0,r^*_{\mathbf{R}}(p))<0$ and $\nabla^2_{13}\Pi_{\mathbf{M}}(p,0,r^*_{\mathbf{R}}(p))<0$, allows us to conclude that $-1<\frac{dr^*_{\mathbf{R}}(p)}{dp}<0$. Recall that we assumed p is such that $r^*_{\mathbf{R}}(p)>0$. For some p', we will have $r^*_{\mathbf{R}}(p')=0$, and $r^*_{\mathbf{R}}(p)$ will remain zero for all $p>p'$, and hence $\frac{dr^*_{\mathbf{R}}(p)}{dp}$ will be zero for all $p>p'$. (If $r^*_{\mathbf{R}}(p)$ were to become positive for some $p''>p'$, this would be a contradiction to the result that $\frac{dr^*_{\mathbf{R}}(p)}{dp}<0$ when $r^*_{\mathbf{R}}(p)>0$.)

Proof of (c) The existence of the Nash equilibrium follows from Lemma 5(a), Proposition 1 and Proposition 2(b). The uniqueness of the Nash equilibrium follows from Lemma 5(b) and parts (a) and (b) of this proposition. (Note that, in order to apply Lemma 5, we need upper bounds on the decision variables of the retailer and the manufacturer, p and $r_{\mathbf{R}}$, respectively. We could satisfy this requirement by picking arbitrarily large numbers to bound the feasible choices for p and $r_{\mathbf{R}}$.)

Proof of Proposition 5

Throughout the proof, let $p^*(r_{\mathbf{R}})$ denote the optimal retail price chosen by the retailer at a given $r_{\mathbf{R}}$ when $r_{\mathbf{C}} = 0$.

Proof of (a) Note that $p_o = p^*(0)$ whereas $\tilde{p} = p^*(\tilde{r}_{\mathbf{R}})$. Therefore, $\tilde{p} - p_o = \int_0^{\tilde{r}_{\mathbf{R}}} \frac{dp^*(r_{\mathbf{R}})}{dr_{\mathbf{R}}} dr_{\mathbf{R}}$. By Proposition 4(a), $-1 < \frac{dp^*(r_{\mathbf{R}})}{dr_{\mathbf{R}}} < 0$. The desired result follows.

Proof of (b) Note that $y_o = y^*(p^*(0), 0, 0)$ whereas $\tilde{y} = y^*(p^*(\tilde{r}_{\mathbf{R}}), 0, \tilde{r}_{\mathbf{R}})$. Now, $\tilde{y} - y_o = \int_0^{\tilde{r}_{\mathbf{R}}} \frac{dy^*(p^*(r_{\mathbf{R}}),0,r_{\mathbf{R}})}{dr_{\mathbf{R}}} dr_{\mathbf{R}}$. Therefore, we will conclude the proof if we can show that $\frac{dy^*(p^*(r_{\mathbf{R}}),0,r_{\mathbf{R}})}{dr_{\mathbf{R}}} > 0$. Note that $\frac{dy^*(p^*(r_{\mathbf{R}}),0,r_{\mathbf{R}})}{dr_{\mathbf{R}}} = \nabla_3 y^*(p^*(r_{\mathbf{R}}), 0, r_{\mathbf{R}}) + \frac{dp^*(r_{\mathbf{R}})}{dr_{\mathbf{R}}} \nabla_1 y^*(p^*(r_{\mathbf{R}}), 0, r_{\mathbf{R}})$. Now, $\nabla_3 y^*(p^*(r_{\mathbf{R}}), 0, r_{\mathbf{R}}) > 0$ from Lemma 2(c), $\frac{dp^*(r_{\mathbf{R}})}{dr_{\mathbf{R}}} < 0$ from of Proposition 4(a) and $\nabla_1 y^*(p^*(r_{\mathbf{R}}), 0, r_{\mathbf{R}}) < 0$ from Lemma 3. (To see why Lemma 3 can be applied here, recall that $\nabla_1 \Pi_{\mathbf{R}}(p^*(r_{\mathbf{R}}), 0, r_{\mathbf{R}}) = 0$ by Proposition 1 since $p^*(r_{\mathbf{R}})$ optimizes $\Pi_{\mathbf{R}}$.) These observations imply that $\frac{dy^*(p^*(r_{\mathbf{R}}),0,r_{\mathbf{R}})}{dr_{\mathbf{R}}} > 0$, which yields the desired result.

Proof of (c) Note that $\Pi_{\mathrm{SC}}(\tilde{p}, 0, \tilde{r}_{\mathbf{R}}) = \Pi_{\mathrm{SC}}(p^*(\tilde{r}_{\mathbf{R}}), 0, \tilde{r}_{\mathbf{R}})$ and $\Pi_{\mathrm{SC}}(p_o, 0, 0) = \Pi_{\mathrm{SC}}(p^*(0), 0, 0)$. Therefore, $\Pi_{\mathrm{SC}}(\tilde{p}, 0, \tilde{r}_{\mathbf{R}}) - \Pi_{\mathrm{SC}}(p_o, 0, 0) = \int_0^{\tilde{r}_{\mathbf{R}}} \frac{d\Pi_{\mathrm{SC}}(p^*(r_{\mathbf{R}}),0,r_{\mathbf{R}}))}{dr_{\mathbf{R}}} dr_{\mathbf{R}}$. Hence, if we can show that $\Pi_{\mathrm{SC}}(p^*(r_{\mathbf{R}}), 0, r_{\mathbf{R}})$ is increasing in $r_{\mathbf{R}}$ for $r_{\mathbf{R}} \leq w - c$, then the desired result will follow. Hence, we want to show that

$$\frac{d\Pi_{\mathrm{SC}}(p^*(r_{\mathbf{R}}), 0, r_{\mathbf{R}})}{dr_{\mathbf{R}}} = \frac{dp^*(r_{\mathbf{R}})}{dr_{\mathbf{R}}} \nabla_1 \Pi_{\mathrm{SC}}(p^*(r_{\mathbf{R}}), 0, r_{\mathbf{R}}) + \nabla_3 \Pi_{\mathrm{SC}}(p^*(r_{\mathbf{R}}), 0, r_{\mathbf{R}})$$

is positive. The following equalities can be verified using (5) and (6):

$$\begin{aligned}
\nabla_1 \Pi_{\mathrm{SC}}(p, r_{\mathbf{C}}, r_{\mathbf{R}}) =& \nabla_1 \Pi_{\mathbf{R}}(p, r_{\mathbf{C}}, r_{\mathbf{R}}) + \nabla_1 \Pi_{\mathbf{M}}(p, r_{\mathbf{C}}, r_{\mathbf{R}}) \\
=& \nabla_1 \Pi_{\mathbf{R}}(p, r_{\mathbf{C}}, r_{\mathbf{R}}) + \left[w - c - (r_{\mathbf{R}} + \beta r_{\mathbf{C}}) \frac{w - v}{p + r_{\mathbf{R}} - v} \right] \nabla_1 y^* \\
& - (r_{\mathbf{R}} + \beta r_{\mathbf{C}}) \int_0^{\frac{p + r_{\mathbf{R}} - w}{p + r_{\mathbf{R}} - v}} \nabla_1 z d\xi \\
\nabla_3 \Pi_{\mathrm{SC}}(p, r_{\mathbf{C}}, r_{\mathbf{R}}) =& \nabla_3 \Pi_{\mathbf{R}}(p, r_{\mathbf{C}}, r_{\mathbf{R}}) + \nabla_3 \Pi_{\mathbf{M}}(p, r_{\mathbf{C}}, r_{\mathbf{R}}) \\
=& \left[w - c - (r_{\mathbf{R}} + \beta r_{\mathbf{C}}) \frac{w - v}{p + r_{\mathbf{R}} - v} \right] \nabla_3 y^*
\end{aligned}$$

Note that $\nabla_1 \Pi_{\mathbf{R}}(p^*(r_{\mathbf{R}}), 0, r_{\mathbf{R}}) = 0$ by definition of $p^*(r_{\mathbf{R}})$ and Proposition 1. Thus, after substitution and rearranging terms, we get

$$\frac{d\Pi_{SC}(p^*(r_R),0,r_R)}{dr_R}=\left(1+\frac{dp^*(r_R)}{dr_R}\right)\left[w-c-r_R\frac{w-v}{p^*(r_R)+r_R-v}\right]\nabla_3 y^*$$
$$+\frac{dp^*(r_R)}{dr_R}\left\{\left[w-c-r_R\frac{w-v}{p^*(r_R)+r_R-v}\right](\nabla_1 y^*-\nabla_3 y^*)\right.$$
$$\left.-r_R\int_0^{\frac{p^*(r_R)+r_R-w}{p^*(r_R)+r_R-v}}\nabla_1 z d\xi\right\}$$

By Lemma 2(c) and Proposition 4(a), the first term above is positive. We show that the second term is also positive, to conclude the proof. Since $\frac{dp^*(r_R)}{dr_R}<0$, all we need to show is

$$\left[w-c-r_R\frac{w-v}{p^*(r_R)+r_R-v}\right](\nabla_1 y^*-\nabla_3 y^*)-r_R\int_0^{\frac{p^*(r_R)+r_R-w}{p^*(r_R)+r_R-v}}\nabla_1 z d\xi<0. \quad (13)$$

Now, for $\xi\leq\frac{p^*(r_R)+r_R-w}{p^*(r_R)+r_R-v}$,

$$\begin{aligned}\nabla_1 y^*-\nabla_3 y^* &= -\gamma y^*\\ &= \nabla_1 z\left(p^*(r_R),0,\frac{p^*(r_R)+r_R-w}{p+r_R-v}\right)\\ &<\nabla_1 z(p^*(r_R),0,\xi)\end{aligned}$$

The first equality follows from Lemma 2(a) and (c), the second from Lemma 1(a), and the inequality holds because $\xi\leq\frac{p^*(r_R)+r_R-w}{p^*(r_R)+r_R-v}$. Thus, using $r_R\leq w-c$, (13) holds.

Proof of Proposition 6

Proof of (a) Note that $p^*(r_C)$ will satisfy $\nabla_1\Pi_R(p^*(r_C),r_C,0)=0$ at any given r_C (by Proposition 1). By implicit differentiation of this identity with respect to r_C, we obtain $\frac{dp^*(r_C)}{dr_C}=-\frac{\nabla^2_{12}\Pi_R(p^*(r_C),r_C,0)}{\nabla^2_{11}\Pi_R(p^*(r_C),r_C,0)}$. We know from Lemma 6(d) and (j) that

$$\nabla^2_{11}\Pi_R(p^*(r_C),r_C,0)<0 \; and \; \nabla^2_{12}\Pi_R(p^*(r_C),r_C,0)>0.$$

Therefore, it follows that $\frac{dp^*(r_C)}{dr_C}>0$. Furthermore, from Lemma 6(d) and (j), we can write:

$$\nabla^2_{12}\Pi_R(p,r_C,r_R)|_{\nabla_1\Pi_R=0}=-\alpha\nabla^2_{11}\Pi_R(p,r_C,r_R)|_{\nabla_1\Pi_R=0}$$
$$+\alpha\frac{w-v}{p+r_R-v}\nabla_1 y^*-\alpha\gamma\int_0^{\frac{p+r_R-w}{p+r_R-v}}zd\xi$$

From the equality above, since $\nabla_1 y^*<0$ when $\nabla_1\Pi_R=0$ (from Lemma 3) and $\gamma>0$ (by ($A3$)), we have $\nabla^2_{12}\Pi_R(p,r_C,r_R)|_{\nabla_1\Pi_R=0}<-\alpha\nabla^2_{11}\Pi_R(p,r_C,r_R)|_{\nabla_1\Pi_R=0}$. Therefore, we have

$$\nabla_{12}\Pi_R(p^*(r_C), r_C, 0) < -\alpha\nabla_{11}\Pi_R(p^*(r_C), r_C, 0).$$

This observation yields $\frac{dp^*(r_C)}{dr_C} < \alpha$.
Proof of (b) The existence of the Nash equilibrium follows from Lemma 5(a), Proposition 1 and Proposition 2(a). (Note that, in order to apply Lemma 5, we need upper bounds on the decision variables of the retailer and the manufacturer, p and r_C, respectively. We could satisfy this requirement by picking arbitrarily large numbers to bound the feasible choices for p and r_C.)

Proof of Proposition 7
Throughout the proof, let $p^*(r_C)$ denote the optimal retail price chosen by the retailer at a given r_C when $r_R = 0$.
Proof of (a) Note that $p_o = p^*(0)$ whereas $\tilde{p} = p^*(\tilde{r}_C)$. Therefore, $\tilde{p} - p_o = \int_0^{\tilde{r}_C} \frac{dp^*(r_C)}{dr_C} dr_C$. By Proposition 6, $0 < \frac{dp^*(r_C)}{dr_C} < \alpha$. The desired result follows.
Proof of (b) Note that $y_o = y^*(p^*(0), 0, 0)$ whereas $\tilde{y} = y^*(p^*(\tilde{r}_C), \tilde{r}_C, 0)$. Now, $\tilde{y} - y_o = \int_0^{\tilde{r}_C} \frac{dy^*(p^*(r_C), r_C, 0)}{dr_C} dr_C$. We will conclude the proof if we can show that $\frac{dy^*(p^*(r_C), r_C, 0)}{dr_C} > 0$. Note that $\frac{dy^*(p^*(r_C), r_C, 0)}{dr_C} = \nabla_2 y^*(p^*(r_C), r_C, 0) + \frac{dp^*(r_C)}{dr_C}\nabla_1 y^*(p^*(r_C), r_C, 0)$. Since $0 < \frac{dp(r_C)}{dr_C} < \alpha$ by Proposition 6 and $\nabla_1 y^*(p^*(r_C), r_C, 0) < 0$ by Lemma 3, we obtain $\frac{dy^*(p^*(r_C), r_C, 0)}{dr_C} > \nabla_2 y^*(p^*(r_C), r_C, 0) + \alpha\nabla_1 y^*(p^*(r_C), r_C, 0)$. Using this last inequality and substituting for $\nabla_1 y^*(p^*(r_C), r_C, 0)$ from Lemma 2(a) and for $\nabla_2 y^*(p^*(r_C), r_C, 0)$ from Lemma 2(b), we can deduce that $\frac{dy^*(p^*(r_C), r_C, 0)}{dr_C} > 0$, which concludes the proof of this part.
Proof of (c) As in the proof of part (c) of Proposition 5, we will show that $\Pi_{SC}(p^*(r_C), r_C, 0)$ is increasing in r_C for $r_C \le w - c$ when $\alpha \ge \beta$. The desired result would then follow. Now, the following equalities can be verified by partial differentiation of (5) and (6):

$$\begin{aligned}
\frac{d\Pi_{SC}(p^*(r_C), r_C, 0)}{dr_C} &= \frac{dp^*(r_C)}{dr_C}\nabla_1\Pi_{SC}(p^*(r_C), r_C, 0) + \nabla_2\Pi_{SC}(p^*(r_C), r_C, 0) \\
&= \frac{dp^*(r_C)}{dr_C}\nabla_1\Pi_R(p^*(r_C), r_C, 0) \\
&\quad + \left(w - c - \beta r_C\frac{w - v}{p^*(r_C) - v}\right)\left(\nabla_2 y^* + \frac{dp^*(r_C)}{dr_C}\nabla_1 y^*\right) \\
&\quad + \beta r_C\left(\int_0^{\frac{p^*(r_C)-w}{p^*(r_C)-v}} \nabla_2 z d\xi - \frac{dp^*(r_C)}{dr_C}\int_0^{\frac{p^*(r_C)-w}{p^*(r_C)-v}} \nabla_1 z d\xi\right) \\
&\quad + p^*(r_C)\int_0^{\frac{p^*(r_C)-w}{p^*(r_C)-v}} \nabla_2 z d\xi \\
&\quad - \beta\left(\int_0^{\frac{p^*(r_C)-w}{p^*(r_C)-v}} z d\xi + \frac{w - v}{p^*(r_C) - v}y^*\right)
\end{aligned}$$

Now, the first term is zero, by definition of $p^*(r_C)$. The second term is positive, because, as in the proof of part (b), $\nabla_2 y^*(p^*(r_C), r_C, 0) + \frac{dp^*(r_C)}{dr_C}\nabla_1 y^*(p^*(r_C), r_C, 0) > 0$, and $r_C \le w - c$. The third term is positive by virtue of Lemma 1(a)-(b) and Proposition 6(a). Using Lemma 6 (a), Lemma 1(a)-(b) and the fact that $\nabla_1 \Pi_R(p^*(r_C), r_C, 0) = 0$ we get that

$$p^*(r_C)\int_0^{\frac{p^*(r_C)-w-v}{p^*(r_C)-v}} \nabla_2 z d\xi = \alpha\left(\int_0^{\frac{p^*(r_C)-w}{p^*(r_C)-v}} z d\xi + \frac{w-v}{p^*(r_C)-v} y^*\right).$$

Thus, the sum of the last two terms can be written as

$$(\alpha - \beta)\left(\int_0^{\frac{p^*(r_C)-w}{p^*(r_C)-v}} z d\xi + \frac{w-v}{p^*(r_C)-v} y^*\right),$$

which is positive because $\alpha \ge \beta$.

Proof of (d) The proof of this part is almost identical to the analogous result in Proposition 3. Set $\tilde{r}_R = 0$ and the proof follows the same line of argument.

REFERENCES

Aydin, G. Porteus, E. L. (2006), Joint inventory and pricing decisions for an assortment, To appear in *Operations Research.*

Barlow, R. E. Proschan, F. (1965), *Mathematical Theory of Reliability*, Wiley, New York.

Bulkeley, W. M. (10 February 1998), 'Rebates' secret appeal to manufacturers: Few consumers actually redeem them', *The Wall Street Journal.*

Cachon, G. Netessine, S. (2003), Game theory in supply chain analysis, *in* D. Simchi-Levi, D. Wu Z.-J. Shen, eds, 'Handbook of Quantitative Supply Chain Analysis: Modeling in the eBusiness Era', Springer.

Chen, X., Li, C.-L., Rhee, B.-D. & Simchi-Levi, D. (2007), 'The impact of manufacturer rebates on supply chain profits', *Naval Research Logistics* **54**, 667–80.

Dreze, X. Bell, D. R. (2003), 'Creating win-win trade promotions: Theory and empirical analysis of scan-back trade deals', *Marketing Science* **22**, 16–39.

Gerstner, E. Hess, J. D. (1991), 'A theory of channel price promotions', *The American Economic Review* **81**, 872–86.

Gerstner, E. Hess, J. D. (1995), 'Pull promotions and channel coordination', *Marketing Science* **14**, 43–60.

Kalyanam, K. (1996), 'Pricing decisions under demand uncertainty: A bayesian mixture model approach', *Marketing Science* **15**, 207–21.

Krishnan, H., Kapuscinski, R. Butz, D. A. (2004), 'Coordinating contracts for decentralized channels with retailer promotional effort', *Management Science* **50**, 48–63.

Lal, R. (1990), 'Price promotions: Limiting competitive encroachment', *Marketing Science* **9**, 247–62.

Lal, R., Little, J. D. C. Villas-Boas, J. M. (1996), 'A theory of forward buying, merchandising and trade deals', *Marketing Science* **15**, 21–37.

Menzies, D. (September 12 2005), 'Mail-in rebates rip', *Marketing.*

Millman, H. (17 April 2003), 'Customers tire of excuses for rebates that never arrive', *The New York Times*.
Narasimhan, C. (1984), 'A price discrimination theory of coupons', *Marketing Science* **3**, 128–47.
Olenick, D. (14 October 2002), 'emachines drops rebate program', *TWICE*.
Petruzzi, N. C. Dada, M. (1999), 'Pricing and the newsvendor problem: A review with extensions', *Operations Research* **47**, 183–94.
Porteus, E. L. (2002), *Foundations of Stochastic Inventory Theory*, Stanford University Press, Stanford.
Raju, J. S., Srinivasan, V. Lal, R. (1990), 'The effects of brand loyalty on competitive price promotional strategies', *Management Science* **36**, 276–304.
Rao, R. C. (1991), 'Pricing and promotions in asymmetric duopolies', *Marketing Science* **10**, 131–44.
Ricadela, A. Koenig, S. (September 1998), 'Rebates' pull is divided by hard and soft lines', *Computer Retail Week*.
Taylor, T. A. (2002), 'Supply chain coordination under channel rebates with sales effort effects', *Management Science* **48**, 992–1007.

Chapter 11
CLEARANCE PRICING IN RETAIL CHAINS

Stephen A. Smith
Department of Operations and MIS, Leavey School of Business, Santa Clara University, Santa Clara, CA 95053, USA

1. INTRODUCTION

1.1 Background

As an application of management science, retail clearance pricing has been an outstanding success. Pilot studies conducted in the 1990's (Smith and Achabal 1998), found that installing a computer based clearance pricing algorithm at a major retail chain resulted in 10 to 15% increases in the revenue capture rate during the clearance period. Increases in sell-through and shorter markdown cycle times also freed up capital and floor space for the retailer's follow-on products. Similar revenue gains during the clearance period have been achieved by commercially offered clearance markdown systems (Merrick 2001). Spotlight Systems, Inc.,[1] a seller of clearance markdown software systems, reported in 2002 that the average gain in gross margin dollars for the department and specialty stores that had implemented their system amounted to about 4% of revenue, or $40 million for every $1 billion of sales. Since U.S. department store sales now exceed $500 Billion per year, there is a very large potential dollar impact, if similar results can be obtained across the industry. Major vendors of ERP systems are now making price optimization a cornerstone of their retail applications suites (Sullivan 2005). This background section discusses why clearance pricing is such an attractive application for retailers and what has allowed it to be successfully implemented through computer based models.

1.2 Trends in retail pricing

Retail department and specialty stores are selling an ever increasing fraction of their merchandise on markdowns, which now account for over one third of

[1] Spotlight Systems was acquired by Profit Logic, Inc. in 2003, which was in turn acquired by Oracle Corporation in 2005.

N. Agrawal, S.A. Smith (eds.), *Retail Supply Chain Management*,
DOI: 10.1007/978-0-387-78902-6_11,

all sales.[2] This is a result of four general trends in these retailers' merchandising strategies:

(1) more products in the assortment
(2) a greater proportion of "fashion" merchandise,
(3) shorter seasons and
(4) more private label (store brand) merchandise.

While these trends give customers a wider selection of product choices and are essential for retailers to remain competitive, they also increase the difficulty of managing the retail supply chain. Fashion and private label items tend to have long lead times for orders from the manufacturer and the total order quantity for the season is usually fixed in advance. This decision is based on the initial sales forecasts, which tend to be inaccurate for fashion and seasonal merchandise. Also, well over half of the retailer's total order for seasonal and fashion items is usually sent to the stores at the start of the season to create an attractive presentation of the merchandise. Since inter-store transfers are not economical, it is difficult to rebalance this inventory if the initial allocation is incorrect. When sales at a given store are lower than expected, retailers must find a way to clear the excess merchandise to make way for the new product arrivals of the coming season.

Clearance pricing involves two decisions: when to start clearance markdowns and how "deep" the markdowns should be, both of which depend on the remaining inventory. Traditionally, these decisions have been made by the buyer who originally chose the merchandise and ordered it from the manufacturer. This may create a disincentive for taking markdowns early enough, since an early decision to mark down really amounts to admitting that the product has underperformed. For seasonal items such as swimsuits and winter coats, demand decreases rapidly near the end of the season; thus delaying a markdown can be very costly. For simplicity, buyers have traditionally taken the same markdown at all stores, or for all stores within a region. This is suboptimal when there are significant inventory imbalances across stores. These factors tend to make clearance markdowns a very complex decision that buyers would be happy to delegate to a computer based pricing algorithm. At the same time, retail managers require a clear demonstration of the "payback," i.e., the return on investment, for any newly implemented system. Thus, any clearance markdown pricing system needs to be able to pay for itself through improvements in gross margin dollars during the clearance period.

The computing resources necessary for clearance pricing have only recently been available to retailers. As late as the 1990's many retailers did not retain store level item sales figures for more than 90 days and sales results were often reported only in dollars of revenue. Often, there were no detailed records of how many units of each item were sold at a given price. The economics of data storage was often the deciding factor in these decisions, because a department store retailer with 100,000 SKUs and 1000 stores simply could not afford to

[2] National Retail Federation data for Department and Specialty Stores.

store all this transaction data for all time periods. Computing resources were also limited among retail staff members, because of the high costs of training and support. Since retail staff members tend to change job assignments frequently, it is important to standardize and document all decision making procedures, and to make the results easily understandable by retail personnel who are not technically trained in using computers. The exponential decline in the cost of data storage and the growth in popularity of personal computers that occurred during the 1990's removed these barriers to implementing computer based clearance pricing algorithms.

1.3 Mathematical models for clearance pricing

An analytical approach to clearance markdown management requires the successful implementation of three system components:

(1) a sales forecasting model
(2) a clearance price optimization algorithm that works at the store and item level
(3) financial performance measurement of the effectiveness of the system

This section discusses a number of the models in the literature that relate to these components of the clearance pricing system.

The modeling assumptions in this paper were motivated by discussions with buyers who manage clearance markdowns at several retail department and specialty store chains. The author also assisted three major retailers in designing computer-based systems that incorporated these models. One unique aspect of this chapter's pricing model is that sales depend explicitly on the retailer's on-hand inventory. The pricing analysis implies that when the rate of sale is sensitive to the inventory level, it is optimal to have higher prices early in the season, followed by deeper markdowns later in the clearance period. Furthermore, inventory sensitivity in the demand makes it optimal to have some amount of leftover merchandise at the end of the clearance period. This leftover inventory, which is typically found in department store chains, may be sold to a discounter, transferred to other channels operated by the retailer or possibly donated to charity. Many retailers recognize the advantage of setting clearance prices at the store level to account for the variation in inventory levels and sales rates across stores. Due to the complexity and time consuming nature of localized pricing, computer-based clearance pricing algorithms are required to implement these store level markdown decisions.

2. RELATED RESEARCH

In general, optimal clearance pricing for retailers involves some type of dynamic pricing. Two excellent surveys on dynamic pricing policies have recently appeared: Elmaghraby and Keskinocak (2003) and Bitran and Caldenty (2003).

The surveyed papers include a variety of factors such as seasonally varying or declining demand, varying customer response to price changes, demand uncertainty, inventory dependent demand and simultaneous pricing and inventory decisions. Since no tractable model can incorporate all of these factors simultaneously, the choice of modeling assumptions requires tradeoffs. The literature summary below focuses on specific subsets of the pricing literature in marketing, economics and inventory management that are relevant to the retail clearance pricing application.

Intertemporal pricing issues similar to those found in clearance markdowns are studied in a deterministic setting by Stokey (1979), Kalish (1983), Dhebar and Oren (1985), Rajan, Rakesh and Steinberg (1992), Braden and Oren (1994). Stokey's analysis considered a family of customer utility functions that decline with time and identified conditions under which the optimal price trajectory is constant or decreasing. Kalish (1983) considered sales rates that vary with both price and cumulative sales-to-date and obtained conditions on sales rate and production cost that determine whether the optimal price trajectories are increasing or decreasing. Dhebar and Oren (1985) determined the optimal price trajectory when there is a positive network externality and decreasing supply cost. The other two papers are discussed below.

Demand uncertainty has been included in dynamic pricing models in a variety of ways. Lazear (1986) and Pashigian (1988) considered clearance markdowns for a single item sold to heterogeneous customers who have a time invariant probability distribution of reservation prices. Gallego and van Ryzin (1994) developed a continuous time optimal pricing model in which demand is generated by Poisson arrivals. Feng and Gallego (1995) develop a continuous time Markov process formulation with stochastic demand that determines the optimal timing and duration of a single price reduction. Bitran, *et al.* (1998), Bitran and Mondschein (1997) and Zhao and Zheng (2000) generalize this by modeling customer demand as Poisson arrivals whose reservation prices change over time. The net result is a nonhomogeneoues Poisson process multiplied by a price sensitivity function. While these models capture demand uncertainty, they do not include the influence of inventory level on demand, which we found was often significant in retail sales. Significant positive correlation between inventory levels and retail sales was also found by Wolfe(1968) and Bhat(1985).

Learning can play a role in dynamic pricing for either the buyer or the seller. Lazear(1986) allowed the seller to infer customers' reservation prices through their responses to a decreasing sequence of discrete prices. Besanko and Winston(1990) investigated the role of customers' knowledge of future prices in intertemporal pricing. Braden and Oren(1994) derive an optimal nonlinear price structure that improves the seller's information about the distribution of heterogeneous customers' price sensitivities. Lariviere and Porteus(1999) considered a multi-period pricing and inventory model with learning, in which the seller uses varying inventory levels as opposed to price changes to obtain information.

The marketing literature on price promotions provides a number of empirically tested functional forms for price response. This paper adopts a multiplicative form with exponential price sensitivity, which has been analyzed and empirically tested by Narasimhan(1984), Russell and Bolton(1988), Bolton(1989), Achabal *et al.* (1990), Smith *et al.* (1994) and Kalyanam (1995). Exponential sensitivity is also applicable for modeling how price influences purchases of consumer durables; Kalish(1985) compared several variations.

There are a number of related papers that develop combined strategies for pricing and inventory management. Eliashberg and Steinberg(1987) considered pricing, inventory and production management policies for a marketing channel subject to seasonal variations. Rajan, *et al.* (1992) considered dynamic pricing and inventory decisions with a variable time horizon and shrinkage costs. Bitran, *et al.* (1998) consider the coordination of prices and inventories across multiple retail outlets in which there are initial allocations of inventories and a further reallocation to rebalance inventories in response to sales. This formulation includes many of the aspects of retail markdown pricing, but the result is a dynamic programming problem with such a large state space that it is likely to be intractable. The authors propose and test some myopic heuristics for approximate solutions. Mantrala and Rao (2001) discuss a decision support system called MARK, which determines discrete prices and inventory levels based on a time varying elasticity demand model. Monahan, *et al.* (2004) analyze a newsvendor model with combined pricing and inventory decisions at discrete time points. Cheng and Sethi (1999) develop a Markov decision model to determine promotion and inventory decisions in a discrete periodic review system. Ray, *et al.* (2005) develop a combined pricing and inventory management model for a two echelon serial supply chain using a demand function with an additive uncertainty term and random delivery times. Netessine (2004) models price and inventory changes at discrete time points, considering the optimization of both prices and the discrete timing of the price changes.

In summary, the model in this chapter differs from those discussed above in that it includes seasonal variations and demand dependence on inventory level, in addition to price sensitivity. At the same time, this paper's model requires the time horizon to be fixed, and ignores time dependent inventory costs and discounting. It allows a single inventory level adjustment, while a number of the previous papers on combined dynamic policies consider more general inventory strategies. Also, this chapter's pricing model does not explicitly include demand uncertainty. However, the updating of the clearance price at discrete time points, as discussed in the last section, provides an approximate myopic solution to the dynamic pricing problem with demand changes. Also, the deterministic optimization formulation allows a closed form pricing solution to be obtained from optimal control theory. For the retail clearance markdown application, it appears that these modeling assumptions are a good compromise that results in a workable clearance pricing model.

This chapter extends the specific results in Smith and Achabal (1998) in several ways. First, it discusses the highly successful application results that have been achieved by commercially available clearance markdown systems since the publication of the original paper. Second, it extends the earlier model to obtain FONC and approximate solutions for the case in which prices change only at pre-assigned discrete time points. An approximate discrete pricing solution is developed, and the continuous solution is used to obtain bounds on the maximum error associated with the approximation. Finally, it obtains closed form expressions for the maximum profit function and presents illustrative numerical analyses for the discrete pricing case.

3. MODEL SPECIFICATIONS AND OPTIMALITY CONDITIONS

In developing a decision making framework for clearance markdowns, it is important to note three ways in which clearance prices differ from other types of retail pricing decisions: (1) clearance markdowns are permanent, i.e., prices are not permitted to increase later, (2) demand tends to decrease at the end of the clearance period due to items becoming "out of season," as well as incomplete assortments and reduced merchandise selection, (3) optimal clearance prices typically differ by location due to inventory imbalances.

Motivated by these observations, the modeling assumptions are as follows:

- Sales rate depends explicitly on price, seasonal variations and inventory level.
- Competition, demand uncertainty, time discounting and time dependent holding costs are not explicitly included in the model.

These modeling choices can be explained as follows. Price dependence specifies the change in sales rate as a function of the percentage markdown. Seasonal variations capture the increase in sales rate that tends to occur during certain prime shopping periods such as Christmas and back-to-school, and the decrease that occurs at the end of the product's season. When the on-hand inventory is too low at a given store, the sales rate may also drop. This is especially true for apparel when there is an incomplete selection of sizes and colors. Additionally, for some items, it is important to have sufficient inventory to create an attractive in-store display to draw customers' attention to the product.

Retailers tend to intentionally schedule larger deliveries during periods with high sales forecasts, e.g., during promotions. In analyzing the corresponding sales data after the fact, this may sometimes seem to imply a false "causality," in that the higher sales during promotions should not be attributed to higher inventories, even though a positive correlation exists. On the other hand, most buyers seem to feel that low inventories do reduce sales, which was supported by our regression results. Retailers often define a minimum on-hand inventory for each product, sometimes called "fixture fill," which is the quantity required for adequate presentation. This is used as a reference level in defining the inventory effect in the model.

Competition and demand uncertainty are not explicitly captured in the sales rate model. However, sales lost to competitors are implicitly reflected in the retailer's seasonally adjusted rate of sale. This is appropriate as long as the competitors do not react directly to the retailer's price changes. For clearance markdowns taken at the store level, competitive reactions seem unlikely, given that most retail chains have hundreds of stores, each with different local competitive environments.

Demand uncertainty clearly exists, but modeling it complicates the analysis to a great extent. Optimal clearance pricing in the presence of gradually decreasing demand uncertainty would require multistage pricing decisions, which would need to be jointly optimized by stochastic dynamic programming. The state space for this problem is extremely large, because it must capture all the possible changes in the states of information that influence each update of the pricing policy. Because the clearance period is relatively short and sales rates are declining, the early clearance markdowns tend to be the dominant decisions economically, thus reducing the importance of multi-stage optimization. The short clearance period also justifies the lack of time discounting and time dependent inventory costs in this model. We therefore develop a deterministic pricing formulation without discounting.

3.1 Model formulation

The model is specified as a continuous function of time with the following parameters

t_0 = current time of the season
t_e = end of the season, sometimes known as the "outdate"
t = an arbitrary time $t_0 \leq t \leq t_e$
I_0 = on hand inventory at time t_0
$p(t)$ = price trajectory at time
$s(t)$ = cumulative sales from time t_0 to time t
$I(t)$ = $I_0 - s(t)$ = the on-hand inventory at time t
s_e = total units sold by the outdate t_e
$x(p,I,t)$ = the sales rate at time t, with price p and on-hand inventory I.
c_e = salvage value per unit at the end of the season
$c(I_0)$ = cost of adjusting I_0, if changes are permitted
$R(I_0)$ = total revenue obtained from the I_0 units

The total sales $s(t)$ up to time t clearly satisfies

$$s(t) = I_0 - I(t) = \int_{t_0}^{t} x(p(\tau), I(\tau), \tau) d\tau, \tag{1}$$

which implies the differential equation

$$I'(t) = -x(p(t), I(t), t) \text{ for each } t. \tag{2}$$

It is also required that $s_e < I_0$, where the unsold units $I_0 - s_e = I(t_e)$ are salvaged.

In general, the retailer's objective is to maximize total revenue during the clearance period, since the cost of ordering I_0 is a sunk cost. However, changes in I_0 with costs captured by the function $c(I_0)$ may be permitted in some cases. The net profit can then be expressed as:

$$R(I_0) - c(I_0) = \int_{t_0}^{t_e} p(t)x(p(t), I(t), t)dt + c_e(I_0 - s_e) - c(I_0),$$

$$\text{subject to } I_0 \geq s_e = \int_{t_0}^{t_e} x(p(t), I(t), t)dt. \tag{3}$$

This objective function can be optimized using optimal control methods, as discussed in detail in Smith and Achabal (1998). These results will be summarized below and then extended to develop exact and approximate solutions for the discrete pricing case.

First order necessary conditions (FONC) for maximizing (3) with respect to $p(t)$, subject to the stated constraints can be obtained by forming the Hamiltonian $H = (p - \lambda)x$ and treating $I(t)$ as the state variable and $p(t)$ as the control (see, e.g., Kamien and Schwartz 1981, pp. 143–8). The Lagrange multipliers are

$$\theta = \text{the Lagrange multiplier for the constraint } I_0 - s_e \geq 0$$

$$\lambda(t) = \text{the Lagrange multiplier for } I'(t) = -x(p(t), I(t), t) \text{ at time } t.$$

The FONC for the optimal control $p(t)$ and the corresponding state variable $I(t)$ are:[3]

$$\partial H/\partial I = [p - \lambda]x_I = -\lambda' \qquad \partial H/\partial p = [p - \lambda]x_p + x = 0 \tag{4}$$

$$\text{with the boundary condition } \lambda(t_e) = c_e + \theta \tag{5}$$

Eliminating $p - \lambda$ from the two partial derivative equations gives

$$\lambda' = xx_I/x_p \qquad \text{and} \qquad p + x/x_p = \lambda. \tag{6}$$

Evaluating (6) at $t = t_e$ and combining with (5) yields the boundary condition for θ

$$(p + x/x_p)_{t=t_e} = c_e + \theta. \tag{7}$$

[3] Subscripts p and I denote partial derivatives and the independent variable t has been suppressed for notational compactness.

3.1.1 The separable sales rate case

Specific assumptions concerning the functional form of the sales rate allow (6) and (7) to be solved explicitly for the optimal price trajectory. For this paper, a multiplicative, separable function with exponential price sensitivity is assumed,

$$x(p, I, t) = k(t)y(I)e^{-\gamma p}, \tag{8}$$

where $k(t)$ = the seasonal demand at time t
$y(I)$ = the inventory effect when on-hand inventory is I
γ = the price sensitivity parameter for demand.

Although much of this paper's development can be carried through for a more general demand function, a closed form solution can be obtained only for a separable demand function like (8). A slightly different closed form solution can also be obtained for constant elasticity price dependence of the form $p^{-\gamma}$. Both exponential price sensitivity and constant elasticity demand functions have been widely studied in marketing. These have generally been found to be superior to linear price sensitivity in empirical studies. [See, e.g., Kalyanam (1995) and Smith, *et al.* (1994) for references.]

For the separable form (8), we have that $x/x_p = -1/\gamma$ is a constant. From (6), it therefore follows that $p'(t) = \lambda'(t)$. Thus, (6) yields an ordinary differential equation that can be solved for $p(t)$

$$p'(t) = xx_I/x_p = -\frac{1}{\gamma}k(t)y'(I(t))e^{-\gamma p(t)}. \tag{9}$$

Mathematically similar formulations have been studied in other contexts. Kalish (1983), Dhebar and Oren (1985) and Mahajan, Muller and Bass (1990) developed formulations that are sensitive to experience effects rather than inventory, which lead to similar necessary conditions for the optimal price trajectories. Rajan, Rakesh and Steinberg (1992) obtained optimal price solutions for a separable demand form that is analogous to (8), but with a time varying γ. Gallego and van Ryzin(1994) obtained an optimal price trajectory for the case of exponential price sensitivity and Poisson demand arrivals. These formulations do not consider the dependence of sales on the current inventory level or seasonal variations, however.

Rajan, *et al.* allow a variable cycle length and they explicitly consider shrinkage and other inventory costs. They obtain closed form optimal price trajectories for the cases of linear and exponential price sensitivities. Variable cycle length is used for clearance pricing of some discontinued non-seasonal items, but seasonal items, which constitute the bulk of retail clearance items, have a fixed clearance calendar to coincide with the planned arrival of new merchandise.

3.1.2 Compensating prices

Equation (9) can be solved by proving that the optimal $p(t)$ adjusts the sales rate so as to exactly compensate for any reduction in sales due to $y(I(t))$. This result is stated as the following lemma.

Lemma 1 *For the multiplicatively separable sales rate function given by (8), equation* (9) *implies that the optimal policy is to adjust* $p(t)$ *so that sales remain proportional to* $k(t)$.

Proof We wish to show that for the optimal $p(t)$

$$\frac{x(p(t), I(t), t)}{k(t)} = y(I(t))e^{-\gamma p(t)} \text{ is constant in } t. \tag{10}$$

Suppressing the dependence on t and I and differentiating, we have

$$\frac{d}{dt}(ye^{-\gamma p}) = [I' y' - \gamma y p']e^{-\gamma p} = [-ky' e^{-\gamma p} - \gamma p']ye^{-\gamma p} = 0,$$

from (9), after substituting $I' = -kye^{-\gamma p}$ from (2). *QED.*

Lemma 1 implies that the price $p(t)$ at any time t can be expressed in terms of the final price $p(t_e)$ and the ending inventory $I(t_e)$) as follows

$$y(I(t))e^{-\gamma p(t)} = y(I(t_e))e^{-\gamma p(t_e)} \text{ for all } t, \tag{11}$$

Equation (11) also shows that the optimal price depends upon $I(t)$ but not upon t. Therefore, by defining a new function $P(I(t)) = p(t)$, (9) can be solved for the price trajectory as a function of the inventory level

$$P(I) = p(t_e) + \frac{1}{\gamma}\ln\left(\frac{y(I)}{y(I(t_e))}\right). \tag{12}$$

The total sales s_e must satisfy from (1)

$$s_e = \int_{t_0}^{t_e} k(t)y(I(t))e^{-\gamma p(t)}dt = y(t_e)e^{-\gamma p(t_e)}K,$$
$$\text{where } K = K(t_e) = \int_{t_0}^{t_e} k(t)dt. \tag{13}$$

One of two possible cases must hold at time t_e. Either $\theta > 0$ and $s_e = I_0$, or $\theta = 0$ and thus $p(t_e) = c_e + 1/\gamma$ from (7). If $\theta = 0$, we determine s_e from the relationship

$$s_e = y(I_0 - s_e)Ke^{-\gamma c_e - 1}. \tag{14}$$

This has a unique solution since $y(I_o - s_e)$ is decreasing in s_e.

3.1.3 Determining optimal inventory and maximum profit

We can use the change of variable $I = I(t)$ and the price function $P(I)$ to rewrite the integral in the total revenue as

$$\int_{t_0}^{t_e} p(t)x(p(t), I(t), t)dt = \int_{t_0}^{t_e} p(t)(-I'(t))dt = \int_{I_0 - s_e}^{I_0} P(I)dI. \tag{15}$$

Substituting for $P(I)$ from (12), we have

$$R(I_0) = s_e p(t_e) + \frac{1}{\gamma} \int_{I_0 - s_e}^{I_0} \ln\left(\frac{y(I)}{y(I_0 - s_e)}\right) dI + c_e(I_0 - s_e). \tag{16}$$

This allows us to compute the revenue that will be obtained by using the optimal pricing policy.

Equation (16) can also be used to solve for the optimal I_0, if it is a decision variable, subject to the relationships between I_0 and s_e specified above. For the case in which $y(0) = 0$, FONC can be obtained by maximizing $R(I_0 - C(I_0)$ with respect to I_0 and s_e, subject to (14). Letting η be the Lagrange multiplier for (14), it can be shown that the FONC imply that $\eta = -1/\gamma$ and that

$$p(I_0) = c_e + \frac{1}{\gamma}\left\{1 + \ln\left(\frac{y(I_0)}{y(I_0 - s_e)}\right)\right\} = c'(I_0) + 1/\gamma. \tag{17}$$

This can be solved simultaneously with (14) to obtain the optimal I_0 and s_e.

Conceptually, it is also possible to use the solution of (17) and (14) to optimize the initial inventory purchase at the beginning of the season. However, there are practical reasons why this is generally not advisable. Expanding the size of the time interval $[t_0, t_e]$ to include the whole season implies that the same exponential price sensitivity must hold for the demand during the entire time interval. Intuitively, it seems unlikely that this will be true, since price sensitivity may increase or decrease or even require different functional forms during different parts of the season. Thus, it does not seem appropriate to include the original inventory purchase as a decision variable in the context of the clearance pricing model. Smith and Achabal (1998) discuss some adjustments in on-hand inventory that may be possible during the clearance period.

4. DISCRETE PRICE CHANGES

In practice, retailers change prices at discrete points in time, rather than continuously. In this section, optimal discrete pricing will be derived and compared to the results for continuous pricing. The discrete pricing case is considerably more complex to solve than the continuous case. However, an approximate discrete solution and error bound can be derived.

An approximate solution for the discrete case can be obtained by choosing prices in each time interval that yield the same unit sales as the continuous case for that time interval. It is shown that the typical revenue losses from this approximation are no more than 1% to 2% for two or more price points. The continuous solution is used to bound the maximum error for the approximate discrete solution, since the exact discrete solution can never be better than the continuous solution.

Suppose the retailer may change prices at n previously set times, e.g., once per week. Let

$t_i =$ time of the ith price change
p_i = price for time period i
$s_i(t)$ = cumulative sales up to time t for $t_{i-1} < t < t_i$
$s_i = s_i(t)$ = cumulative sales to the end of period i.

The continuous functions $s_i(t)$, $i=1, \ldots, n$ satisfy the differential equation

$$s_i'(t) = k(t)y(I_0 - s_i(t))e^{-\gamma p_i} \text{ for } t_{i-1} \leq t \leq t_i, \tag{18}$$

with boundary conditions $s_i(t) = s_i$ for $i=1, \ldots, n$. The discrete optimization problem is

$$\max_{p_1,\ldots p_n} \sum_{i=1}^{n} p_i(s_i - s_{i-1}), \tag{19}$$

subject to (18) and its boundary conditions. The variables separate in (18), to yield

$$\frac{ds_i}{y(I_0 - s_i)} = e^{-\gamma p_i}k(t)dt. \tag{20}$$

The differential equation in (20) can be solved for a specific function $y(I)$, if the left hand side can be integrated. The optimization problem can then be solved by a discrete search over the vector of prices $p_1, \ldots, p_n$, subject to the functions $s_i(t)$ obtained from (20).

4.1 Solution for the power function form

In this section, we will solve the special case in which the inventory sensitivity follows a power function[4] of the form

$$y(I) = (I/I_r)^{\alpha}, \text{for a fixed reference value } I_r. \tag{21}$$

[4] In Smith and Achabal (1998), additional solution details are given for the general function $y(I)$ and numerical analyses are performed for a linear function $y(I)$.

This form gives considerable flexibility since for various choices of α, it can be either convex, concave or a linear function of the on-hand inventory. This form has $y(0)=0$, which implies that $\theta=0$ in (7) and s_e is determined from (14). Thus, there will be left over inventory to be salvaged at the end of the season in this case. In practice, this occurs for virtually all clearance items. Also, $p(t_e) = p_e = c_e + 1/\gamma$ from (7) for $\theta = 0$.

Sometimes in (21) the effect of inventory dependence can be truncated at $I = I_r$. This assumes that inventory larger than I_r, does not affect sales. This may often be an appropriate assumption because, as noted previously, higher inventories may sometimes falsely appear to cause higher sales. Thus, whether or not to truncate the inventory effect is really a judgment call, based on the nature of the sales environment that is being analyzed.

For the power function (21), the fraction of units sold

$$f_e = s_e/I_0 \tag{22}$$

is related to I_0 from (14) as follows

$$f_e = \left(\frac{I_0}{I_r}\right)^{-\alpha} \frac{K}{I_0} e^{-\gamma(c_e+1/\gamma)}(1-f_e)^{\alpha}. \tag{23}$$

The price and total revenue equations then can be written as

$$P(I) = p_e + \frac{\alpha}{\gamma}\ln\left(\frac{I/I_0}{1-f_e}\right) \tag{24}$$

$$R(I_0) = I_0\left[c_e + f_e/\gamma - \frac{\alpha}{\gamma}\{f_e + \ln(1-f_e)\}\right] \tag{25}$$

Note that in (24) and (25) I_r and K do not appear, but f_e depends on I_0/I_r and K/I_0 through (23).

Some of the characteristics of these functions can be summarized as follows:

Lemma 2 *The fraction f_e of the inventory sold is decreasing in I_0 for $\alpha < 1$, increasing in I_0 for $\alpha > 1$ and constant for $\alpha = 1$. For $\alpha = 1$, we have*

$$f_e = \frac{a}{1+a}, \text{ where } a = \frac{Ke^{-\gamma p_e}}{I_r}. \tag{26}$$

Thus, the revenue $R(I_0)$ is linear in I_0 for $\alpha = 1$.

Proof Taking the total derivative of (23) and rearranging terms, we obtain

$$\frac{df_e}{dI_0} = \frac{(1-f_e)(\alpha-1)}{I_0^{2-\alpha}(1-f_e)^{1-\alpha} + \alpha I_0 a}. \tag{27}$$

This shows the behavior of f_e, with respect to changes in I_0, based on the term $\alpha - 1$ in the numerator. QED.

4.1.1 Optimal discrete pricing

The differential equation (18) can be solved by integration for the special case (21) to obtain

$$-\frac{I_r^{\alpha}\{I_0 - s_i(t)\}^{1-\alpha}}{1-\alpha} = e^{-\gamma p_i}[K(t) - K(t_{i-1})] + Z_i, \tag{28}$$

where Z_i is the constant for the function $s_i(t)$ and $K(t)$ is the cumulative seasonal coefficient function from (13). At the initial condition $t = t_{i-1}$ in (28) $s_i(t_{i-1}) = s_{i-1}$ and we obtain the constant term

$$Z_i = -\frac{I_r^{\alpha}\{I_0 - s_{i-1}\}^{1-\alpha}}{1-\alpha}.$$

Equation (28) then acts as a constraint in solving the optimization problem (19). No closed form solution can be obtained, but the optimal $p_1, \ldots p_n$ can be determined by numerical methods.

4.1.2 Discrete pricing to match the optimal continuous sales

An approximate pricing solution can be obtained by choosing p_i so that the sales in period i match those obtained for the continuous pricing case. That is, we calculate the cumulative sales obtained up to time t_i in the continuous case

$$s_i = y(t_e)e^{-\gamma p_e}K(t_i), \text{ for } i = 1, \ldots, n. \tag{29}$$

Using these s_i, we determine the corresponding prices by solving the relationships

$$e^{-\gamma p_i}[K(t_i) - K(t_{i-1})] = \frac{\{I_0 - s_{i-1}\}^{1-\alpha} - \{I_0 - s_i\}^{1-\alpha}}{(1-\alpha)I_r^{-\alpha}} \tag{30}$$

from (28) for $p_1, \ldots, p_n$. Here it is convenient to express the p_i in terms of

$$f_i = \text{the fraction of units sold up to time } t_i.$$

Because of the compensating price property, it follows that

$$f_i = \frac{s_i}{I_0} = f_e\frac{K(t_i)}{K}, \tag{31}$$

when the optimal price trajectory $P(I)$ is used. Thus, once f_e is determined from (23), the f_i follow immediately from (31). Therefore

$$p_i = -\frac{1}{\gamma}\ln\left(\frac{I_0}{1-\alpha}\left(\frac{I_r}{I_0}\right)^{\alpha}\frac{\{1-f_{i-1}\}^{1-\alpha}-\{1-f_i\}^{1-\alpha}}{K(t_i)-K(t_{i-1})}\right). \tag{32}$$

The total revenue obtained using this discrete pricing is then given by

$$\overline{R}(I_0) = I_0\left[\sum_{i=1}^{n} p_i[f_i - f_{i-1}] + (1-f_e)c_e\right]. \tag{33}$$

Since the maximum revenue $R(I_0)$ obtained with the optimal continuous pricing solution is greater than or equal to the revenue that can be obtained with any discrete solution, it bounds the maximum discrete revenue obtained from (19) as well as the revenue obtained with the approximate solution in (33). Thus we have proved the following lemma.

Lemma 3. The percentage profit loss from using the approximate discrete prices obtained from (30) in place of the exact discrete price solution obtained from (28) is bounded as follows

$$\text{Profit Loss\%} \leq \frac{R(I_0) - \overline{R}(I_0)}{R(I_0)}. \tag{34}$$

Furthermore, the profit loss from using optimal discrete pricing obtained from (19) instead of optimal continuous pricing from (12) has this same upper bound. It is illustrated in the next section that this percentage loss is less than 1% to 2% for typical parameter values.

5. NUMERICAL EXAMPLES

In this section, we compute the price trajectories, total sales and total revenue for some parameter values to gain insights about the sensitivity of the results to the various input parameters. We will also compare the continuous and discrete pricing solutions.

To reduce the number of variables, all cases use the values

$$I_0 = I_r = 1000 \text{ units}, \quad t_0 = 0, \quad t_e = 1 \quad \text{and} \quad K(t) = tK.$$

That is, we assume that there are no seasonal variations and the on hand inventory exactly equals I_r. The solutions can be extended to other I_0 values from (24) and (25). The time unit scale can be chosen arbitrarily, since all time variations can be expressed as functions of the inventory level I. Solutions are obtained by solving (23) for s_e by a one dimensional search, e.g., the Excel Goal Seek function, and then computing the prices and total revenues from (24) and (25).

Different demand rates can be tested by changing K or by changing the ratio $K/\ I_0$. Since K is difficult to interpret intuitively, we define the Base demand parameter

$$\text{Base (demand)} = Ke^{-\gamma p_e}, \tag{35}$$

which corresponds to the total unit demand at the minimum price p_e with no inventory effect ($\alpha = 0$). Also note that $p_e = c_e + 1/\gamma$ is the optimal price when $\alpha = 0$ and inventory can be salvaged at a unit cost c_e. We will use a retail price of $p_0 = \$10.00$ as a reference value and write all other costs and revenues as multiples of p_0. For these graphs, I_0 is not a decision variable, so $c(I_0)$ is a sunk cost that can be omitted from this numerical analysis.

For the first set of graphs, we use the following parameter values, which represent typical numbers for an apparel item

$c_e = 20\%$, $\gamma = 3.33$, $\alpha = 0.5$, 1.0 or 1.5 and Base = 500 to 1500.

Let us first consider the total sales $s_e = f_e\, I_0$ in Figure 11-1 as a function of the Base values and $\alpha = 0.5$, 1.0 and 1.5. These curves are concave increasing, as one might expect, and the smaller values of α give the largest total sales in every case. This is because the negative effects of inventory on sales are less for smaller values of α.

Now let us consider Figure 11-2, the optimal price trajectory for the single fixed Base Demand = 1000. From Figure 11-1, the total sales for $\alpha = 0.5$, 1.0 and 1.5 are 838, 677 and 578, respectively. Each curve in Figure 11-2 shows the compensating behavior of the optimal price trajectory, as more inventory is sold. Also, we know from Figure 11-1 that $\alpha = 1.5$ corresponds to the least total inventory sold. In all cases, it is best to price higher initially and then gradually decrease the price to compensate for the increasing inventory effect, as

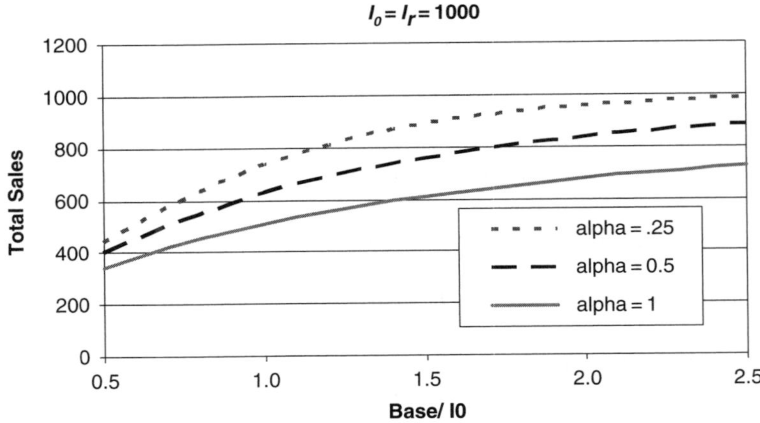

Figure 11-1. Total sales $f_e\, I_0$ versus Base/ 10 $I_0 = I_r = 1000$

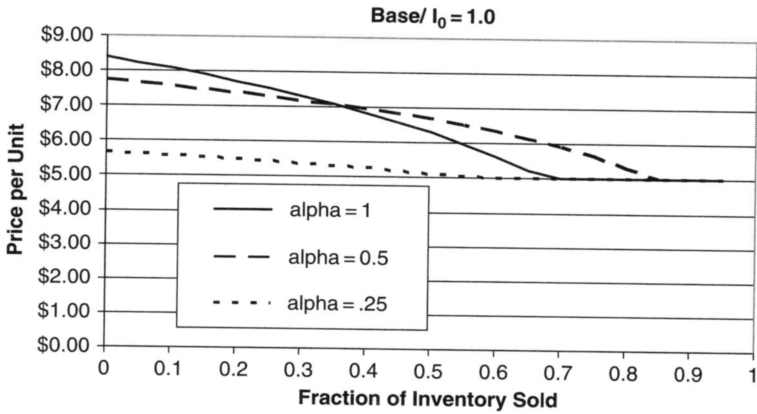

Figure 11-2. Optimal Price Trajectories Base/ I_0 = 1.0

described by (24). The crossing patterns of the price curves in Figure 11-2 can be explained as follows. We know that $\alpha = 1.5$ must have the steepest drop, because it compensates for the largest inventory effect, while $\alpha = 0.5$ must yield the flattest curve. All curves must have the same terminal price p_e. The highest initial price therefore occurs for $\alpha = 1.5$. Figure 11-3 shows the behavior of the optimal initial price $p(I_0)$ for other values of Base Demand.

Figure 11-4 shows the total revenue obtained by using the optimal price trajectory (24) in each case. It is interesting to note in Figure 11-4 that the revenues generated for the three values of α are fairly close to each other. This implies that if inventory effects are modeled correctly, then the almost the same revenue can be obtained through appropriate pricing. For larger α values, higher prices maximize the profit by selling fewer units.

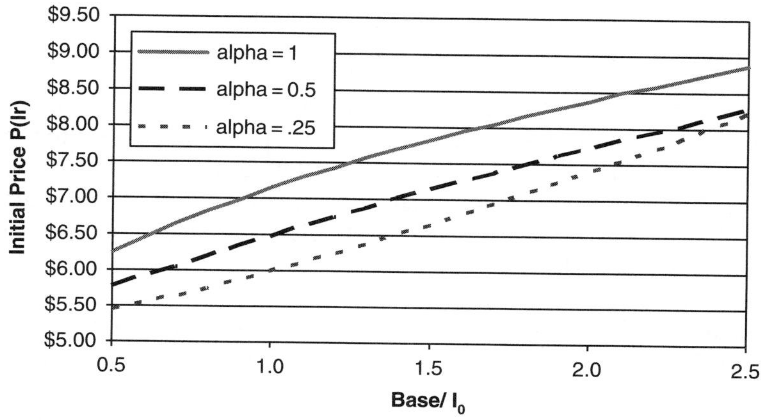

Figure 11-3. Optimal Initial Price

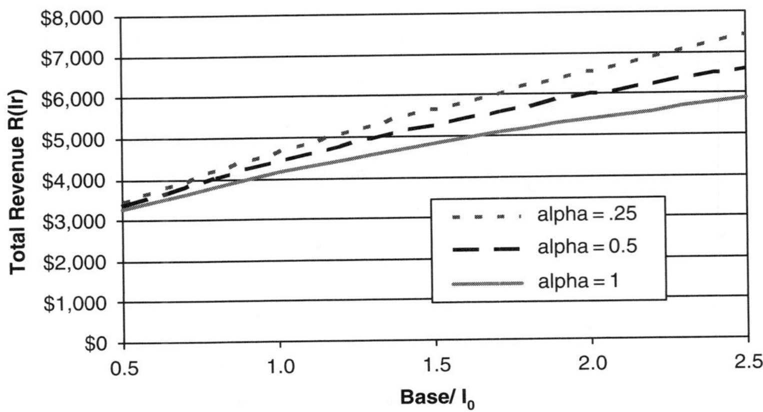

Figure 11-4. Total Revenue versus Base/ I_0

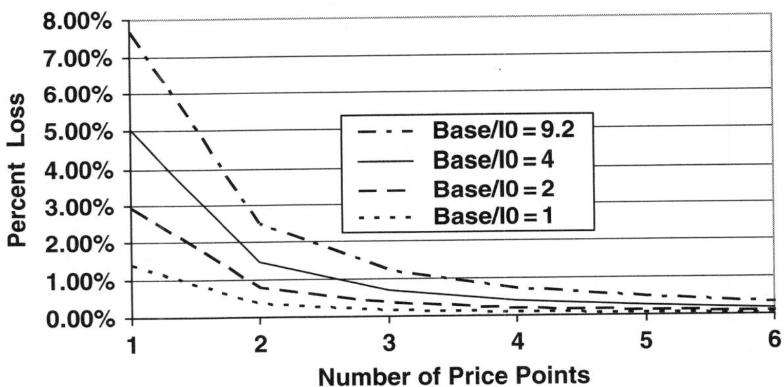

Figure 11-5. Profit Loss Bounds for Approximate Discrete Prices for $c_e \geq 0, \alpha \leq 1$ & $\gamma > 0$

Figure 11-5 shows the bound on the profit loss as a result of approximating the optimal continuous price trajectory with the discrete prices (32) that match the continuous sales at the discrete points. That is, the percentage losses in Figure 11-5 are obtained from (34). The other assumptions behind Figure 11-5 are as follows. For $\alpha \leq 1$, it is intuitively clear that $\alpha = 1$ yields the worst percentage loss, since the price drops more rapidly for higher α. This was also consistent with extensive calculations. Second, $c_e = 0$ is also the worst case for percentage loss, because with no salvage value the price trajectory drops must achieve all the profits. But with $c_e = 0$, we see that the factor I_0/γ appears in both (25) and (33), and so I_0/γ cancels out in (34). Thus, the curve in Figure 11-5 holds for all I_0 and γ as well. It is clear from Figure 11-5 that errors are always generally less than 1% or 2% if at least two price points are used. The worst case occurs for

Base/ $I_0 = 9.2$, which corresponds to the lowest demand level that requires an optimal price higher than the base price of p_0 = \$10.00.

6. CONCLUSIONS

Both practical and theoretical insights can be drawn from the experiences with the clearance markdown methodology described in this paper. From a practical standpoint, improvements in clearance markdown policies have had major financial impacts on a number of firms because clearance sales volumes are substantial and any increased revenues from improved clearance policies go directly to the bottom line. Clearance markdown algorithms are now a key component of merchandise pricing for many retail chains, which are part of a sector with sales exceeding \$500 billion per year.

The markdown response model in this chapter differs from other dynamic pricing models in that it includes a dependence on inventory level. Retail buyers in the initial studies, particularly for apparel products, felt that having adequate inventory for presentation strongly affects sales. Regression analyses have also found that low inventories are highly correlated with reduced sales. Adopting a multiplicative, exponential price response function, which has previously been successful in modeling the response to promotional markdowns, leads to an optimal clearance price trajectory that exactly compensates for the effects of reduced inventory, independent of the form of the inventory sensitivity.

General properties of the optimal pricing policy for merchandise that is sensitive to inventory level can provide guidelines for developing corporate strategies for these products. Inventory sensitivity implies that prices should be set higher before the clearance period begins, and then reduced gradually during the clearance period. For many products, it is optimal to leave some quantity of merchandise unsold at the end of the season, especially if it has a salvage value. At the same time, our pricing studies indicated that the initial clearance markdowns should be deeper than buyers were accustomed to taking, while excessive markdowns at the end of the season should be avoided in favor of salvaging, or even discarding, unsold merchandise.

One of the implementation requirements is parameter estimation. Smith and Achabal (1998) discuss some regression based approaches for estimating the parameters for sales forecasting and markdown response models. These methods have often been combined with subjective estimation of certain response parameters, or use of seasonal variations that were computed at a higher level of aggregation. While these estimation methods are based partially on subjective choices, they have been sufficiently accurate to achieve significant improvements in operating results at a number of retailers.

This model can also provide a basis for further research in pricing policies that include dependence on inventory effects. Possible enhancements, which have been considered in other related research, include time discounted cash flows and time dependent inventory holding costs. When the clearance markdown period is longer, these time dependent aspects become more important. Another interesting generalization is the use of initial clearance prices to elicit information about the customer markdown response parameters. When combined with the sensitivity of sales to inventory, this remains an unsolved problem to the author's knowledge. Finally, these successful practical applications should encourage others to apply management science models in situations that require a combination of regression analysis and subjective parameters choices.

Acknowledgment The author is especially indebted to Professor Dale Achabal, Director of the Retail Management Institute at Santa Clara University, for initiating the clearance markdown research, and for leading the projects that resulted in the successful implementations of clearance pricing.

REFERENCES

Achabal, Dale, Shelby McIntyre and Stephen Smith (1990), "Maximizing Profits from Department Store Promotions," *Journal of Retailing*, Vol. 66, No. 4 (Winter), 383–407.

Agrawal, V. and S. Seshadri, (2000) "Impact of Uncertainty and Risk Aversion on Price and Order Quantity in the Newsvendor Problem," *Manufacturing and Service Operations Management*, Vol. 2, No. 4, 410–423.

Besanko, David and Wayne L. Winston (1990), "Optimal Price Skimming by a Monopolist facing Rational Consumers," *Management Science*, Vol. 36, No. 5 (May), 555–567.

Bhat, Rajendra R. (1985), Managing the Demand for Fashion Items, UMI Research Press, Ann Arbor Michigan.

Bitran, Gabriel R. and Rene Caldenty (2003), "An Overview of Pricing Models for Revenue Management," *Manufacturing and Service Operations Management*, Vol. 5, No. 3, 203–229.

Bitran, Gabriel, Rene Caldenty and Susana Mondschein (1998) "Coordinating Clearance Makrdown Sales of Seasonal Products in Retail Chains," *Operations Research*, Vol. 46, 609–624.

Bitran, Gabriel R. and Susana V. Mondschein (1997), "Periodic Pricing of Seasonal Products in Retailing," *Management Science*, Vol. 43, No. 1 (January), 64–79.

Bolton, Ruth N. (1989), "The Relationship Between Market Characteristics and Promotion Price Elasticities," *Marketing Science*, Vol. 8, No. 2 (Spring), 153–169.

Braden, David J. and Shmuel S. Oren (1994), "Nonlinear Pricing to Produce Information," *Marketing Science*, Vol. 13, No. 3 (Summer), 310–326.

Cheng, Feng and Suresh. P. Sethi. (1999). "A periodic review inventory model with demand influenced by promotions decisions," *Management Science*, 45(11) 1510–1523.

Dhebar, Anirudh and Shmuel Oren (1985), "Optimal Dynamic Pricing for Expanding Networks," *Marketing Science*, Vol. 4 (Fall), 336–351.

Eliashberg, Jehoshua and Richard Steinberg (1987), "Marketing-Production Decisions in an Industrial Channel of Distribution," *Management Science*, Vol. 33 (August), 981–1000.

Elmaghraby, Wedad and Pinar Keskinocak (2003), "Dynamic Pricing in the Presence of Inventory Considerations," *Management Science*, Vol. 49, No. 10 (October) 1287–1309.

Federgruen, Awi and A. Heching (2002) "Multi-location Combined Pricing and Inventory Control," *Manufacturing and Service Operations Management*," Vol. 4, No. 4, 275–295.

Federgruen, Awi and A. Heching (1999) "Combined Pricing and Inventory Control Under Uncertainty," *Operations Research*, Vol. 47, 454–475.

Feng, Youyi and Guillermo Gallego (1995), "Optimal Starting Times for End-of-Season Sales and Optimal Stopping Times for Promotional Fares," *Management Science*, Vol. 41, No. 8, (August), 1371–1391.

Gallego, Guillermo and Garrett van Ryzin (1994) "Optimal Dynamic Pricing of Inventories with Stochastic Demand," *Management Science*, Vol. 40, No. 8, (August), 999–1020.

Kalish, Shlomo (1983), "Monopolistic Pricing with Dynamic Demand and Production Cost," *Marketing Science*, Vol. 2, No. 2 (Spring), 135–159.

—(1985), "A New Product Adoption Model with Price, Advertising and Uncertainty," *Management Science*, Vol. 31, No. 12 (December), 1569–1585.

Kalyanam, Kirthi (1996), "Pricing Decisions Under Demand Uncertainty: A Bayesian Mixture Model Approach," *Marketing Science*, Vol. 15, No. 3, 207–221.

Kamien, Morton I. and Nancy Schwartz (1981), Dynamic Optimization, North Holland, New York.

Karlin, Samuel, (1960), "Dynamic Inventory with Varying Stochastic Demands," *Management Science*, Vol. 6, No. 3 (April), 231–258.

Lariviere, Martin and Evan Porteus (1999) "Stalking Information: Bayesian Inventory Management with Unobserved Lost Sales," *Management Science*, 45, 3 (March), 346–363.

Lazear, E. P.(1986), "Retail Pricing and Clearance Sales," *The American Economic Review*, Vol. 76 (March), 14–32.

Mahajan, Vijay, Eitan Muller and Frank Bass (1990), "New Product Diffusion Models in Marketing: A Review and Directions for Research," *Journal of Marketing*, Vol. 54 (January), 1–26.

Mantrala, M. K. and S. Rao (2001) "A Decision Support Systems that Helps Retailers Decide Order Quantities and Markdowns for Fashion Goods," *Interfaces*, Vol. 31, No. 3, S146–S165.

Merrick, Amy (2001) "Priced to Move: Retails Attempt to get a Leg Up on Markdowns with New Software," *Wall Street Journal*, (August 7) A1, A6.

Narasimhan, Chakravarthi (1984), "A Price Discrimination Theory of Coupons," *Marketing Science*, Vol. 3, No. 2 (Spring), 128–147.

Netessine, Serguei (2004), "Dynamic Pricing of Inventory/ Capacity with Infrequent Price Changes," forthcoming in *European Journal of Operations Research.*

Pashigian, B. Peter (1988), "Demand Uncertainty and Sales: A Study of Fashion and Markdown Pricing," *The American Economic Review*, Vol. 78, No. 5 (December), 936–953.

Rajan, Arvind, Rakesh and Richard Steinberg (1992), "Dynamic Pricing and Ordering Decisions by a Monopolist," *Management Science*, Vol. 38, No. 2 (February), 240–262.

Ray, Saibal, Shanling Li and Yuyue Song (2005), "Tailored Supply Chain Decision Making Under Price Sensitive Demand and Delivery Uncertainty," *Management Science*, Vol. 51, No. 12, 1873–1891.

Russell, Gary J. and Ruth N. Bolton (1988), "Implications of Market Structure for Elasticity Structure," *Journal of Marketing Research*, Vol. 25, No. 3 (August), 229–241.

Smith, Stephen A., Dale Achabal and Shelby McIntyre (1994), "A Two Stage Sales Forecasting Procedure Using Discounted Least Squares," *Journal of Marketing Research*, Vol. 31 (February), 44–56.

Smith, Stephen A. and Dale D. Achabal, (1998) "Clearance Pricing and Inventory Policies for Retail Chains, *Management Science*, Vol. 44, No. 3, 285–300.

Stokey, Nancy (1979), "Inter-temporal Price Discrimination," *Quarterly Journal of Economics*, Vol. 93 (August), 355–371.

Sullivan, Laurie (2005) "Getting the Price Right: SAP Goes Shopping," *Information Week*, November 28, 2005.

Wolfe, Harry B. (1968), "A Model for Control of Style Merchandise," *Industrial Management Review*, [now Sloan Mgt. Rev.], Vol. 9 (Winter), 69–82.

Zhao, W. and Y-S Zheng (2000) Optimal Dynamic Pricing for Perishable Assets with Nonhomogeneous Demand," *Management Science*, Vol. 46 No. 3, 375–388.

Chapter 12
MARKDOWN COMPETITION

Seungjin Whang
Graduate School of Business, Stanford University, Standford, CA, USA

Abstract We present a stylized model of markdown competition. We consider two retailers who compete in a market with a fixed level of initial inventory. The initial inventory level is only known to the retailer, and not to the other. To maximize the profit, each retailer would mark down at a time of his individual choice. The model assumes deterministic demands, a single chance of price change, and a prefixed set of prices. We consider a two-parameter strategy set where a retailer chooses the timing of markdown as a function of the current time, his inventory level and the other's move so far. We characterize the equilibrium of the game and derive managerial insights.

1. INTRODUCTION

Dynamic price optimization, as a branch of revenue management, investigates the price as a key decision variable in a dynamic business environment. In particular, it studies how to "operationalize" pricing decisions by considering additional dimensions like time and inventories. Perhaps the most canonical example is the ground-breaking work by Gallego and van Ryzin (1993) who study the optimal price trajectory based on the actual realization of sales and the length of remaining sales period. Since then, a wide variety of dynamic pricing models came into existence. In those models, demands may be deterministic or *stochastic* (Gallego and van Ryzin, 1993), the set of prices *predetermined* or arbitrary (Feng and Xiao, 2000), the number of price changes *limited* or unlimited (Feng and Gallego, 1995), time continuous or *discrete* (Dudey, 1992), customers *strategic* or myopic (Aviv and Pazgal, 2003), the setting of the game *completely known* or revealing over time (Lazear, 1986), and sellers *monopolistic* or competing (Belobaba, 1987). See Talluri and van Ryzin (2004) or Bitran and Caldency (2003) for an extensive review of the literature.

Competition, although present in almost every real setting, has not received enough attention in the dynamic pricing literature, compared to other aspects. This paper attempts to fill the gap by presenting a stylized model of dynamic markdown competition. We consider two retailers who compete in a market

N. Agrawal, S.A. Smith (eds.), *Retail Supply Chain Management*,
DOI: 10.1007/978-0-387-78902-6_12,

with a fixed level of initial inventory. The initial inventory level is only known to the corresponding retailer, and not to the other. To maximize the profit, each retailer would permanently mark down once at a time of his individual choice. The model assumes deterministic demands, a single chance of price change, and a predetermined set of prices. We consider a two-parameter strategy set where a retailer chooses the timing of markdown as a function of the current time, his inventory level, and the other's move so far. We characterize the equilibrium of the game and derive managerial insights.

Dynamic markdown competition – where a retailer marks down as a counter to the competitor's move – is a familiar facet of business practice. Consider, for example, the cut-throat competition in the game device market:

> "Microsoft cut the price of its Xbox game console by about a third in the U.S. and Canada and announced a similar price cut for Japan Wednesday. The move had been expected by market watchers and comes on the heels of Sony Computer Entertainment America's price reduction for the PlayStation 2 on Tuesday. Effective immediately, Xbox consoles will cost $199.99 in the U.S., down from $299.99, Microsoft says in a statement. Xbox, Sony's PlayStation 2, and Nintendo's GameCube now all cost about $200 in the U.S. In Japan, where Xbox sales have been sluggish since its launch late February, the Xbox will be cut to $193 from $270 effective May 22, Microsoft says." (Evers, IDG News Service, 2002).

Our model extracts two elements of the business practice captured in the article – the timing of markdown in response to the competitor's move and based on its own inventory position.

This is not the first research work on dynamic price competition. For example, Dudey (1992) studies a model where two duopolistic firms face multiple customers, one at a time in sequence. For each customer, the two firms simultaneously submit their price quotes, and the customer would take the lower offer so far as the price is lower than her reservation price. Each firm starts with a fixed quantity of inventory, so that the price quote is a function of the time, her own inventory level and the other firm's inventory level, as well as the customer's reservation price. Assuming that both firms have complete information of the game (including the evolution of inventory positions), the paper characterizes the equilibrium strategy of each firm.

Varian (1980) and Lal (1990) interpret price promotions as a mixed equilibrium strategy among competing retailers. Lal (1990), for example, considers three retailers, two national brands and one local brand, in a market consisting of switchers and loyals. Loyals are loyal to their preferred national brand, while switchers always buy the cheapest available. The dilemma facing a national brand is that he cannot extract all the surplus from his loyals *and* win switchers' market segment, too, due to the threat coming from the local brand. Thus, implicit collusion is supported as a non-cooperative equilibrium, where the two national brands take turns lowering the price in the form of promotion. Hence, the regular price extracts loyals' surplus, and the promotional price attracts switchers. In a similar market setting, Rao (1991) also studies two retailers – a national brand and a local brand – competing in promotion. Each firm makes a three-stage sequential decision of regular price, promotion depth and promotion frequency. Two firms simultaneously take actions at each stage, and the outcome of the previous stages is jointly observed before moving on to the next stage. They characterize the equilibrium of the multi-stage, multi-decision game with complete information. In the above line of work the players in this game are allowed to change prices, but not as an *ex-post* counter to the other's decisions.

Netessine and Shumsky (2004) study horizontal competition in which two airlines compete over "overflow passengers." Each airline has a fixed capacity and offers two classes, high-fare and low-fare, of seats at two different prices. Each airline faces a random demand to each class, which is exogenously given. Each airline sets a "booking limit" to the number of low-fare seats, so the overflow customers denied tickets at one airline attempt to purchase tickets at the other airline. The paper investigates the strategy of each airline in choosing the booking limit in this non-cooperative game with complete information.

Our model differs from the above work in that it is set up as a non-cooperative game with incomplete information, and players' strategy is the timing of markdown. The rest of the paper is organized as follows. In §2 we provide the details of the model. §3 analyzes the problem of a monopolistic retailer who would choose the time of markdown in the base model. §4 forms the core of the paper where we demonstrate the equilibrium strategies of two duopolistic retailers in choosing the markdown time. The last section concludes with a summary and managerial implications.

2. THE MODEL

Consider a pair of retailers (denoted by $i = 1, 2$) competing in a seasonal or fashion product market. At time 0, each retailer, facing uncertain demand, orders a fixed quantity of the product, based on his individual forecast. The order arrives before the selling season starts. The two retailers are symmetric in terms of market power and cost structure, but may differ in their forecasts and order quantities. The forecast as well as the order quantity is privately known to the respective retailer. The order quantity by one retailer is viewed to the other as a random variable drawn from a common distribution F over $[0, \infty)$. At time 1 the selling season starts, and the demand rate at each possible pair of retail prices is revealed to both retailers. Retailers have no chance to replenish the stock even if they realize the demand is larger than initially forecasted.

In standard microeconomics, the demand function defines the 'total' demand level at each price. It does not capture how the demand materializes across time. To fix this, we introduce a 'demand trajectory' that shows the distribution of demand over time. In the present paper we assume a specific demand trajectory in the form of $e^{-\tau/\beta}$ over time $\tau \in [0, \infty)$, where $\beta(> 0)$ is the 'demand rate' defining the demand intensity. Thus, the demand arriving in the time interval [0,t] is here given by $\int_0^t e^{-\tau/\beta} d\tau$ or $\beta[1 - e^{-t/\beta}]$, and the total demand over the entire season is β. This particular demand trajectory assumes that the demand of the product peaks upon its introduction and exponentially declines over time. Even if the selling season is infinitely long in this setup, the exponential decay (with the right choice of β) will ensure that the demand fades away fast in time, thereby approximating the demand pattern of a seasonal or fashion product. Further, note that the demand realization process has no uncertainties once the demand parameter is revealed. Obviously, it is a strong assumption, but it keeps the analysis tractable. In addition, the deterministic model will serve as an anchor case to stochastic models in developing a heuristic or an upper bound (see Gallego and van Ryzin (1993)).

Note that the higher the demand rate β, the slower the demand decays over time and the larger the total demand. β is determined by the prices set by the retailers. Each retailer starts the season with the price set at p_0, but may choose to mark down to $p_1(< p_0)$ at a time of his individual choice. p_0 and p_1 are prefixed prior to the season. This price change would change the demand rates for both retailers. To simplify the notation, let β_{ij} $(i, j \in \{0, 1\})$ denote the demand rates β facing the retailer whose own price is p_i and the other's is p_j. For example, if his price is p_0 and hers is p_1, he faces β_{01} and she faces β_{10} as the demand rate. We assume that $\beta_{10} > \beta_{11} > \beta_{00} > \beta_{01}$. In case he marks down and she does not, for example, his demand rate β_{10} will be the highest of the four cases (due to the combination of a larger market and bigger market share), and hers β_{01} will be the lowest. If both mark down, the demand rate β_{11} facing each retailer falls somewhere between the two extremes, but will be higher than β_{00} the initial demand rate, due to a larger market.

We assume that sales are permanently lost from the market if the retailer visited stocks out. One scenario that supports this assumption is the following: If a potential customer visits a retailer who is out of stock, she will not learn about the existence of the product, so she will not search for it at the other retailer's. More generally, we assume that stockouts at one retailer's do not affect the sales at the other retailer's. This adds another strong assumption that if one stocks out, the current demand intensity continues to hold at the other retailer.

Compared to the existing literature, the present model imposes a series of simplifying assumptions of deterministic demands, a single chance of price change, and a prefixed set of prices. Further, we do not discount cash flow for simplicity, and assume that any unsold items at the end of the season are thrown away at zero salvage value and zero cost. In return, the model highlights the timing of competitive markdowns under asymmetric information (about the initial stock level).

3. THE CASE OF A MONOPOLISTIC RETAILER

Before we study the case of competition, we first consider a monopolistic retailer who starts the season at price p_0 with the stock level S. Assume that the demand parameter at price p_i is β_i for $i = 0, 1$, where $p_0 > p_1$ and $\beta_0 < \beta_1$.

Suppose now that the retailer would choose the time to mark down. The demand trajectory enables us to evaluate the impact of a price change on the season's overall profit to each retailer and to formulate the markdown-timing problem as follows.

$$\max_{t\geq 0}\int_0^t p_0 e^{-\tau/\beta_0}d\tau+\int_t^T p_1 e^{-\tau/\beta_1}d\tau=p_0\beta_0(1-e^{-t/\beta_0})$$
$$+p_1\beta_1(e^{-t/\beta_1}-e^{-T/\beta_1}), \tag{P1}$$

where

$$\beta_0(1-e^{-t/\beta_0})+\beta_1(e^{-t/\beta_1}-e^{-T/\beta_1})\leq S. \tag{1}$$

Inequality (1) is the capacity constraint that ensures that total sales do not exceed the initial inventory, where T denotes the time of running out of stock. We assume that T can take the value of infinity, which happens when S is large enough.

We form the Lagrangian function:

$$\mathcal{L}(t,T,\lambda)=p_0\beta_0(1-e^{-t/\beta_0})+p_1\beta_1(e^{-t/\beta_1}-e^{-T/\beta_1})-\lambda[\beta_0(1-e^{-t/\beta_0})$$
$$+\beta_1(e^{-t/\beta_1}-e^{-T/\beta_1})-S], \tag{P2}$$

where λ is the Lagrangian multiplier associated with the capacity constraint. After straightforward manipulation, the Kuhn-Tucker theorem yields the following result.

Theorem 1 *To the monopolistic retailer with a starting inventory* S, *the optimal time* $t^*(S)$ *to mark down is given by*

$$t^*(S)=\begin{cases}\infty, & \text{if } S<\beta_0;\\ \frac{\beta_0\beta_1}{\beta_1-\beta_0}\ln\frac{p_0-\lambda(S)}{p_1-\lambda(S)} & \text{if } \beta_0\leq S\leq S^*;\\ \frac{\beta_0\beta_1}{\beta_1-\beta_0}\ln\frac{p_0}{p_1} & \text{if } S>S^*,\end{cases}$$

where $\lambda(S)$, *the (non-negative) Lagrangian multiplier to the capacity constraint, satisfies*

$$S=\beta_0\left[1-\left(\frac{p_1-\lambda(S)}{p_0-\lambda(S)}\right)^{\frac{\beta_1}{\beta_1-\beta_0}}\right]+\beta_1\left(\frac{p_1-\lambda(S)}{p_0-\lambda(S)}\right)^{\frac{\beta_0}{\beta_1-\beta_0}}, \tag{2}$$

and S^* *is the smallest value of* S *with* $\lambda(S)=0$; *that is,*

$$S^*=\beta_0\left[1-\left(\frac{p_1}{p_0}\right)^{\frac{\beta_1}{\beta_1-\beta_0}}\right]+\beta_1\left(\frac{p_1}{p_0}\right)^{\frac{\beta_0}{\beta_1-\beta_0}}. \tag{3}$$

Also, $\beta_0<S^*<\beta_1$.

If the retailer has tight supply, he will never mark down, or equivalently, his optimal markdown time will be infinity. This is because in the absence of cash flow discounting, he has no incentive to mark down if he can sell everything he has even if it takes a long time. The cutoff inventory level is β_0, which is the quantity he can sell without a markdown. Here the choice of the value ∞ is somewhat arbitrary. To be exact, the solution to (P2) in this range of S is $t^*(S) = T^*$, where $T = T^*$ satisfies (1) in equality. This means that the retailer marks down at the time he runs out of stock. This is equivalent to the event of no markdown ever (especially as observed by the other retailer if she exists as in later sections), hence comes our choice of infinity. In the other extreme case (i.e., an ample inventory), he cannot sell all he has, so he will maximize his profit by lowering the price at time $\frac{\beta_0\beta_1}{\beta_1-\beta_0}\ln\frac{p_0}{p_1}$, which remains constant to any retailer whose inventory level is larger than S^*. In the middle range of the inventory, the timing of his markdown will depend on the inventory level. The higher the inventory level, the quicker comes the markdown. In this case, the retailer will time the markdown to sell all his inventory. Loosely speaking, $t^*(S)$ is decreasing in $S \in [0, \infty)$.[1] The monopolist with a high inventory will be more anxious, so he will rush to cut the price to move the volume.

4. MARKDOWN COMPETITION

We now turn to the case of two retailers competing in the choice of markdown timing. The strategy for each retailer is the choice of its markdown time, taking the other retailer's strategy as given. More specifically, retailer $i\,(=1,2)$ (he) will choose the time $\sigma_i(S_i, \mathcal{H}_t)$ to mark down, where σ_i is not only a function of his private inventory level S_i, but also of the history $\mathcal{H}_t$ of the game up until his decision time t. In our model that has assumed away demand uncertainties (after the season starts), the relevant information contained in $\mathcal{H}_t$ is the actions taken by the other retailer j (she) and the current time. The strategy determines in advance what to do in each contingency, as the game evolves and uncertainties are resolved. The strategy will maximize the expected profit at each time point for the rest of the game based on the realized path.

Retailer i's expected profit depends on his own inventory level S_i, as well as retailer j's strategy σ_j that depends on her inventory level S_j. To derive his optimal strategy, retailer i must take into account the uncertainties about S_j to predict her strategy and develop his own strategy. Our equilibrium concept is similar to Bayesian subgame-perfect equilibrium (Kreps, 1990). Further, we restrict our attention to 'symmetric' equilibrium in which the two retailers use the same strategy function and play with different arguments.

Now consider the set $\mathcal{S} = \{\tilde{\sigma}(t_a, t_b, \mathcal{H}_t) | 0 \leq t_a \leq t_b\}$ (or $\{\tilde{\sigma}(t_a, t_b)\}$ for short) of two-parameter strategies for each retailer that operate as follows: "Wait and

---

[1] This statement is not mathematically accurate since the function $t^*(S)$ is not well defined in the interval $[0, \beta_0]$, but the meaning is clear in the present context.

see if the other retailer marks down; if the latter does before t_b, then mark down either immediately or at t_a, whichever comes later. If the other does not mark down until t_b, then don't wait any longer and mark down now before the other." When both retailers play strategies in $\mathcal{S}$, retailer i faces three alternative scenarios depending on retailer j's markdown time τ. τ may fall in one of the three time intervals $I_a := [0, t_a)$, $I_b := [t_a, t_b)$, and $I_c := [t_b, \infty]$. If it falls in I_a, retailer i is not "ready" yet, so he will wait and mark down later at t_a. If in I_b, he will immediately match retailer j's markdown. In I_c, retailer i will move first without further waiting for retailer j's move.

While this strategy set appears to contain a wide set of plausible actions, it is not exhaustive by any means. For example, one can consider a three-parameter strategy like "Wait and see if the other retailer marks down; if the latter does before t_a, then mark down at $t'_a(> t_a)$. If the latter does after t_a but before t_b, then mark down at t_b. If the other does not mark down until t_b, then don't wait any longer and mark down before the other." Clearly, this example, although not so convincing on its own, alludes to an infinite number of possible strategy sets, underscoring the fact that $\mathcal{S}$ is just one of them.

Now retailer i's decision is to find a pair $(t^*_a(S_1), t^*_b(S_1))$, or simply (t^*_a, t^*_b), that determine his optimal strategy in $\mathcal{S}$. To derive t^*_a first, suppose that the game started at time 0, and soon retailer j marked down at time t in I_a. The current demand rate for retailer i is β_{01}, but his markdown decision would change it to β_{11}. We now solve

$$\max_{t_a \geq t} \int_t^{t_a} p_0 e^{-\tau/\beta_{01}} d\tau + \int_{t_a}^{T} p_1 e^{-\tau/\beta_{11}} d\tau = p_0\beta_{01}(1 - e^{-t_a/\beta_{01}}) + p_1\beta_{11}(e^{-t_a/\beta_{11}} - e^{-T/\beta_{11}}) \quad \text{(P3)}$$

subject to

$$\beta_{01}(e^{-t/\beta_{01}} - e^{-t_a/\beta_{01}}) + \beta_{11}(e^{-t_a/\beta_{11}} - e^{-T/\beta_{11}}) \leq S_i - \beta_{00}(1 - e^{-t/\beta_{00}}).$$

After adding a constant $\int_0^t p_0 e^{-\tau/\beta_{01}} d\tau$ to the objective and slight modification of the constraint, we have:

$$\max_{t_a \geq t} \int_0^{t_a} p_0 e^{-\tau/\beta_{01}} d\tau + \int_{t_a}^{T} p_1 e^{-\tau/\beta_{11}} d\tau = p_0\beta_{01}(1 - e^{-t_a/\beta_{01}}) + p_1\beta_{11}(e^{-t_a/\beta_{11}} - e^{-T/\beta_{11}}) \quad \text{(P3}')$$

subject to

$$\beta_{01}(1 - e^{-t_a/\beta_{01}}) + \beta_{11}(e^{-t_a/\beta_{11}} - e^{-T/\beta_{11}}) \leq S_{it},$$

where $S_{it} := S_i - [\beta_{00}(1 - e^{-t/\beta_{00}}) - \beta_{01}(1 - e^{-t/\beta_{01}})] := S_i - \Delta_t$. It is easy to verify that Δ_t is positive and monotone increasing in t.

This problem has the same structure as (P1), with β_0, β_1 and S_i replaced by β_{01}, β_{11} and S_{it}. Hence, we have the following solution from Theorem 1.

$$t_a^*(S_{it}) = \begin{cases} \infty, & \text{if } S_{it} \le \beta_{01}; \\ \frac{\beta_{01}\beta_{11}}{\beta_{11}-\beta_{01}} \ln \frac{p_0-\lambda(S_{it})}{p_1-\lambda(S_{it})}, & \text{if } \beta_{01} < S_{it} < S^\circ; \\ \frac{\beta_{01}\beta_{11}}{\beta_{11}-\beta_{01}} \ln \frac{p_0}{p_1}, & \text{if } S_{it} \ge S^\circ, \end{cases} \tag{4}$$

where $\lambda(S_{it})$, the (non-negative) Lagrangian multiplier to the capacity constraint, satisfies

$$S_{it} = \beta_{01}\left[1 - \left(\frac{p_1 - \lambda(S_{it})}{p_0 - \lambda(S_{it})}\right)^{\frac{\beta_{11}}{\beta_{11}-\beta_{01}}}\right] + \beta_{11}\left(\frac{p_1 - \lambda(S_{it})}{p_0 - \lambda(S_{it})}\right)^{\frac{\beta_{01}}{\beta_{11}-\beta_{01}}}, \tag{5}$$

and

$$S^\circ = \beta_{01}\left[1 - \left(\frac{p_1}{p_0}\right)^{\frac{\beta_{11}}{\beta_{11}-\beta_{01}}}\right] + \beta_{11}\left(\frac{p_1}{p_0}\right)^{\frac{\beta_{01}}{\beta_{11}-\beta_{01}}}. \tag{6}$$

Also, note that $\beta_{01} < S^\circ < \beta_{11}$.

Suppose now that the time point t_a^* has passed without retailer j's move. The new time interval I_b starts, so retailer i will immediately adopt if the other marks down. But if she does not, retailer i cannot wait forever for her move, so he faces the problem of choosing "the preemptive markdown time" t_b, i.e., the time to stop waiting and mark down first.

To find the optimal t_b^*, we first introduce some notation. For the moment, assume that $t_b^*(\cdot)$ is monotone decreasing. Let $G(\tau)$ denote the probability of the other retailer marking down by time τ, with $\bar{G}(\tau) := 1 - G(\tau)$ and $g(\tau) := G'(\tau)$. Also let $\bar{G}^o(\tau|t)$ denote the probability that retailer j will mark down later than time τ on the condition that she has not marked down until time t; i.e., $\bar{G}^o(\tau|t) := 1 - G^o(\tau|t) = \bar{G}(\tau)/\bar{G}(t)$, for $\tau \ge t$. Let g^o and g respectively denote the probability density (or frequency) function of G^o and G.

At time $t(> t_a^*)$, retailer i will choose t_b^* by solving the following (P4):

$$\begin{aligned} &\max_{t_b \ge t} \int_t^{t_b^-} [p_0\beta_{00}(e^{-t/\beta_{00}} - e^{-\tau/\beta_{00}}) + p_1\beta_{11}(e^{-\tau/\beta_{11}} - e^{-T_1(\tau)/\beta_{11}})]dG^\circ(\tau|t) \\ &+ [p_0\beta_{00}(e^{-t/\beta_{00}} - e^{-t_b/\beta_{00}}) + p_1\beta_{11}(e^{-t_b/\beta_{11}} - e^{-T_2/\beta_{11}})]g^\circ(t_b|t) \\ &+ \int_{t_b^+}^{\infty} [p_0\beta_{00}(e^{-t/\beta_{00}} - e^{-t_b/\beta_{00}}) + p_1\beta_{10}(e^{-t_b/\beta_{10}} - e^{-\tau/\beta_{10}}) + p_1\beta_{11}(e^{-\tau/\beta_{11}} - e^{-T_3(\tau)/\beta_{11}})]dG^\circ(\tau|t) \\ &+ [p_0\beta_{00}(e^{-t/\beta_{00}} - e^{-t_b/\beta_{00}}) + p_1\beta_{10}(e^{-t_b/\beta_{10}} - e^{-T_4/\beta_{10}})]g^\circ(\infty|t), \end{aligned} \tag{P4}$$

subject to the following capacity constraints

$$\beta_{00}(1-e^{-\tau/\beta_{00}})+\beta_{11}(e^{-\tau/\beta_{11}}-e^{-T_1(\tau)/\beta_{11}})\le S_i, \forall\tau\in[t,t_b)$$

$$\beta_{00}(1-e^{-t_b/\beta_{00}})+\beta_{11}(e^{-t_b/\beta_{11}}-e^{-T_2/\beta_{11}})\le S_i$$

$$\beta_{00}(1-e^{-t_b/\beta_{00}})+\beta_{10}(e^{-t_b/\beta_{10}}-e^{-\tau/\beta_{10}})+\beta_{11}(e^{-\tau/\beta_{11}}-e^{-T_3(\tau)/\beta_{11}})\le S_i,\forall\tau\in(t_b,\infty)$$

$$\beta_{00}(1-e^{-t_b/\beta_{00}})+\beta_{10}(e^{-t_b/\beta_{10}}-e^{-T_4/\beta_{10}})\le S_i.$$

In the above, T_i $(i=1,2,3,4)$ represents the time to run out of inventory under four different scenarios; $T_1(\tau)$ is the time to run out of stock when both retailers mark down at time $\tau\in[0,\tau_b)$, T_2 when both mark down at t_b, $T_3(\tau)$ when i first marks down at t_b and j follows at $\tau\in(t_b,\infty)$, and T_4 when i first marks down at t_b and j does not follow. The objective function in (P4) represents the expected profit to retailer i when he plays $\tilde{\sigma}_i(t_a^*,t_b)$ while retailer j plays $\tilde{\sigma}_j(t_a^*,t_b^*)$.

Note that G can be derived from the distribution of random variables S_j via $t_a^*(\cdot)$ and $t_b^*(\cdot)$, and is a mixed (i.e., continuous and discrete) distribution. Regrettably, (P4) is very difficult to solve. One way to tackle the problem is to form a Lagrangian and obtain its saddle point (Luenberger, 1969). To derive the equilibrium strategy, we obtain the FOC of the Lagrangian for (P4), and then invoke the symmetric equilibrium assumption, so retailer i's choice of t_b should be equal to retailer j's optimal t_b^*, hence $t_b^{*-1}(t_b)=t_b^{*-1}(t_b^*(S_i))=S_i$. Then, we have (see the details in the Appendix):

$$\begin{aligned}
&(p_0e^{-t_b^*/\beta_{00}}-p_1e^{-t_b^*/\beta_{11}})F(S_i)+p_1(e^{-t_b^*/\beta_{11}}-e^{-t_b^*/\beta_{10}})F(t_a^{*-1}(t_b^*)+\Delta_{t_b})\\
&-\lambda_1(t_b^*)[\beta_{00}(1-e^{-t_b^*/\beta_{10}})-\beta_{11}(e^{-t_b^*/\beta_{11}}-e^{-T_1(t_b^*)/\beta_{11}})-S_i]\\
&-\lambda_2'(t_b^*)[\beta_{10}(1-e^{-t_b^*/\beta_{10}})-\beta_{11}(e^{-t_b^*/\beta_{11}}-e^{-T_2/\beta_{11}})-S_i]-\lambda_2(t_b^*)(e^{-t_b^*/\beta_{10}}-e^{-T_2/\beta_{11}})\\
&+\lambda_3(t_b^*)[\beta_{00}(1-e^{-t_b^*/\beta_{10}})+\beta_{11}(e^{-t_b^*/\beta_{11}}-e^{-T_3(t_b^*)/\beta_{11}})]-\bar{\Lambda}_3(t_b^*)(e^{-t_b^*/\beta_{00}}-e^{-t_b^*/\beta_{10}})\\
&+(p_0e^{-t_b^*/\beta_{00}}-p_1e^{-t_b^*/\beta_{10}})F(\beta_{01})\\
&+\lambda_4(t_b^*)[\beta_{00}(1-e^{-t_b^*/\beta_{00}})+\beta_{10}(e^{-t_b^*/\beta_{10}}-e^{-T_4/\beta_{10}})-S_i]=0.
\end{aligned}\tag{7}$$

A corner solution to (P4) occurs when retailer i has an initial inventory less than β_{01}. He would ultimately sell out even at the regular price, so he would never mark down, or his markdown time will be infinity.

Hence, the following theorem summarizes the equilibrium.

Theorem 2 *Consider the set $\mathcal{S}=\{\tilde{\sigma}(t_a,t_b,\mathcal{H}_t)|0\le t_a\le t_b\}$ of two-parameter strategies for each retailer that operate as follows: "Wait and see if the other retailer marks down; if the latter does before t_b, then mark down either immediately or at t_a, whichever comes later. If the other does not mark down until t_b, then don't wait any longer and mark down before the other." Let*

$$t_a^*(S_{it})=\begin{cases}\infty, & \text{if } S_{it}\le\beta_{01};\\ \frac{\beta_{01}\beta_{11}}{\beta_{11}-\beta_{01}}\ln\frac{p_0-\lambda(S_{it})}{p_1-\lambda(S_{it})}, & \text{if } \beta_{01}<S_{it}<S^{\circ};\\ \frac{\beta_{01}\beta_{11}}{\beta_{11}-\beta_{01}}\ln\frac{p_0}{p_1}, & \text{if } S_{it}\ge S^{\circ},\end{cases}\tag{4}$$

where $\lambda(S_{it})$, the (non-negative) Lagrangian multiplier to the capacity constraint, satisfies

$$S_{it} = \beta_{01}\left[1 - \left(\frac{p_1 - \lambda(S_{it})}{p_0 - \lambda(S_{it})}\right)^{\frac{\beta_{11}}{\beta_{11}-\beta_{01}}}\right] + \beta_{11}\left(\frac{p_1 - \lambda(S_{it})}{p_0 - \lambda(S_{it})}\right)^{\frac{\beta_{01}}{\beta_{11}-\beta_{01}}},$$

and

$$S^\circ = \beta_{01}[1 - (\frac{p_1}{p_0})^{\frac{\beta_{11}}{\beta_{11}-\beta_{01}}}] + \beta_{11}(\frac{p_1}{p_0})^{\frac{\beta_{01}}{\beta_{11}-\beta_{01}}}.$$

And, let

$$t_b^*(S_i) = \begin{cases} \infty, & \text{if } S_i \leq \beta_{01}; \\ B^*(S_i), & \text{otherwise,} \end{cases}$$

where $B = B^*(S_i)$ satisfies (7). If $B^*(\cdot)$ is monotone decreasing and $t_b^*(S) \geq t_a^*(S - \Delta_t)$ for each $t, S \in [0, \infty)$, then $\tilde{\sigma}_i(t_a^*, t_b^*)$ forms the equilibrium in $\mathcal{S}$ of the markdown game.

The equilibrium is depicted in Figures 12-1 and 12-2. Each instance of initial inventory S determines the two parameters (t_a^*, t_b^*), which in turn define his markdown strategy. As an example, suppose two retailers 1 and 2 start with inventory positions S_1 and S_2, respectively, as shown in Figure 1. At the beginning both retailers sell at the regular price p_0. As time passes (moving up in the Y axis on the Figure), retailer 2 with a higher inventory S_2 reaches the time point $t_b^*(S_2)$ and marks down to price p_1. Let $t := t_b^*(S_2)$. Since

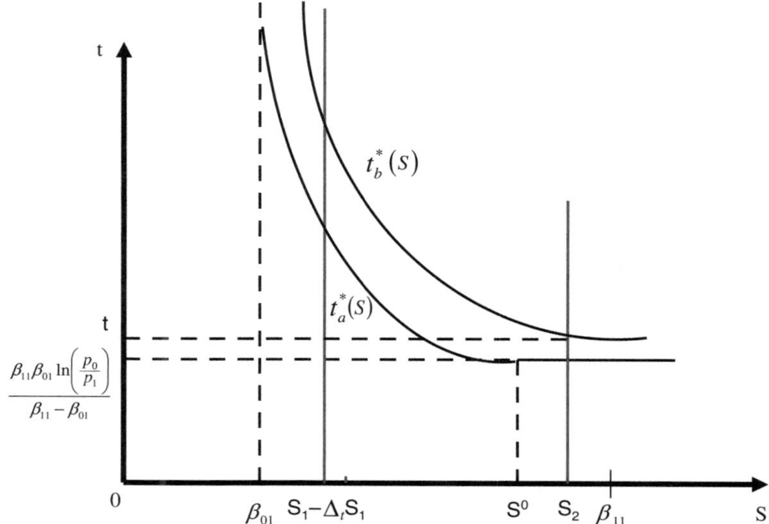

Figure 12-1. Equilibrium for Markdown Competition – Case 1

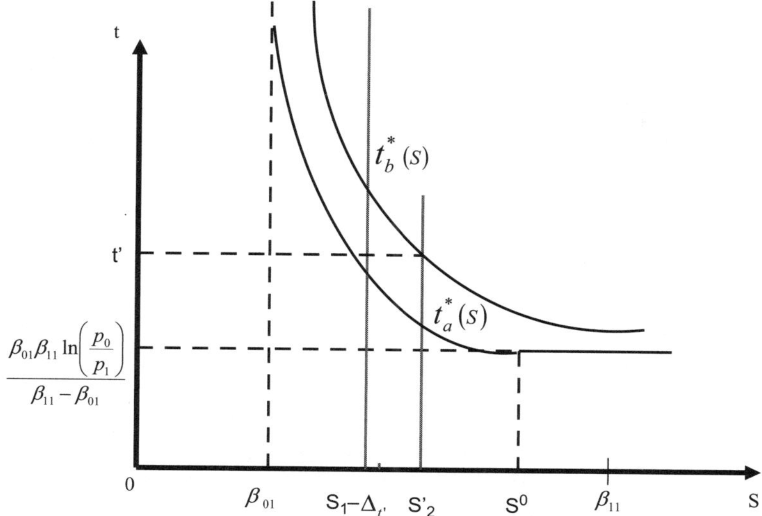

Figure 12-2. Equilibrium for Markdown Competition – Case 2

$t_a^*(S_1 - \Delta_t) > t$ in the Figure, retailer 1 does not immediately match the markdown, but instead waits until $t_a^*(S_1 - \Delta_t)$ and marks down. Thus, the two markdowns will be separated by some time. Now consider another pair of retailers that start with inventory levels S_1 and S_2' as in Figure 2. Again, retailer 2 moves first at time $t_b^*(S_2') := t'$. But this time $t_a^*(S_1 - \Delta_{t'}) < t'$, so retailer 1 will immediately follow the markdown. This is the case where markdowns are "clustered" around the same time. The first mover disturbs the status quo to the other, who is then forced to take a mitigating action.

5. MANAGERIAL IMPLICATIONS AND CONCLUSION

The paper studies how two retailers compete in choosing markdown times. We have restricted our search to a set $\mathcal{S}$ of two-parameter strategies that capture a lot of plausible behaviors. The equilibrium strategy is a function of three elements – the competitor's move so far, the current time (relative to t_a^* and t_b^*), and his own inventory level (captured through $t_a^*(S_i - \Delta_t)$). In our deterministic model, the latter two are overlapping. In equilibrium one retailer's markdown may prompt the other to match instantaneously, especially if the latter has a large inventory and the selling season is almost over (e.g., S_{it} is large and λ is zero). Unfortunately, we could not obtain a closed-form solution to one parameter t_b^*, but the structure of the solution provides several managerial insights.

First, the markdown policy has a direct impact on its preceding inventory decisions. One can view the inventory and markdown decisions together as a bigger sequential game – first considering the subgame (P4) of markdown competition, and then rolling up the solution to the inventory decision. That is, one should solve the following inventory-markdown integrated problem:

$$\max_S \Pi(\tilde{\sigma}(t_a^*(S), t_b^*(S)) - C(S), \tag{P5}$$

where $\Pi(\tilde{\sigma}(t_a^*(S), t_b^*(S))$ is the expected profit under the optimal strategy $\tilde{\sigma}(t_a^*(S), t_b^*(S))$ (solving (P4)), and $C(S)$ is the cost of procuring S. This may lead to larger or smaller inventory levels than the traditional newsvendor solution, depending on model parameters. On the one hand, the unit margin or the underage cost is not as high as the newsvendor operation without a mark down, so the optimal inventory level will be smaller than the newsvendor solution. On the other hand, however, the demand will be higher at a marked-down price. Hence, the retailer who is willing to mark down if necessary may possibly choose a larger-than-newsvendor inventory level if the markdown still grants the retailer a positive margin.

Second, we anticipate that markdowns will be frequently clustered around a certain time. Note in Figure 12-2 that clustering happens when the two retailers start with similar levels of inventory. Since their demand signals are likely to be positively correlated or if they have a uni-modal density function like the normal distribution, they will order similar quantities, so clustering of markdowns will be more likely. See Gul and Lundholm (1995) and its references for other instances of clustering.

Third, the present work proposes an alternative model of price dispersion. Economists have long studied various models of price dispersion as a deviation from the traditional "law of one price" (see Varian (1980) and its references). For example, Varian (1980) (plus its Errata, Varian (1981)) analyzes the competition among n retailers facing two types of customers – informed and uninformed. Informed customers know the price distribution of a certain item and purchase the item at the store with the lowest price. Uninformed customers randomly choose a store and buy the item there if the price is lower than her reservation price. Each store's strategy is the assignment of probabilities to different prices to charge. Varian demonstrates, among others, that no symmetric equilibrium exists where all stores charge the same price, and even strongly, that there would be no point masses in the equilibrium pricing strategies. Thus, price-randomization is the only equilibrium, hence arises price dispersion. Our model presents another possibility of price dispersion. It differs from Varian in two major ways (besides other differences like permanent vs. temporary price changes, and information asymmetry vs. symmetry). First, the model allows a retailer to choose a dynamic strategy of taking, or not taking, an action upon observing the other's move, while each retailer in Varian sets a price randomly drawn from a pre-determined density function. The difference boils down to whether retailers can monitor each other's price. Obviously, it will vary across different markets and products, but given the Internet and the mass media, prices are getting easier to monitor these days.

The other key difference of our model it that it captures the inventory position as a driver of price dispersion. Note from the Figures that the retailer's markdown time is a decreasing function of his initial inventory position. Markdown happens either on its own initiative (due to a high inventory level and a disappointing demand rate) or motivated by the competitor's markdown. In either case, competition redirects the market demand from one retailer with a low inventory to another with a high inventory. On the one hand, it is similar to the behavior of a monopolist who "shapes demands" across different products by dynamically adjusting the prices of two products to shift the demand away from a low-stock product to a high-stock product. But markdown competition would enhance economic efficiency by achieving inventory pooling. Note this happens in a decentralized manner and despite informational asymmetry - as envisioned by Hayek (1945) (who assumed there is "only one price for any commodity" in one market). Unfortunately, for lack of a consumer choice model, our model would be insufficient to formally investigate the efficiency issue.

Fourth and last, note from Figure 1 or Theorem 2 that a markdown will happen only after a certain time A^* $(:= \frac{\beta_{01}\beta_{11}}{\beta_{11}-\beta_{01}} \ln \frac{p_0}{p_1})$. This comes from two observations: (1) the preemptive markdown time $t_b^*(\cdot)$ is a decreasing function of the initial inventory level, and (2) even if a retailer has a lot of inventory (larger than S°), his optimal markdown time remains at A^*. This seems consistent with our perception that markdowns are what we expect towards the end of lifecycle.

The paper deliberately took a minimalist approach, loaded with a series of simplifying assumptions. Relaxation of these assumptions (e.g., deterministic demand) would be desirable. But given that we could not obtain any crisp results from the present simple model, I would rather hope to see a model that is even simpler and yet insightful, or empirical study that would supplement our modeling approach.

APPENDIX: A SKETCHY DERIVATION OF (7)

Note first in (P4) that since $\bar{G}^o(\tau|t) = \bar{G}(\tau)/\bar{G}(t)$ and $g^o(\tau|t) = g(\tau)/\bar{G}(t)$, every $G^\circ(\cdot|t)$ and $g^\circ(\cdot|t)$ can be respectively replaced by $G(\cdot)$ and $g(\cdot)$. Note also that G can be derived from the distribution of random variables S_j via $t_a^*(\cdot)$ and $t_b^*(\cdot)$, and is a mixed (i.e., continuous and discrete) distribution. The first term is his expected profit when retailer j first marks down and he follows immediately. Thus, the probability of retailer j's markdown happening no later than τ is given by $G(\tau) = P(t_b^*(S_j) \leq \tau) = P(S_j \geq t_b^{*-1}(\tau)) = 1 - F(t_b^{*-1}(\tau))$. Thus, $g(\tau) = -dF(t_b^{*-1}(\tau))/d\tau$. The second term captures the case where retailer i first marks down at t_b and retailer j immediately follows. In this case $g(\tau)$ has a probability mass at $\tau = t_b$, since any retailer j whose S_{jt_b} (or $S_j + \Delta_{t_b}$) value satisfies $t_a^*(S_{jt_b}) < t_b \le t_b^*(S_j)$ will immediately follow retailer i's move. Thus, $g(t_b) = F(t_b^{*-1}(t_b)) - F(t_a^{*-1}(t_b) + \Delta_{t_b})$. The third captures the case where retailer i first marks down at t_b and retailer j follows later at τ. Retailer i's demand rate changes from β_{00}, to β_{10} (at t_b) and then to β_{11} (at τ). In this case $G(\tau) = P(t_a^*(S_j - \Delta_{t_b}) \le \tau) = P(S_j - \Delta_{t_b} \ge t_a^{*-1}(\tau)) = 1 - F(t_a^{*-1}(\tau) + \Delta_{t_b})$, giving $g(\tau) = -dF(t_a^{*-1}(\tau) + \Delta_{t_b})/d\tau$. The last term covers the case where retailer i first marks down, but retailer j never follows, since her initial inventory is lower than β_{01}, so she can sell all at the regular price even in the worst scenario (i.e., at demand rate β_{01}). This happens with probability $F(\beta_{01})$, which is here denoted by $g(\infty)$. The constraints ensure that sales do not exceed the inventory in each instance of τ.

Regrettably, (P4) is very difficult to solve. One way to tackle the problem is to form a Lagrangian and obtain its saddle point (Luenberger, 1969). The FOC of the Lagrangian, after straightforward manipulation and letting $t=0$ without loss of generality, gives:

$$
\begin{aligned}
&(p_0 e^{-t_b/\beta_{00}} - p_1 e^{-t_b/\beta_{11}})F(t_b^{*-1}(t_b)) + p_1(e^{-t_b/\beta_{11}} - e^{-t_b/\beta_{10}})F(t_a^{*-1}(t_b) + \Delta_{t_b}) \\
&\quad - \lambda_1(t_b)[\beta_{00}(1 - e^{-t_b/\beta_{10}}) - \beta_{11}(e^{-t_b/\beta_{11}} - e^{-T_1(t_b)/\beta_{11}}) - S_i] \\
&\quad - \lambda_2'(t_b)[\beta_{10}(1 - e^{-t_b/\beta_{10}}) - \beta_{11}(e^{-t_b/\beta_{11}} - e^{-T_2/\beta_{11}}) - S_i] - \lambda_2(t_b)(e^{-t_b}/\beta_{10} - e^{-T_2/\beta_{11}}) \\
&\quad + \lambda_3(t_b)[\beta_{00}(1 - e^{-t_b/\beta_{10}}) + \beta_{11}(e^{-t_b/\beta_{11}} - e^{-T_3(t_b)/\beta_{11}})] - \bar{\Lambda}_3(t_b)(e^{-t_b/\beta_{00}} - e^{-t_b/\beta_{10}}) \\
&\quad + (p_0 e^{-t_b/\beta_{00}} - p_1 e^{-t_b/\beta_{10}})F(\beta_{01}) \\
&\quad + \lambda_4(t_b)[\beta_{00}(1 - e^{-t_b/\beta_{00}}) + \beta_{10}(e^{-t_b/\beta_{10}} - e^{-T_4/\beta_{10}}) - S_i] = 0.
\end{aligned} \tag{7}
$$

where $\lambda_1, \lambda_2, \lambda_3, \lambda_4$ are the Lagrangian multipliers to the four constraints of (P4) in that order, and $\bar{\Lambda}_3(t_b) := \int_{t_b^+}^{\infty} \lambda_3(\tau)d\tau$. By definition of symmetric equilibrium, retailer i's choice of t_b should be equal to retailer j's optimal t_b^*, hence $t_b^{*-1}(t_b) = t_b^{*-1}(t_b^*(S_i)) = S_i$. Applying this to (7), we have:

$$
\begin{aligned}
&(p_0 e^{-t_b^*/\beta_{00}} - p_1 e^{-t_b^*/\beta_{11}})F(S_i) + p_1(e^{-t_b^*/\beta_{11}} - e^{-t_b^*/\beta_{10}})F(t_a^{*-1}(t_b^*) + \Delta_{t_b}) \\
&\quad -\lambda_1(t_b^*)[\beta_{00}(1 - e^{-t_b^*/\beta_{10}}) - \beta_{11}(e^{-t_b^*/\beta_{11}} - e^{-T_1(t_b^*)/\beta_{11}}) - S_i] \\
&\quad -\lambda_2'(t_b^*)[\beta_{10}(1 - e^{-t_b^*/\beta_{10}}) - \beta_{11}(e^{-t_b^*/\beta_{11}} - e^{-T_2/\beta_{11}}) - S_i] - \lambda_2(t_b^*)(e^{-t_b^*/\beta_{10}} - e^{-T_2/\beta_{11}}) \\
&\quad + \lambda_3(t_b^*)[\beta_{00}(1 - e^{-t_b^*/\beta_{10}}) + \beta_{11}(e^{-t_b^*/\beta_{11}} - e^{-T_3(t_b^*)/\beta_{11}})] - \bar{\Lambda}(t_b^*)(e^{-t_b^*/\beta_{00}} - e^{-t_b^*/\beta_{10}}) \\
&\quad + (p_0 e^{-t_b^*/\beta_{00}} - p_1 e^{-t_b^*/\beta_{10}})F(\beta_{01}) \\
&\quad + \lambda_4(t_b^*)[\beta_{00}(1 - e^{-t_b^*/\beta_{00}}) + \beta_{10}(e^{-t_b^*/\beta_{10}} - e^{-T_4/\beta_{10}}) - S_i] = 0.
\end{aligned} \tag{8}
$$

Acknowledgment I would like to thank the editor and the referees for offering valuable input to earlier drafts.

REFERENCES

Aviv, W. and A. Pazgal (2003), "Optimal Pricing of Seasonal Products in the Presence of Forward-Looking Consumers," Working Paper, Olin School of Business, Washington University, St. Louis, MO 63141.

Belobaba, P.P. (1987), "Airline Yield Management: An Overview of Seat Inventory Control," *Transportation Science* 29(3), 63–73.

Bitran R. and R. Caldency (2003), "Pricing Models for Revenue Management," *Manufacturing and Service Operations Management* 5(3), 203–229.

Dudey, M. (1992), "Dynamic Edgeworth-Bertrand Competition," *The Quarterly Journal of Economics* 107(4), 1461–1477.

Evers, J. "Microsoft announces Xbox Price Cut," *PCWorld,* May 15, 2002, http://www.pcworld.com/news/article/0,aid,99524,00.asp.

Feng, Y. and G. Gallego (1995), "Optimal Starting Times for End-of-Season Sales and Optimal Stopping Times for Promotional Fares," *Management Science* 41(98), 1371–1391.

Feng, Y. and B. Xiao (2000), "Optimal Policies of Yield Management with Multiple Predetermined Prices," *Operations Research* 48(2), 332–343.

Gallego, G. and G. van Ryzin, (1993), "Optimal Dynamic Pricing of Inventories with Stochastic Demand over Finite Horizons," *Management Science* 40, 999–1020.

Gul, F. and R. Lundholm (1995), "Endogenous Timing and the Clustering of Agents' Decisions," *Journal of Political Economy* 103(5), 1039–1066.

Hayek, F. (1945), "The Use of Knowledge in Society," *American Economic Review* 35, 519–530.

Kreps, D. (1990), *A Course in Microeconomic Theory*, Princeton Book Company.

Lal, R. (1990), "Price Promotions: Limiting Competitive Encroachment," *Marketing Science* 9(3), 247–262.

Lazear, E., (1986), "Retail Pricing and Clearance Sales," *The American Economic Review* 76(1), 14–32.

Luenberger, D. G. (1969), *Optimization by Vector Space Methods,* John Wiley & Sons, Inc., New York.

Netessine, S. and R. Shumsky (2004), "Revenue Management Games: Horizontal and Vertical Competition," Working Paper, Wharton School, The University of Pennsylvania.

Rao, R. (1991) "Pricing and Promotions in Asymmetric Duopolies," *Marketing Science,* 10 (2), pp. 131–144.

Talluri, K. and G. van Ryzin (2004), *Theory and Practice of Revenue Management*, Springer-Verlag, New York.

Varian H., (1980), "A Model of Sales," *American Economic Review,* 70(4), 651–659.

Varian H., (1981), "Errata: A Model of Sales," *American Economic Review* 71(3), 517.

Index

Early Titles in the
INTERNATIONAL SERIES IN OPERATIONS RESEARCH & MANAGEMENT SCIENCE
(*page ii*)

Axsäter, S. / *INVENTORY CONTROL*

Wolkowicz, H., Saigal, R., & Vandenberghe, L. / *HANDBOOK OF SEMI-DEFINITE PROGRAMMING: Theory, Algorithms, and Applications*

Hobbs, B.F. & Meier, P. / *ENERGY DECISIONS AND THE ENVIRONMENT: A Guide*
to the Use of Multicriteria Methods

Dar-El, E. / *HUMAN LEARNING: From Learning Curves to Learning Organizations*

Armstrong, J.S. / *PRINCIPLES OF FORECASTING: A Handbook for Researchers and*
Practitioners

Balsamo, S., Personé, V., & Onvural, R./ *ANALYSIS OF QUEUEING NETWORKS WITHBLOCKING*

Bouyssou, D. et al. / *EVALUATION AND DECISION MODELS: A Critical Perspective*

Hanne, T. / *INTELLIGENT STRATEGIES FOR META MULTIPLE CRITERIA DECISION MAKING*

Saaty, T. & Vargas, L. / *MODELS, METHODS, CONCEPTS and APPLICATIONS OF THE ANALYTIC HIERARCHY PROCESS*

Chatterjee, K. & Samuelson, W. / *GAME THEORY AND BUSINESS APPLICATIONS*

Hobbs, B. et al. / *THE NEXT GENERATION OF ELECTRIC POWER UNIT COMMITMENT MODELS*

Vanderbei, R.J. / *LINEAR PROGRAMMING: Foundations and Extensions, 2nd Ed.*

Kimms, A. / *MATHEMATICAL PROGRAMMING AND FINANCIAL OBJECTIVES FOR SCHEDULING PROJECTS*

Baptiste, P., Le Pape, C. & Nuijten, W. / *CONSTRAINT-BASED SCHEDULING*

Feinberg, E. & Shwartz, A. / *HANDBOOK OF MARKOV DECISION PROCESSES: Methods*
and Applications

Ramík, J. & Vlach, M. / *GENERALIZED CONCAVITY IN FUZZY OPTIMIZATION* AND DECISION ANALYSIS

Song, J. & Yao, D. / *SUPPLY CHAIN STRUCTURES: Coordination, Information and* Optimization

Kozan, E. & Ohuchi, A. / *OPERATIONS RESEARCH/ MANAGEMENT SCIENCE AT WORK*

Bouyssou et al. / *AIDING DECISIONS WITH MULTIPLE CRITERIA: Essays in* Honor of Bernard Roy

Cox, Louis Anthony, Jr. / *RISK ANALYSIS: Foundations, Models and Methods*
Dror, M., L'Ecuyer, P. & Szidarovszky, F. / *MODELING UNCERTAINTY: An Examination* of Stochastic Theory, Methods, and Applications
Dokuchaev, N. / *DYNAMIC PORTFOLIO STRATEGIES: Quantitative Methods and Empirical Rules* for Incomplete Information
Sarker, R., Mohammadian, M. & Yao, X. / *EVOLUTIONARY OPTIMIZATION*
Demeulemeester, R. & Herroelen, W. / *PROJECT SCHEDULING: A Research Handbook*
Gazis, D.C. / *TRAFFIC THEORY*
Zhu/ *QUANTITATIVE MODELS FOR PERFORMANCE EVALUATION AND BENCHMARKING Ehrgott & Gandibleux*/ MULTIPLE CRITERIA OPTIMIZATION: State of the Art Annotated Bibliographical Surveys
Bienstock/ *Potential Function Methods for Approx. Solving Linear Programming Problems*
Matsatsinis & Siskos/ *INTELLIGENT SUPPORT SYSTEMS FOR MARKETING* DECISIONS
Alpern & Gal/ *THE THEORY OF SEARCH GAMES AND RENDEZVOUS*
Hall/*HANDBOOK OF TRANSPORTATION SCIENCE - 2nd* Ed.
Glover & Kochenberger/ *HANDBOOK OF METAHEURISTICS*
Graves & Ringuest/ *MODELS AND METHODS FOR PROJECT SELECTION:* Concepts from Management Science, Finance and Information Technology
Hassin & Haviv/ *TO QUEUE OR NOT TO QUEUE: Equilibrium Behavior in Queueing Systems*
Gershwin et al/ *ANALYSIS & MODELING OF MANUFACTURING SYSTEMS*
Maros/ *COMPUTATIONAL TECHNIQUES OF THE SIMPLEX METHOD*
Harrison, Lee & Neale/ *THE PRACTICE OF SUPPLY CHAIN MANAGEMENT: Where Theory and Application Converge*
Shanthikumar, Yao & Zijm/ *STOCHASTIC MODELING AND OPTIMIZATION OF MANUFACTURING SYSTEMS AND SUPPLY CHAINS*
Nabrzyski, Schopf & Weglarz/ *GRID RESOURCE MANAGEMENT: State of the Art and Future Trends*
Thissen & Herder/ *CRITICAL INFRASTRUCTURES: State of the Art in Research and Application*
Carlsson, Fedrizzi, & Fullér/ *FUZZY LOGIC IN MANAGEMENT*
Soyer, Mazzuchi & Singpurwalla/ *MATHEMATICAL RELIABILITY: An Expository Perspective*
Chakravarty & Eliashberg/ *MANAGING BUSINESS INTERFACES: Marketing, Engineering, and Manufacturing Perspectives*
Talluri & van Ryzin/ *THE THEORY AND PRACTICE OF REVENUE MANAGEMENT*

Kavadias & Loch/*PROJECT SELECTION UNDER UNCERTAINTY: Dynamically Allocating Resources to Maximize Value*

Brandeau, Sainfort & Pierskalla/ *OPERATIONS RESEARCH AND HEALTH CARE: A Handbook of Methods and Applications*

Cooper, Seiford & Zhu/ *HANDBOOK OF DATA ENVELOPMENT ANALYSIS: Models and Methods*

Luenberger/ *LINEAR AND NONLINEAR PROGRAMMING, 2nd* Ed.

Sherbrooke/ *OPTIMAL INVENTORY MODELING OF SYSTEMS: Multi-Echelon Techniques,* Second Edition

Chu, Leung, Hui & Cheung/ *4th PARTY CYBER LOGISTICS FOR AIR CARGO*

Simchi-Levi, Wu & Shen/ *HANDBOOK OF QUANTITATIVE SUPPLY CHAIN ANALYSIS: Modeling in the E-Business Era*

Gass & Assad/ *AN ANNOTATED TIMELINE OF OPERATIONS RESEARCH: An Informal History*

Greenberg/ *TUTORIALS ON EMERGING METHODOLOGIES AND APPLICATIONS IN OPERATIONS RESEARCH*

Weber/ *UNCERTAINTY IN THE ELECTRIC POWER INDUSTRY: Methods and Models for Decision Support*

Figueira, Greco & Ehrgott/ *MULTIPLE CRITERIA DECISION ANALYSIS: State of the Art Surveys*

Reveliotis/ *REAL-TIME MANAGEMENT OF RESOURCE ALLOCATIONS SYSTEMS: A Discrete Event Systems Approach*

Kall & Mayer/ *STOCHASTIC LINEAR PROGRAMMING: Models, Theory, and Computation*

Sethi, Yan & Zhang/ *INVENTORY AND SUPPLY CHAIN MANAGEMENT WITH FORECAST UPDATES*

Cox/ *QUANTITATIVE HEALTH RISK ANALYSIS METHODS: Modeling the Human Health Impacts of Antibiotics Used in Food Animals*

Ching & Ng/ *MARKOV CHAINS: Models, Algorithms and Applications*

Li & Sun/ *NONLINEAR INTEGER PROGRAMMING*

Kaliszewski/ *SOFT COMPUTING FOR COMPLEX MULTIPLE CRITERIA DECISION MAKING*

* A list of the more recent publications in the series is at the front of the book *

Printed in the United States of America